Dairy Herd Hea
Management

A guide for veterinarians and dairy professionals

By Jos Noordhuizen

Edited by Helen Warren

About the author

Jos Noordhuizen was born on April 9th, 1947 in the Netherlands. He got his DVM diploma from Utrecht Veterinary Faculty in January 1975, after which he was appointed for 3 years in the Clinic of Obstetrics & Gynaecology of that Faculty, where he practiced Herd Fertility Schemes on dairy farms. After having spent some time in private food animal practice, he went back to Utrecht where he got his PhD on the subject Herd Health & Production Management on Dairy Farms in 1984.

In 1988 he was appointed professor in animal husbandry at the Wageningen Agricultural University. Later In 1997 he was invited for the chair of professor in ruminant health at the Utrecht Faculty where he stayed until 2005. He was guest professor in veterinary schools in Gent (Belgium), Nantes (France) and Lyon (France).

He is currently adjoint professor at Charles Sturt University, School of Animal & Veterinary Science (Australia) and consultant at VACQA-international (Portugal).

He has supervised over 35 PhD projects, published over 300 papers in scientific and practice journals, edited 4 books on various veterinary and dairy farming topics, organized several international courses, was invited for giving seminars all over the world, and participated in joint development projects in Costa Rica, Thailand, Vietnam and Sweden.

Jos was member of the Dutch National Health Council, appointed by HM The Queen, member of the scientific committee on animal health and welfare of the European Commission in Brussels, founder and president of the Dutch Association for Veterinary Epidemiology & Economics, cofounder of the European College of Bovine Health Management, and member of several international associations and congress organisation committees. He currently lives with his wife in Normandy (France).

Bibliographic details

British Library Cataloguing in Publication Data
The Dairy Herd Health And Management Book. A gude for veterinarians and dairy professionals.

I. Noordhuizen J.
ISBN 978-1-899043-36-1

Edited By Warren H.

For more information on content/references please contact the author.

Context Products Ltd, 53 Mill Street, Packington, Ashby de la Zouch, Leicestershire. LE65 1WN England, Tel: +44 (0) 1530 411337
Email: admin@contextbookshop.com
www.ContextBookshop.com

About the Book
This publication takes a new approach to the subject of Dairy Herd Health and Management. The author's 35 plus years experience in the area has lead to the conceptualisation and collection of different topics to better support dairy herd health and management programmes.

About Context
Context is a specialist publisher based in England producing targeted practical technical publications.. Our expertlse allows us to plan, design and produce publications that are colourful and easy to read yet providing you with the essential information you need quickly. We can also offer versions in different languages if required.

Sections

Herd health and productivity management

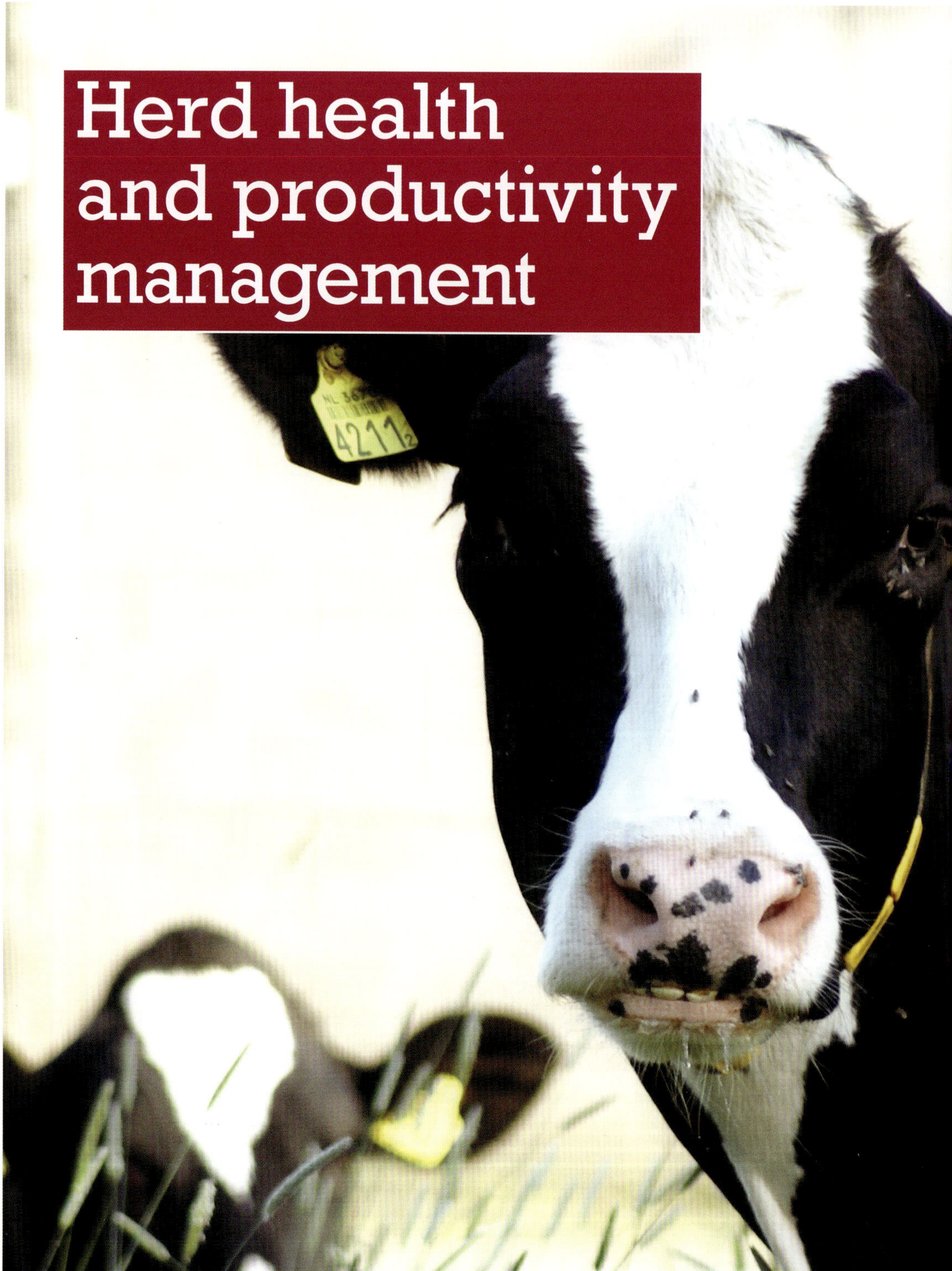

“Herd health and productivity management (HHPM) has become a core business of modern veterinary practices in developed countries. The unit of concern is the herd as opposed to the sick cow. This is because the farmer earns his income through healthy cows and loses money via sick cows. HHPM comprises the various, most important farming areas. For each area, a basic monitoring protocol exists. However, at the same time, individual area specificities should be taken into account. This chapter provides various practical methods used to detect strong points and points for improvement, organise a farm visit and interpret farm performance. Problem analysis requires a different type of protocol, where a stepwise procedure helps the farmer to understand where he stands in relation to average benchmarks and other units. Performance parameters are always compared with reference values.”

Satisfaction inventory

1= not satisfied at all
2= slightly satisfied
3= generally satisfied
4= satisfied
5= very satisfied

Table 1.1 Satisfaction inventory for different farming areas

Name Address						Date
Milk Production	1	2	3	4	5	Important remarks
Litres of milk per cow per year (mean)						
Milk fat %						
Milk protein %						
Milk production in first lactation						
Milk production at start of lactation (0 – 50 days)						
Udder health	1	2	3	4	5	Important remarks
% cows with clinical mastitis per year						
% cows with high cell counts per year						
% cows with subclinical mastitis per year						
Mastitis cases during the dry period						
Outcome of mastitis treatments						
Claw health	1	2	3	4	5	Important remarks
% lame cows per year						
Outcome of routine herd claw trimming						
% cows with foot rot per year						
% cows with laminitis per year						
% cows with Mortellaro disease per year						
% cows with hock lesions per year						
Overall health	1	2	3	4	5	Important remarks
% cows with rumen acidosis						
% cows with ketosis						
% cows with infectious diseases						
% cows with skin lesions						

Satisfaction inventory

Health of calves and maiden heifers	1	2	3	4	5	Important remarks
% diarrhoea in neonate calves						
% diarrhoea in older calves						
% respiratory disease in calves/heifers						
Reproduction and fertility	**1**	**2**	**3**	**4**	**5**	**Important remarks**
Calving interval						
Age of heifers at first calving						
Interval between calving–first insemination						
Interval between calving–conception						
Detection of heat among cows						
Detection of heat among maiden heifers						
% cows with cystic ovarian disease per year						
% cows with retained afterbirth per year						
% cows with endometritis per year						
Number of inseminations per pregnant cow						
Quality of milk (bulk tank)	**1**	**2**	**3**	**4**	**5**	**Important remarks**
Mean cell count (x 1000/ml)						
Mean bacteria count (x 1000/ml)						
Milk fat %						
Milk protein %						
Butyric acid spores						
Antibiotic residues						
Other observations eg nutrition, hygiene, other						

Satisfaction inventory

1= Identify
2= Prioritise
3= Plan

Please answer the following questions
Circle which is applicable

Question 1
On the basis of the above 'satisfaction inventory', do you think that there is sufficient support to address problems with scores of 1 or 2?

YES | NO

If YES arrange visit to discuss action plan.
What are your priorities with regards to addressing the problems?

[1]...

[2]...

[3]...

If NO why is there reluctance to address the problems?

....

Question 2
If you answered **YES** to Question 1, do you feel that you can work together with your vet to analyse and try to solve the herd problems?

YES | NO

If **YES,** there is the opportunity for a visit to discuss an Integrated Herd Health & Productivity Management advisory programme, which would be tailored to your particular issues, highlighted priorities and management practice, as well as current farming objectives. Would you be interested in this?

If NO, what is the main reason for declining this opportunity?

....

Question 3
If you have answered YES to all the previous questions, would you prefer us to visit in the immediate future or later on?

Right away | Later

Question 4
Would you prefer to have an individual meeting on your own farm or as part of a group of other farmers?

Individual visit | In a group

Thank you for having taken the time to fill in and return this inventory.

General farm information relating to HHPM

Table 1.2 General information about an individual farm under an HHPM programme

Name Address	Farm Code Number Phone E-mail
Frequency of HHPM: 1 per 1 / 2 / 3 / 4 / 5 / 6 weeks	Boots and overall are present: Yes / No
Overview Breed Number of lactating cows Number of calves/maiden heifers Type of farm (open / closed) Type of farm house Number of feeding places at feed rack Number of cubicles Foot bath present Yes / No Quota of milk (kg or tons) Quota leased (kg or tons) Area of pastures Area of grass in total Area of maize Area of other types of land/crop AI or running bull Other species present Density of cattle per hectare	
Nutrition Mixer wagon present Maize supplied Additional products Concentrate feeder present Number of silos Source of water Mineral mix is supplied	
Udder health Milking hours Number of machines/claws Type of milking parlour Udder preparation practice Disinfection of teats Yes / No Type of cubicle & size Type of bedding (material) After milking cows stand at feed rack Yes / No Milk Recording Somatic cell counts (+ frequency) Dry-off treatment with antibiotics/sealer	
Herd health situation Status of herd for IBR Status of herd for BVD Status of herd for *paratuberculosis* Status of herd for *leptospirosis* Status of herd for *salmonellosis*	

SWOT analysis

The SWOT analysis follows a preset protocol in all dairy farms. More information can be found at www.vacqa-international.com.

For Veterinary Colleges or Faculties, the access to and use of this website is free. Follow the instructions on this website for subscription. Only one password will be issued so other users have to follow the instructions on the first page of the website in order to subscribe. Each dairy farm is identified by a farm code and a password; in this way one can save the entered data of several farm visits and compare them.

The following farm areas can be addressed while conducting a SWOT analysis:

Dairy cows
- Reproduction at herd level
- Claw health
- Udder health
- Production & nutrition

Calves & maiden heifers
- Performance

Cattle welfare and cow comfort public health
- Public health
- Food safety

The SWOT scoring sheets can be printed out and used as guidance during farm visits to ensure all aspects of the farm are taken into account. This is especially relevant on farms with a herd size of fewer than 40 cows. Data can then be entered for larger herds (>40 cows) directly onto the website to obtain score sheets for the subsequent visit. It is preferable that users make a translation of the English text into their own language to allow the farmer a better understanding of the process.

The SWOT scoring sheets can be used during the farm visit for:
- General diagnostic farm evaluation (routine inspection)
- More in-depth problem analysis
- Integrated herd health & productivity management programmes.

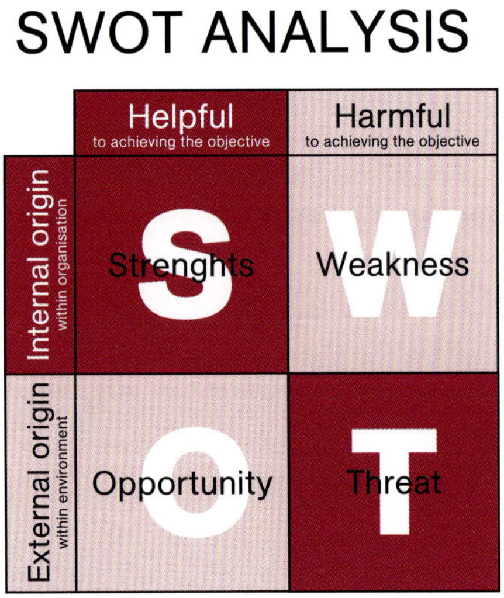

Figure 1.1 SWOT analysis

Transition period score sheet

Adapted from GD Deventer NL 2008

Table 1.3 General score sheet to assess the level of disease resistance in cows during the transition period

Farmer name	Name of veterinarian	Date of visit	
		YES	NO
Nutrition	The average BCS of dry cows is > 3.5		
	A significant decrease in dry matter consumption by dry cows is observed		
Early lactation	Loss of >1 BCS point in the first 6 weeks after calving		
	Rumen Fill score after calving is < 3		
Minerals, Vitamins	Dry cows and/or heifers receive a total mixed ration (TMR)		
		SCORE	
Stress factors	Most calvings take place isolated and outside the herd (score 4)		
	Forage component at the end of the dry period or the day of calving differs by > 50% of the forage type in early lactation (e.g. low energy/ high fibre versus high energy + fibre)(score 2)		
	At the end of the dry period, the cow has not been adapted to at least 2 kg of concentrates (score 2)		
	The neonate calf is not separated from its dam within 6 hrs of birth (score 2)		
	There are too few cubicles for all cows in the herd (score 6)		
	There is not enough space at the feed table (score 4)		
	There are cow comfort problems on this dairy farm (score 4)		
	TOTAL		
		YES	NO
Herd health	Prevalence of infectious diseases is rather high (> 15% of cows in the herd are affected		
	Prevalence of other disorders (eg. mastitis, lameness, metabolic disorders) is rather high (> 15% of cows in the herd)		

If "**YES**" has been entered under one or more of the given situations above, this can be indicative of a reduction in disease resistance. In this case, a more in-depth analysis should be carried out to try to find the cause(s) of this reduction.

A score of **6 or higher under 'Stress factors'** indicates reduced **disease resistance**.

High prevalence of the diseases and disorders noted above can indicate a lowered disease resistance in the herd.

Key factors for large dairy farms

From Dossier Grands Tropeau.
December 2007
L'Eleveur Laitier,
153. Jean-Luc Ménard

Factor 1 - Buildings

- Barn/shed climate = ventilation without obstacles
- Ventilated walls or half-open walls are okay
- Side doors standing open is rather bad for optimal ventilation
- Mobile wall sections over the whole length are fine for flexibility

The milking parlour can be placed:

1 Parallel to the cow houses
2 Perpendicular to cow houses
3 At the end of the cow house

Factor 2 - Specialisation tasks for animal groups on the farm

- Cows - dry
- Cows - in lactation
- Heifers
- Calves
- Combinations of the preceding

} Organisation & specialisation of tasks
Herd divided in groups:
2 x lactation
1 x dry group
(at least part of the year)

Factor 3 - Cow Comfort

- Straw yard
- Cubicles+mattress
- Combinations

} Workload and -time
(Cows forced to pass crossings need time)
Availability of straw
Cow circulation & traffic
Risk of mastitis

Waiting area

1.2 m^2 /cowL = 12 m^2 / 10 cow
When cows enter in different, subsequent, smaller groups 6 m^2 / 10 cow are only needed for the waiting area.

Provide one calving pen / 20 cows (when calvings are concentrated).

Cubicle size

To calculate the correct cubicle size, use the largest 30% of the herd or group as a guideline.

Size needed at the shoulder rail = 0.5 m width /cow
Size needed at the feed table = 0.75 m width / cow

Feed bunk for ad lib consumption

1 place / 2 - 3 cows
Push the forage to the cows or use a mobile feed table.

60
85

Herd health check list - a working example

Table 1.4 Farm calendar in Herd Health Performance Management

	Jan	Feb	Mar	Apr	May	Jun	July	Aug	Sep	Oct	Nov	Dec
CMT herd	X											X
Milking machine check/ calibration	X	X										
Routine claw trimming			X	X						X	X	
Use of foot bath			X	X						X	X	
Heifers inseminated								X	X			
Vaccination		X							X			
Weighing of calves											X	X
Liverfluke check												X
Nematodes - check faeces 5 wks after start of calves in pasture						X						
Ration change - barn to pasture		X	X									
Calves out to pasture				X	X							
Evaluation of pasture										X		
Buffer ration given at pasture								X	X			
Ration change - pasture to barn									X	X		
Analysis of forages											X	X
Evaluation of ration fed in the barn												X
Concentrate feeder check										X		
Evaluation of housing	X	X									X	
Evaluation of climate/environment	X	X									X	
Hygiene check			X								X	
Body condition score	X		X		X		X		X		X	
Rumen fill score	X	X	X	X	X				X	X	X	X
Faecal scores (viscosity and undigested particles)	X	X	X	X	X		X		X	X	X	X
Teat-end callosity score				X						X		
Mobility / locomotion score			X	X							X	X

Indicator cows and casuality

Table 1.5 Indicator cows - casuality and areas to investigate	
Area	Investigate
Milk fever	Rations too rich at the end of lactation; poor transition
Ketosis + negative energy balance	Too little energy in the ration at start of lactation; too little dry matter intake around calving
Hock lesions :abscess	Poor cubicle quality or poor bedding; claw lesions
Teat-end callosity	Poor milking machine function
Abomasal displacement	Deviations in feed or feeding management; infections; stress
Poor BCS scores	Deviations in feeding; high production; health problems
Teat lesions	Housing issues; lameness
Deviations in kg milk produced	See section on Feed & Feeding Management. Other areas of concern are health and stress
Deviations in milk fat	See section of Feed & Feeding Management. Other area of concern is health
Deviations in milk protein	See section of Feed & Feeding Management. Other area of concern is health
Diarrhoea in calves	Poor birth management or poor colostrum management; poor hygiene; infections
Poor calf growth rate	Incorrect feeding management; health problems; poor management

Pooled sample technique for herd metabolic profile testing

Random sampling of 8 to 12 cows - preferably in the close-up group just prior to calving & just after calving.

Table 1.6 Pooled sample technique for herd metabolic profile testing			
BUN	Total proteins	Alcalic Phosphatase	The herd accounts for 73% of variation ($p<0.003$)
Créatinine	Albumine	CK	
Glucose	Total bilirubine	gamma GT	
Na	Total cholesterol	AST	
K	Triglycerides	SDH	
Cl	BHB	CO2 total	
Ca	P	Mg	Anion gap

van Saun (2006) Proceedings of the 12th ICPD, Michigan, USA pp47

Issues facing dairy farms

Stray voltage problems

Stray voltage problems in dairy cattle refers to an AC current flow of > 4 mA (measured as 2 Volt AC with a 500 Ohm resistor). The neutral is linked to the earth. Electromagnetic fields of > 10 kV/m can be found.

Consequences are, for example, a decrease in melatonin levels; increase of IGF-1; rise in blood electrolyte concentration.

Checking for stray voltage should take place in areas where cows: Walk; Eat; Stand/lay down; Are being milked.

If stray voltage problems are suspected, take the following actions

1 Then check for it with a hand-held meter.
If testing negative, then repeat it for 3 or 4 consecutive days.
Contact the Power Company!

2 Let an electrician or expert check all wire system on the farm.

3 Evaluate the milking system during milking

4 Review the housing (barns), feeding, health care and breeding procedures, because very often it is stated that there are stray voltage problems, while the cause for poor performance is in other farming domains.

Ref Robert E.Graves. Hooked on Stray Voltage.
www.das.psu.edu/research-extension/dairy/August 2011
www.pq.goeke.net

Stress & non-esterified fatty acids (NEFA) & betahydroxybutyrate (BHB)

NEFA's can be detected in the blood when a cow mobilises her fat reserves for energy reasons. Stress conditions put high(er) demands on energy.

Stress if NEFA concentrations increase above the threshold of 0.5 mMol/L → prevalence of abomasal displacement increases (with an Odds Ratio= 3.5). However, from about 2 weeks before to 2 weeks after calving a peak in NEFA concentration does occur physiologically.
At 0.5 mMol/L and higher, pathologies are already identifiable.

At a BHB level

- > 1200 µMol/L → Odds Ratio of abomasal displacement is 7.0
- > 1200 µMol/L → there is already a situation of subclinical ketosis
- > 200 µMol/L → Odds Ratio of abomasal displacement is 3.0

The parameters NEFA + Urea before calving give a fair indication about the cow's nutritional and energy status.

The parameters NEFA + BHB + Urea after calving give a fair indication about the nutritional and energy status of the cow.

Ref: LeBlanc *et al.*, (2004) Proceedings of the 12th ICPD, Michigan USA pp43; Counotte G, GD Deventer, Holland, August 2011)

Novel Ingredients (nutraceuticals)

1 Anti-oxidants*[1]
2 Anti-viral*[1]
3 Stimulation of appetite and performance*[1]
4 Stimulation of animal health - spices, herbs, plant extracts, botanicals, essential oils. *[1]
5 Probiotics = DFM = direct-fed microbials *[1]
6 Management and the environment*[2]
7 Novel ingredients*[2]
Ref *[1] Wang *et al.*, 1998. Nottingham University Press (Lyons & Jacques, editors), pp 273)
Ref *[2]: Risley & Lopez (2006) *Proceedings of the 12th ICPD, Michigan USA* pp166-177)

Protocol

Preparation of a HHPM visit
Target time to complete 30 minutes
max suggested is 1 hr

Three phases to a farm visit

HHPM Document Checklist

In order to be well-prepared for the visit, the following documents should be made available:

- A The milk recording sheets from last month
- B The forms from the dairy processing plant
- C Information on rations actually fed
- D The health log of the farm
- E The farm diary for daily events
- F The previous farm visit reports
- G The reports of any analyses carried out
- H The yearly performance reports for reproduction (if available)
- I The yearly performance reports for milk production (if available)
- J Other documents that may be of interest

These documents serve several purposes:

Regarding the history of the farm:

- To get an overall idea about the farm: trends and performances
- To get a picture of the current performances
- To get an idea about the current problems and the suggested action plans, as well as their effect over time

<u>This results in a written report (telegramme style ie no more than 2 pages of A4)</u>

Regarding the current farm situation:

- Find the cows and the farming areas that may represent a problem
- Compose a list of cows and areas to examine during the visit (routine inspection; which individual cows to examine for what reasons; points to discuss with the farmer; any additional examinations to conduct (milk or blood samples)

<u>This results in an examination list for the farm visit.</u>

Both reports/listings should be discussed with the farmer at the start of the farm visit.

See also Protocol – ' analysis of data'
Protocol - 'The farm visit report'

- **Preparation**
- **Execution**
- **Follow-up**

Preparation

For the routine diagnostic evaluation visits and for problem analysis visits, there is no real preparation, unless one desires to collect some documents beforehand. In situations where previous visits have been made, the reports from each of those visits should be taken (see also the SWOT sheets on the website www.VACQA-International.com).

Where a programme of veterinary herd health & production management is ongoing, several documents can be consulted beforehand. Among these, the five most important are:

- The milk recordings from preceding dates (litres of milk; milk fat %; milk protein %; stage of lactation; cell counts; herd means versus group means versus individual records)
- The milk quality data from the bulk tank milk (cell counts; bacteria counts; milk fat% ; milk protein %; other quality features)
- Any recent events in the herd (calvings; reproduction parameters; udder health; claw health; digestive disorders; metabolic disorders; disease in the young stock)
- Any recent events regarding husbandry and management (changes in quality and quantity of feeds; changes in the ration; milking machine)
- The report from the preceding farm visit with routine inspection results and evaluation of proposed interventions and treatment.

It is advisable to check these forms in detail before making the farm visit. it may be necessary to ask the farmer to deliver those documents to the clinic one or two days before the planned visit.

After reviewing these documents, a list of actions to follow during the visit should be prepared. Finally, any other issues which may need thorough discussion with the farmer should also be noted (see also Farm calendar in this series).

Three phases to a farm visit

Execution

It is essential to arrive at the farm on time if you do not want to denigrate the farmer - he must feel that he is taken seriously. On arrival, act according to the principles of Good Veterinary Practice (hygiene measures for boots and overall; parking the car outside the farm premises; washing hands upon arrival).

When the farmer arrives, ask him whether there have been recent events, which require attention. If so, and they are a priority, put them on the action list. If not, continue with the routine diagnostic evaluation protocol:

- First an inspection of the data regarding the herd performance (wall card with all reproductive data on cows in chronological order of their calving date in the year)
  - cows needing a check up are written on an examination list.
- General inspection of the farm (animals; environment; data).
  - record the findings of this inspection.
- Clinical inspection of cows according to the examination list, e.g. reproduction (anoestrus ; gestation), udder/teats, claws, BCS, etc.
  - record the results of this inspection on the examination list.
- Treatment of cows if needed; cows are released afterwards.
  - record treatments on the examination list.
- Inspection of calves, e.g. health, growth, BCS, etc.
  -record the results on the examination list.
- Particular inspection of a certain production issue according to the priority given by the farmer (e.g. udder, claws, housing, climate, milking machine, hygiene, nutrition)

Subsequently, make an inspection tour of the farm together with the farmer. During this tour, check the general hygiene, the silage clamp/silo, the pasture plots, houses, climatic conditions in the barns. During this, discuss certain issues with the farmer, especially if he attempts to deviate during the tour.

Finally, sit down with the farmer in his office or in the kitchen to finalise an interpretation and draw conclusions. This would be the preliminary farm visit report and comprises the first advice for various issues to be discussed with the farmer as he will need to implement them. For example, the prediction of oestrus dates or the measures to be taken in the case of lameness problems or mastitis. During the discussion, adjust the report to make it easier for the farmer to implement and more practical towards his management style. The final written report should be sent by post or delivered directly to the farm door; this must be done within seven days (to allow the consultation of the veterinarian with other specialists, such as a nutritionist). The farm visit is now finished with a date and time arranged for the next appointment.

Follow-up

Often, a additional analysis may have to be carried out, laboratory results obtained or another specialist consulted. This can be done at the clinic. The analysis procedure and results are written down in an analysis report; this report must also be discussed with the farmer. If needed, an appointment must be made between planned visits to discuss the report and any advice. The report should not exceed 5 pages of A4 and should be sent to the farmer within seven days of the visit. Laboratory results should be included in the report.

Note

Be sure to have discussed costs related to farm visits and follow-up activities before any action is taken! Usually, the costs of farm visits are calculated by the hour; the same is valid for other activities like the follow-up. Of course, you are free to decide to offer the farmer a price reduction; for example not to charge the follow-up costs during the first three or six months or the first year.

Practical guide for the farm inspection visit

1 Upon arrival at the farm, try to get an impression of the level of organization and general hygiene standards.

2 Follow the rules of Good Veterinary Practice with regard to changing boots and clothes; sometimes the farmer may provide these. If so, check whether they are clean.

3 Contact the farmer, his family and the employees if present. Ask the farmer about the specific herd performances with which he is most and least satisfied.

4 Start the animal inspection with the young stock. Use the field scoring sheets to record your findings; discuss them with the farmer (show him the instruction cards if needed). Take the equipment for measuring heart girth in young stock if required.

5 Assess the housing, barn climate, hygiene and rations for the calves and maiden heifers; look for good points and identify areas for improvement.

6 Now conduct an inspection of the cows in groups, such as dry cows, fresh cows, mid-lactation cows and end-lactation cows. Do not forget to check cows in the sick pen or calving pen.

7 Inspect the housing, barn climate, hygiene. Take measurements when needed. Look for good points and identify areas for improvement.

8 Do not forget to look at the silage clamp/ silo (grass; maize) and other products or by-products. Also, ensure the store of bedding material is checked.

9 If animals are grazing, check the condition of walkways (soil type; gravel; concrete; mud), as well as the condition of the pasture in the fields and the number of days animals spend in each field.

10 Finally, discuss your questions and findings derived from the inspection with the farmer. Upon request by the farmer, it may be appropriate to give them some brief conclusions before you complete the overall report.

11 Before you leave, change your boots and clothes and wash your hands. This often provides a good opportunity to informally discuss with the farmer any specific findings together with any conclusions you have drawn. In this respect, you may offer the first suggestions or advice that they may adopt on farm to make improvements. Ensure you arrange a date and time for next farm visit. Now you are able to leave the farm.

12 Back at the clinic, complete your interpretation of your visit findings (good points; areas for improvement) and prepare a detailed conclusion. Then, draw up a plan of action with actions (< 5) for the short term and mid-long term. From these prepare a written report of 1 page of A4. Deliver this report to the farmer within 5 days of the visit. If needed, discuss the report with the farmer.

Decision points in the lactation cycle

Calving	35 D	50 D	70 D	90 D	120 D	>>	305 Days

Heat 1 (often seen)

Heat 2 (missed)

Heat 3 (missed)

Try to limit the losses in fertility and the culling of cows with > 3 inseminations. If not, economic losses will be doubled!!

> 85% of the cows must be pregnant !!

Stage at which fertility is at its lowest !!

Interval during which a cow must be inseminated > try to keep the cows within this interval (energy level in ration; prostaglandins)

Check ovary function (cyclicity?) > try to approximately predict the dates of oestrus (in heat in 1 week, in 2 weeks)

Uterine function (metritis?) > treatment needed
? Other disorders present (ovarian cysts)?

Routine herd diagnostic inspection -BCS, RP, claws, teat-end callosity, hygiene, etc

Figure 1.2 Lactation curve decision points

Cluster removal

Automatic detachment of the milking clusters

Traditionally, the threshold for detachment is set at 200 ml milk/min. However, in 50 % of observed cases, detachment did not function well and occurred too late. As a consequence, duration of milking may be unnecessarily extended and there may also be a rise in clinical mastitis cases.

The threshold for detachment must be set at 400 or even 600 ml milk/min. in high yielding cows.

Farm visits

The structure of farm visits for DHE and for PA is similar. The main difference being that a general evaluation protocol is followed for a DHE, while for PA there is always a specific problem to investigate.

The following text may be used as a guide for herd evaluation and analysis. Diagnostic Herd Evaluation is used as the basis; however, italics will be used where specific details for PA are required.

In all situations, restraint of the cows in yolks or at the feed table is highly recommended to facilitate the evaluation, saving time and money, as well as promoting a safe working environment.

Upon arrival at the farm

Park away from animal sheds. Change boots and clothes to those suitable for the farm – these may be provided by the farmer or you use your own but they must be clean. Prior to the animal inspection, ensure you have washed your hands.

When greeting the farmer, explain the format and objectives of your visit. It is usually necessary that the farmer participates in the visit in order to discuss their own findings with yours.

Problem Analysis, *The key issues for the farm will have already been highlighted and actions to address them, defined in earlier discussions with the farmer or may have shown up in the monthly herd performance assessment. As a basis for further activities it is recommended that you start with the following procedure.*

Execution of the visit - order of animal inspection

1 Adult cows

2 Maiden heifers

3 Calves

In the case of a DHE, begin with the animal inspections in the order listed (refer to Table on Monitoring in this series for further details):

If deemed necessary, ask the farmer for further information on these groups.

Following the animal inspection, carry out an inspection of the animals' environment (refer to Table 1.8 on Monitoring). List all the findings on a field scoring sheet and review the farm data - they can provide additional information, as well as confirmation of initial your conclusions (e.g. acidosis or ketosis in fresh cows). Finally, collate your findings and draw initial conclusions. These are then discussed with the farmer, together with actions to address them. There also needs to be discussion regarding whether the farmer is willing to adopt these actions.

Problem Analysis *Following the above, proceed with the detailed inspection of the farm area in which the original issue(s) occurred. Start with an* anamnesis *on this area, followed by an* in-depth inspection *(for example, of affected animals). The specific farm area will influence the exact nature of the inspection: in case of udder health problems, this may involve the milking machine or the milking practices, the treatment of mastitis, the management of high cell count cows; etc. See specific analysis protocols elsewhere in this series.*

Farm visits

If the farmer is happy and in agreement with the conclusions drawn, the advice then needs to be refined into short-term (priorities) and mid-long term. Discuss with the farmer the various options, including costs and potential benefits - they will determine what would be feasible in relation to their management practices. Finally, a report detailing the visit should be drawn up.

Problem Analysis *The report will mainly comprise the results of the* in-depth *analysis, which are recorded along with those of the* DHE.*As with a DHE, the advice for a PA should be divided into that for the short-term and mid-long term. The report should not exceed 5 pages of A4 – a report of this length is more likely to be read by the farmer. See elsewhere in this series for an example of this report.*

Evaluation of advice and interventions

Finally, the effects of the advice and various interventions should be monitored during subsequent farm visits. The next visit should be no later than 1 month after the current visit in order to remain familiar with facts and events.

Effects can be evaluated by conducting again a DHE of animals, environment and farm data (looking for strong points and areas for improvement). This is the most effective and rapid way to ascertain what developments have taken place.

It is equally important to determine whether the farmer has carried out suggested interventions and, if they have not, why not.

For evaluation purposes, inspection data can be stored and edited on the website www.vacqa-international.com .

If necessary, advice and suggested actions may be adjusted and detailed in a new written report for the farmer.

Selecting cows for examination

It is NOT necessary to always find the last individual cows in the herd with a certain disorder on a certain date; it is the HERD LEVEL that counts!!

Option A- Inventory at the start of an HHPM programme. All findings during a farm visit must be recorded on paper (field scoring sheet).

Reproduction

All lactating cows at the feed rack are to be examined. Use cattle markers.

1. Make
2. Check
3. Make a distinction between pregnant and non-pregnant cows.
4. Check among the non-pregnant cows those which are and those which are not inseminated.
5. Check the interval since the last AI of the inseminated cows (maybe too short to do pregnancy check)
6. Check those non-inseminated cows for whether they are cycling or not (not seen? Too soon after calving?)
7. Identify those non-cycling cows that are healthy and those that have a health issue (set diagnosis; treat them)

Health & Production

- Look for the cows (or ask the farmer) that have recently had mastitis, a high cell count, lameness or metabolic disorder.
- Look in the milk records for cows in early lactation (< 120 days) showing a drop in milk yield, fat and/or protein of > 10% since the last recording date (Note that a drop of less than 8% between two dates is acceptable). Try to find out the possible causes of this deviation (NEB? Ketosis? Acidosis? Milk fever?).
- The above cows should be, for the time being, suspected of health issues and recorded!
- Conduct an inspection (diagnostic herd evaluation) of the cows at the feed table and record the deviating cows (poor posture; hock lesions; poor hygiene score; malformation of claws; teat-end-eversions; teat-end callosity; BCS; RF; FC; UFF, etc).

Option B- HHPM programme already in place

The preparation of the farm visit is similar to that mentioned above under 'Health & Production'

In this case, the Herd Overview Card on Reproduction & Health and the Farm Diary are available. The farmer can assist in selecting the appropriate cows for examination. For Reproduction select cows from the Herd Overview Wall Card that fall into one of the following categories:

1. Inseminated more than 35 days ago (or 40 or 45 or 50 days ago) à pregnant or not?
2. Not seen in heat after calving, or observed only once (suspect NEB? Ketosis? Cyclic? Cysts on the ovary?).
3. Irregular inter-oestrus or inter-AI intervals (10 days were prostaglandins applied? 4–5–6–9 weeks after AI).
4. Around 30–40 days after calving, to know whether they are healthy, cyclic and hence fit for AI.
5. Cows which show or have shown problems (retained afterbirth; endometritis; > 3 AI).

For Health (when mastitis, lameness, metabolic disorders etc have been noted on the Wall Card or in the Farm Diary). For Production, use the milk recording sheets from different dates to check for deviations in kg milk, milk fat and milk protein, especially in early lactation (< 120 days); for acidosis, also check cows around 150 days of lactation.

Parameters

Save time by making a selection of named activities. Tailor-make for each farm according to its situation

Table 1.7 Parameters to monitor during a farm visit as part of a Diagnostic Herd Evaluation

Cows, calves, maiden heifers	Animal environment & management	Farm data
Body condition score, BCS	General hygiene on the farm	Milk recording (L; fat; protein)
Rumen Fill score, RF	Milking hygiene	Somatic cell counts
Faecal Consistency score, FC	Milking practice	Milk quality issues (Lab parameters)
Undigested Fibre in Faeces score, UFF	Barn climate (humidity; ventilation; draughts)	Quality of silages
Teat-End Callosity score, TEC	Housing (passageways; cubicles; feed bunk; feed table)	Quality of grass
Teat lesions (and mastitis)	Cow comfort (feed; climate; housing; behaviour)	Quality of pastures
Skin lesions	Manure scraping (system; frequency)	Quality of soil
Cows with poor leg posture	Claw trimming & foot bath	Quality of water (source)
Lame cows (Mobility/locomotion score, LS)	Ration composition (calculations; frequency)	Herd performance parameters
Hock lesions (depilation; abcess)	Concentrate dispenser (check dose)	Laboratory reports
Cow hygiene score, HYG (thighs/ udder)	Exercise areas	Sanitary herd book
Typical behaviour (agonistic/ antagonistic)	Roads to pasture (quality and risks)	Herd Disease-free Certificates
Growth pattern in calves	Management, hygiene & quality of colostrum	Milk urea (values and trends)
Diarrhoea in calves	Housing & climate for calves	Culling rate (% and reasons)
Respiratory disease in calves	Housing, stress for maiden heifers (par 1)	Good Dairy Farming code of practice
Results of rectal palpation	Quality of feed bunk and hygiene	
Clinical disease cases	Floor design and maintenance in houses	
Clinical indicator cases		
Condition	Indicator	
Milk fever	BCS at the end of lactation and in dry period too high; ration (composition) in dry period deviation.	
Ketosis	BCS in dry period too high + severe negative energy balance (NEB) around calving; ration; diseases.	
Rumen Acidosis	NEB; BCS before/after calving; ration (composition; fibres); poor rumen fill.	
Calf Diarrhoea	Birth history; hygiene; poor colostrum management; poor colostrum quality.	

Metritis scoring monitoring chart

After Laurent GOBY, Boehringer Ingelheim Animal Health, Germany; in Internatl Dairy Topics vol. 10, nr.4, 2011

Vaginal discharge score 0

Clear or no discharge

=HEALTHY

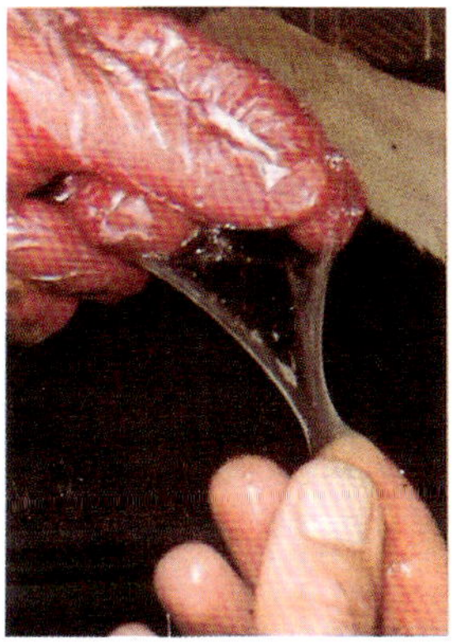

Vaginal discharge score 2

Less than 50% pus, bad smell with or without fever

=MILD METRITIS

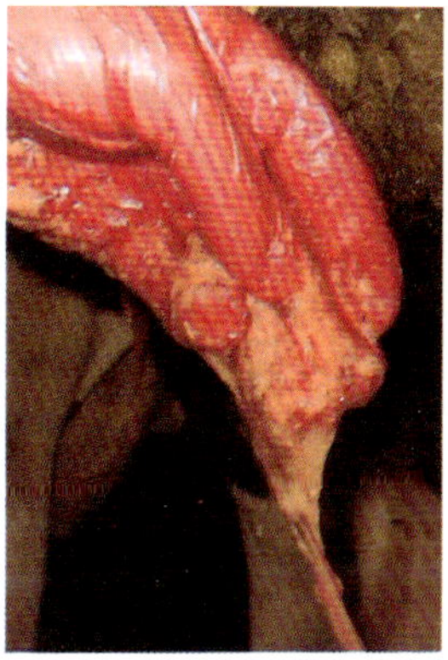

Vaginal discharge score 3

More than 50% pus, and bad smell r

=MILD METRITIS

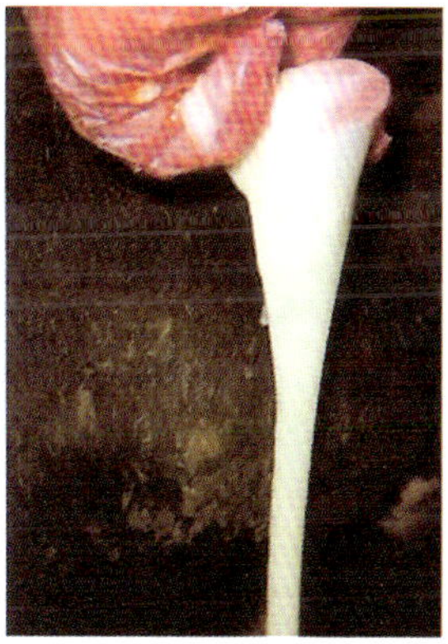

Vaginal discharge score 4

Red brown watery discharge, rotting flesh, putrified smell and fever

=SEVERE METRITIS

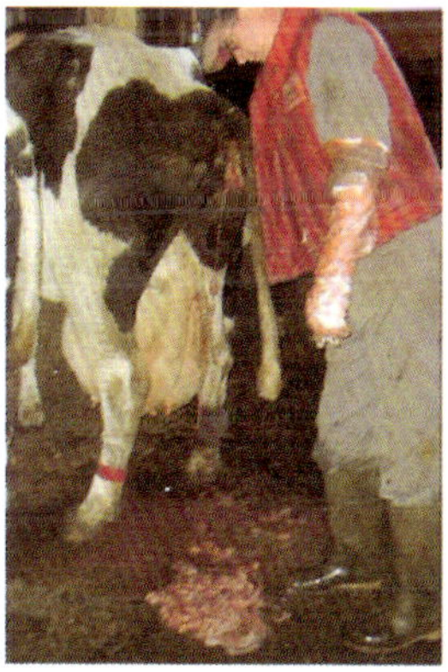

➲ intra-uterine treatment is indicated from score 2 onward, possibly paralleled by intramuscular treatment at scores 3 and 4.

➲ In accordance with other scoring methods, this scoring ranges from 0 to 4 (5 points scale).

Figure 1.3 Metritis scoring monitoring chart

Health monitoring

The compulsory periodical farm visit for assessing (the **prevalence** of) defined health disorders is a component of Dairy Quality Management.

Farm visits are made by the practitioner four times a year and farm visit reports are sent to the Dairy Company. If no special remarks regarding diseases are made, the farm will be designated a 'good animal health' farm. However, prevalence surveys are never representative for the true disease situation on a farm given the fact that there are always some diseases (e.g. of short duration, which do not appear in prevalence surveys) present.

Therefore, a programme for continuous animal health monitoring (CAHM) was designed to replace the quarterly prevalence surveys. Continuous Animal Health Monitoring focuses on incidences of disease.

CAHM comprises the following elements (together with the standards and score points):

CAHM is calculated each quarter on the basis of the four preceding average scores (rolling average).

Interpretaion of CAHM results

* ≥ 70 points Status A
  The animal health state is good.
* 60-69 points Status B
  The animal health situation is intermediate. This also applies when the last quarter score is less than 50 points. The farmer is advised to consult his practitioner to tackle prevailing problems.
* <60 points Status C
  The animal health situation is poor and the practitioner must be consulted to improve this situation.If the situation stays below a score of 60, the farmer must provide written project plans regarding control of disease on the farm.

Currently, there are 7000 participating dairy farmers in The Netherlands; the remainder are currently under the quarterly prevalence visit scheme.
In 2010, there were 97% of farmers with an average rolling score of ≥ 60 points.

The other 3% encountered problems with mastitis, somatic cell count, milk production losses and/or disease control programmes.

Table 1.9 Continuous animal health monitoring -together with the standards and score points

Cattle mortality	≤ 2.13% per quarter *A	25
Mortality in young stock	≤ 8.5% per quarter	10
New udder infections	≤12.5% per quarter	20
Bulk tank somatic cell count	≤ 300.000 cells/ml per quarter	15
Decrease in herd milk production	≤ 10% per quarter	10
Farm is closed	no cattle introduced in last 3 months	10
Farm certified or "unsuspected"	for BVD, IBR, lepto, paratbc, salmonella	10*B

*A= on small farms this percentage is set equal to 1 cow
*B= this is based on 2 score points per disease control programme as named

Field kit - basic utensils

A Basic information about the farm (geography; herd size; activities; information on technologies used etc.)
B A car that is clean both inside and outside
C Clean, disinfected boots; clean, washed work clothes
D Examination gloves + lubricant
E Documents; guides; instruction cards; lists; field scoring sheets etc. Writing paper with pencils (portable PC + printer?). The documents needed are dictated by whether you are conducting a Diagnostic Herd Evaluation, a Problem Analysis or a routine programme visit of HHPM
F Material with which to take samples (blood; milk; urine; faeces etc.), labels, waterproof pencils + laboratory documents
G Wax cattle markers of different colours to mark animals
H Tape measure (centimeters) to establish heart girth in calves
I Tape measure (centimeters) to establish cubicle size, silage clamps (can use laser equipment)
J Claw trimming knives + hard brushes for claw check-ups
K Disinfectant + bucket + clean towels
L Smoking powder for ventilation testing
M Two thermometers (< 42°C for animals and < 150 °C for hot water testing and silage)
N Crop analysis guide for feedstuff quality and ration composition
O Concave mirror to observe teat ends (for callosity)
P Usually, one or two boxes are sufficient to contain this material in the car.

The farm visit report

Each farm visit results in a written report to the farmer, the final version of which can be delivered immediately or sent later the same week. The veterinarian should always keep a copy of any report. The written report should not exceed the one page of A4 for reasons of readability and easy overview.

The report should comprise (next to the farm ID and date of visit) the following five chapters:

1. Findings

* Findings in cows (BCS, RF, FC, FF, TEC etc) with the notation + or – after each finding (+ = strong point; - = weak point)
* Findings during the inspection of housing, climate, feeding, etc and with the same + and – notations
* Findings on the evaluation of farm data (milk recording; dairy factory; etc) again with the same + and – notations

2. Conclusions

* A summary interpretation and synthesis of all findings (including mention of any detected problems)

3. Plan of Action

* activities/interventions for the short-term (no more than five to be listed to maintain motivation of the farmer)
* Activities for the mid-long term (which are considered as lower priority)

4. Treatments advised

5. New date/time for the next visit; calculation of effective time spent on the farm

It is often necessary to add an ANNEX to the farm visit report (usually this would be advice regarding any laboratory results, for example the bacteriology of milk samples and subsequent adjustment of the

Evaluation of the plan

Herd Treatment Advisory Plan, or a new composition for the ration). This ensures that the farmer receives immediate and appropriate detailed advice.

Due to the nature of the report (i.e. notations) it is strongly recommended that abbreviations and a 'telegramme style' be employed to keep the report within one page A4. This is the reason why many details will not actually appear in a farm visit report as it designed to provide an overall operational approach).

In contrast, a problem analysis report will be much more detailed. Examples of a farm visit report from the field can be found elsewhere in this series.

Action plan and the evaluation of the plan

Once the discussion and interpretation of the herd data have been carried out and the predominant risk factors evaluated, an action plan must be developed, as well as a procedure for evaluating the effects of this plan.

Action plan

The action plan must be discussed with the farmer at the end of the farm visit before a written report comprising this plan can be sent out two or three days after the visit. This report should not exceed 1 page if it relates to the reporting of a diagnostic herd evaluation and not exceed 5 pages if it relates to the report of a problem or data analysis.

During the following farm visits the effects of implementing the action plan will be evaluated and, if needed, the plan adjusted. Moreover, the level of compliance of the farmer can be checked.

The action plan must focus on the predominant disorders or problems and certainly not elaborate on everything that has been encountered. Once the substantial, important problems have been addressed, then the next issues can be dealt with.

The action plan has 2 phases: plans for the short term and those for the mid-long term:

-The action plan for the short term must comprise the advice/interventions concerning the weak points that are relevant and easily addressed e.g. installation of a foot bath, conducting a functional claw trimming session, increasing the frequency of manure scraping in the barn ...

The maximum number of actions should be 5 for this action plan.

- The action plan for the mid-long term can comprise more in-depth and strategic advice e.g. modification of the building, culling strategy, change to robotic milking.

It should always be accompanied by a prevention plan to avoid re-occurrence of the problem.

Evaluation of the action plan

This evaluation is paramount to the success of an action plan. It comprises:

- Discussion of the plan and its execution by the farmer
- Evaluation of the effects of treatments and the control of risk factors by evaluating the monthly parameters of herd performance
- Regular evaluation of the herd (see under Diagnostic Herd Evaluation)
- A calendar indicating the dates of subsequent farm visits for such evaluations.

Farm visit report

Typical farm visit for Farm A on xx/mm/yy

Positive points/areas

On entering, the farm premises are clean and well organised; bulk tank room and milking parlour are very clean.

Cows
Generally, a good BCS given the milk yield (BCS around 3); BCS at start of lactation is okay (±3, without too much variation). Rumen Fill score is okay in all lactation stages (2 to 3); good scores for Faecal Consistency and Faeces fibres (2-3).

Score of hygiene & cleanness is good (scores 1-2). Teat-end scores are good (1-2).

Mobility/locomotion score (static & dynamic) is ±acceptable; not for # 678 and #955.

There are some hock lesions (4 on 20 cows); so be careful → recheck next visit.

Environment
The Barn is a good size for the number of cows (see reference standards) and is well lit (> 200 lux). Barn climate and cow comfort is good, temperature is somewhat high (ventilation). Cubicles with straw allow sufficient room for head movements

Distribution of feed to cows is okay; concentrate dispenser dose is well gauged.

Farmer is to the point and ambitious.

Farm data
Level of production (L) and milk fat & protein are okay (milk recording)

Bulk milk somatic cell counts were historically okay (but problem in March)

Calving peak is between May and December, so today (8 sep) most of the cows are at the end of lactation. There are about 42% of first lactation cows (= many!) → check culling policy of farmer.

Few mastitis cases (5 on 70 cows in July).

Milk quality parameters do not show any deviation in the past year.

Points for improvement:

Cows
BCS after 50 days: too many thin cows; end lactation okay; BCS in dry cows is too high. Rumen Fill score in dry cows is far from the ideal. Level of rumination in the herd (2 tests) was not more than 50% of the cows (norm is >70%); causes are stress? temperatures? disease?

Too many cows (± 50%) show poor leg posture; possibly claw lesions (Mobility/locomotion score may be slowly shifting to the undesirable side). Some cows have long claws on the front feet; this hampers an appropriate feed intake.

Cow # 810 shows endometritis at 18 days after calving → needs follow up.

Environment/management
Too few drinking places (3 times 2x30 cm)= 180 cm (norm is 45 cm width per 10 cows, so 315 cm in total for 70 cows ; in heat periods 2 times more).

Farm data
Some cows are suspected of having rumen acidosis given the milk recording results of April–May and June–July. These cows must be checked at next recording date. In April several cows showed a substantial drop in milk yield.

In November 2008 there has already been a drop in milk yield → precaution → put to the barn earlier in the season? Manage a smooth transition from pasture to barn.

There is a group of cows with high SCC; they are a risk for their herd-mates with respect to mastitis. How

Farm visit report

to handle them best within the herd? We need to know their bacteriological profile for better treatment options (so milk samples needed).

The yearly culling rate is rather high (38% versus norm of <30%); main reasons are fertility and lameness. Calving interval is 420 days (distribution to be checked); 30% of cows had more than 3 services (20% primiparous, only 7% in maiden heifers).

Discussion and conclusions

- Fertility problem in the herd, needs to be explored in more detail. We may consider an analysis of heat detection after calving and at 3–6–9 weeks after AI. Also, a check on prevalence of endometritis and cysts is needed.
- Subclinical claw lesions are present, but diagnosis is not sufficiently reliable. So we need a sample of 20% of the herd to be trimmed for diagnosis, either during routine claw trimming (inventory of primary diagnoses), or during a specific farm visit.
- High SCC cows are an issue and may pose a risk to other cows.
- BCS scores need to be better controlled at the end of lactation and in the dry period to avoid problems in reproduction and production in the next lactation.
- Adjust the number of drinking places; add new ones if necessary; high yielding cows consume more than 120 L water per day.

Advice for the short-term

- Continue routine claw trimming once or twice per year (do not neglect front claws!); cows that show poor mobility/locomotion score or poor leg posture must be checked. A photo-diagnosis instruction card for the farmer may be of great help for properly diagnosing primary lesions.
- It would be highly desirable for the farmer to write down all claw lesions that he diagnoses. This is the only way of having a rapid insight into the development of claw lesions. Proper diagnoses are needed to tailor a plan of action (risk factors are different for each diagnosis). At the same time, manure scraping must be more frequent (4 to 6 times per day) because the floor must be kept dry and clean.
- Adjust the rations at the end of lactation and in the dry period to better control BCS (at 3.5 instead of 4) and subsequent problems (poor feed intake; poor transition; poor ovulation rate).
- Add drinking places up to a total width of at least 350 cm (double the width in summer).

Advice for the mid-long term

- Analyse in detail the fertility problems (intervals which deviate from normal values, cows with poor performance, cysts, endometritis, etc); action plans will then follow.
- Analyse the cows with high SCC to better understand this particular problem. A bacteriological profile is indispensable; milk samples must be taken. The samples can be stored in the freezer, labeled with the individual cow ID, date and quarter sampled. The samples can be sent to the lab in one batch. On the basis of this profile the Herd Treatment Advisory Plan can be designed (see elsewhere in this series). Other measures include milking high SCC cows last, separating high SCC cows from other cows and consideration given to culling of chronically high SCC cows.

This is an expanded report; "expanded" for reasons of clarity and readability.

NOTE: This report is provisional and can be adapted after the results of the next visit dated -- -- --..

Performance report

Table 1.8 Example of a monthly herd performance report in HPPM

Year= 2008 - 2009					Farm Code=			
Date of visit (dd-mm)	31-10	30-11	24-12	31-01	28-02	....	....	Ref value
# cows present Lact / Dry	76/19	69/16	68/14	81/9	79/18			17% dry
L milk per day	30	28	28	30	32			
Mean milk fat %	41	39	40	41	41			
Mean milk protein%	34	30	33	33	34			
Ratio milk fat to protein	1.20	1.0	1.21	1.24	1.21			
% cows with fat < 3.00	1	2	4	3	5			
% cows with protein < 3.00	3	2	4	8	11			
Urea (bulk tank)	-	-	-	-	-			
Somatic cell count (SCC) Max	398	411	453	822	612			< 250
SCC Min	78	88	92	81	67			< 100
% new cows SCC > 250	6	9	12	19	9			< 10%
% cows with SCC > 250	8	11	13	21	22			< 15%
Bacteria Count in milk	7	26	43	6	13			< 50
# Clinical mastitis cases	3	4	2	5	2			< 25%
# Abortions	0	1	0	0	0			< 4%
# Retained placenta	0	1	1	2	0			< 10%
# Metritis	1	2	0	1	3			< 10%
# Abomasal displacement	0	0	0	0	2			< 5%
# Ketosis	0	0	0	1	2			< 5%
# Acidosis	1	0	0	0	1			< 5%
# Lameness cases	0	3	2	4	3			< 25%
# Stillborn calves	0	0	2	2	1			< 5%
# Calves dead > 3 days age	0	0	1	1	1			< 5%
# Calves with diarrhoea	0	0	1	3	4			< 10%
# Calves with respiratory problems	0	0	0	1	0			< 10%
# Cows culled	0	4	1	2	1			< 25%
Reason: Production			1		1			< 22%
Reproduction		2						< 30%

Performance report

Date of visit (dd-mm)	31-10	30-11	24-12	31-01	28-02	....	....	Ref value
Udder/teats		1		1				< 15%
Claws/legs				1				< 10%
Other		1						< 23%
% pregnant per AI	54	50	46	51	53			> 50%
# AI/cow inseminated	1.6	1.8	1.7	1.6	1.8			< 1.7
Calving interval (actual)	378	369	379	381	383			365-385
Calving interval (expected)	381	379	389	391	392			365-385
% anoestrus on number of cows examined	10	12	17	21	23			< 10%

→ Remarks are written on a separate page.
→ Once per 6 and per 12 month, an evaluation or progress report must be delivered.

Interpretation

There are more cows in lactation in January and February (due to new calvings).

The 17% of dry cows is applicable for calving intervals of 365 days and a dry period length of 60 days. If the dry period is shorter and/or the calving interval is shorter, then this rate can increase.

The mean production per cow per day increases around February due to new cows coming into lactation?

The mean milk fat% decreases in November (transition from pasture to housing poorly organized?) but figures are not too bad . The same comment is valid for milk protein%, with the exception of November where the fat-to-protein ratio is zero Indicating an issue at this point.

Around December—January—February the number of cows with a low milk protein increases → sign of energy shortage in the ration (ketosis?) à is to be kept under surveillance or ration adjusted.

There are large differences between cows regarding their individual SCC. This is also true of the number of mastitis cases; this indicates an issue of concern in the herd.

Cows with a new SCC > 250 and the total SCC indicate an existing udder health problem that will likely explode.

The milk quality (microbiology) is always below the threshold of 50.000/ml but there are time points of increased risk: November and December. Should be kept under surveillance ←

Abortions: no problem

Retained placenta: no problem

Metritis: this figure increases; needs monitoring ←

Abomasal displacement: not a true problem

Ketosis: not a true problem

Acidosis: no problem

Lameness: is always present in the herd - diagnosis? Needs monitoring ←

Stillborn calves: five in three months is too many - needs monitoring ←

Dead calves at > 3 days age: three in three months is too many – needs monitoring ←

Calves with diarrhea: continues to increase - needs monitoring ←

Calves with respiratory problems: none

Culled cows: the total percentage is relatively low; culling strategy should be explored; they maybe culling too few cows.

Reproduction:

The pregnancy rate is acceptable, as is the number of AI per cow. The intervals do not show a problem in either the actual or the expected values, although there is a slight increase in recent months. The rate of cows examined for "not seen in heat" increases – needs monitoring ←

➔ An Action Plan must now developed with measures/interventions for the short- (maximum of five items) and mid-long term!!

Protocol report on the analysis of farm data

Each farm will have several documents that have a certain interest for the veterinarian in an HHPM programme (either as a fully integrated HHPM, routine farm diagnostic evaluation visits or stand-alone problem analysis).

It is essential that veterinarians are able to (1) read these farm documents properly; (2) correctly interpret the results and (3) create a synthesis of the information obtained from these different farm documents.

> ***Farm documents refer to parameters, such as milk recording (monthly sheets; year reports), reports on milk quality parameters from the dairy processor and reports on the reproductive performance of the herd.***

The general approach to data analysis comprises three levels :

1 Evaluation of the average herd situation (e.g. 8000 L milk/cow/yr; or a mean calving interval of 401 days)

2 Evaluation of performance parameters for a group of cows (e.g. in early lactation, or dry cows)

3 Evaluation of performance of individual cows.

All three levels should be addressed in the data analysis. Preferably, each evaluation should be carried out while comparing figures from the current date with those from previous dates, as well as with those from large populations or the region. The latter two comparisons serve as reference values to highlight positive or negative trends. See also the sheet on Reference Values in this series.

Example for the year report on milk production:

Production at herd level (means):

- 7500 L milk, milk fat % 3.9, milk protein % 3.10, SCC 252.000/ml
- Stage of lactation 5th month (150 days)

Production of primiparous cows:

- 28 L milk/cow/day, milk fat % 3.60, milk protein % 2.70, SCC 112.000/ml

Production of parities 2 + 3:

- 33 L milk/cow/day, milk fat % 3.70, milk protein % 3.00, SCC 234.000/ml

Production of parities 4+

- 38 L milk, milk fat % 3.90, milk protein % 3.20, SCC 321.000/ml

Cow 122 Par 1 24 L milk milk fat% 2.80
milk protein% 2.60 SCC 655 88 stage of lactation

Cow 122 should be highlighted or examination during the next visit given her poor milk yield. See also other examples in this series of farm data.

The milk recording forms

The milk production data should be evaluated from the most recent date. Check that this date is close enough to the date of the farm visit.

Secondly, evaluate the milk recording forms from the last two or three dates in parallel in order to detect trends in productivity: L of milk, milk fat and milk protein, as well as somatic cell count values

Remind yourself of the shape of a normal lactation curve - this facilitates the interpretation (peak milk yield; persistency; level etc).

Think about the possible reasons underlying detected trends or problems (feed related? due to disease?)

Protocol

Regarding the herd fertility status

Traditionally, reproductive performance parameters are calculated as 'means'. Statistically speaking, however, this is not correct due to the skewed distribution of observations. For example, the different intervals measured from calving to xxx (e.g. first heat, first AI, conception) are not normally distributed. In these cases, the 'median' should be calculated as opposed to the mean. For evaluating reproductive performance, the box-plot method (box-and-whiskers) can be used to correct present the data. For a box-plot, the median is first calculated followed by the 25% of observations both left and right of the median. Finally the 'tails' left and right are determined (Figure 1.4).

In Figure 1.4, the median is at 408 days and the box is represented by 377–408 days (left of the median) and by 408–466 (right of the median). Note that the two boxes do not necessarily have to be of the same volume!! The 'tails' are 352–377 (left) and 466–568 (right). As can be expected in cattle reproduction, the right tail is much longer than the left one (a few repeat breeder cows with extremely long calving interval).

Hence, we have now five parameters of information instead of only one mean calving interval and improvements can be detected in any of these five parameters (e.g. smaller right tail; smaller box; lower median).

In the above example, it is clear that, although the median is acceptable, the right box is large and the right tail is very long, indicating a reproductive problem in the herd.

Remarks:

- If you are unable to follow the correct method of evaluation, or if you do not succeed in finding the right answers to the questions, it is suggested that you consult colleagues (veterinarians) or other specialists (e.g. nutritionists). This is also an important method of furthering you knowledge.
- Logical, clear and sound reasoning is essential for herd performance evaluation and analysis.

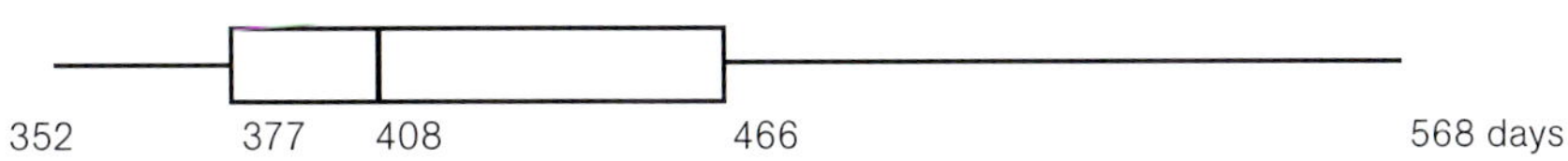

Figure 1.4 Box-plot of calving interval data

Analysis & reporting of farm data

Every farm has different documents available that are of certain interest to the veterinarian conducting a HHPM programme (either in the format of audits, routine inspections or integrated programmes).

It is paramount that the veterinarian is able to (1) read these documents; (2) correctly interpret the information within them and (3) concisely discuss the implications of information.

General farm documents include:

1. Milk recording forms (sheets per date of recording; technical reviews; year reports).
2. Documents from the dairy processing plant comprising milk quality data (milk fat and protein; cell counts; quality deviations; etc.)
3. Reproductive performance reports

The general approach to analysing farm data comprises 3 levels:

1. Evaluation of the average situation of the herd performance (for example: 8500 kg milk per cow per year; mean calving interval of 401 days)
2. Evaluation of the performance at group level (for example the cows between 0 and 100 days lactation, 100 to 200 days, 200 to 300 days, dry cows; % mastitis in early lactation < 100 days)
3. Evaluation of certain individual cows

These 3 levels must be addressed when analysing herd data. Ideally, each evaluation should be carried out by comparing data from one date to the following date using population reference values or regional farmers' group references. The comparison over time allows for the detection of trends.

Example of a review of milk production:

Herd production means:

- 8500 L milk, milk fat% 3.90; milk protein% 3.10; cell count 252.000/ml; herd lactation stage 5 month= 150 days

Production of first lactation animals:

- 29 L/animal/day; milk fat% 3.70; milk protein% 2.70; cell count 112.000/ml

Production of parities 2 & 3:

- 33 L milk/animal/day; milk fat% 3.70; milk protein% 3.00; cell count 234.000/ml

Production of parity 4 and higher:

- 39 L milk/animal/day; milk fat% 3.91; milk protein% 3.22; cell count 321.000/ml

Cow # 122

- Parity 1
- 24 L milk/day, milk fat% 2.80; milk protein% 2.60, cell count 688.000/ml, 88 days in milk. Is this cow in negative energy balance/ketosis? Does she have a mastitis? This cow would be one to examine during the next farm visit. See also other farm and HHPM sheets in this series.

The milk recording forms

Analysis and evaluation start with assessing the milk recording forms from the most recent recording date. Check whether the date is suitably close to current farm visit date.

Next, the milk recording forms of the preceding two and three dates should be compared with the most recent one to detect changes or issues with performance in L milk, milk fat%, milk protein% and cell counts. Both positive and negative trends should be noted. Also, bear in mind the shape of the lactation curve, with peak milk date, persistency and overall production level. Think about possible reasons for any deviation from the expected.

Subsequently, the potential risk factors on the farm contributing to these deviations (diseases; feed changes; stress; genetics) should be explored. It is

Analysis & reporting of farm data

here that knowledge and skills from different farming situations should be used to develop a discussion and find a solution.

The reproductive performance review:

'Means' are most commonly used to express several performance figures (mean calving interval; mean interval from calving to conception; etc.). Statistically speaking, the calculation of such means is incorrect because the data are not normally distributed!

In skewed distributions the "median" should be calculated, that is the observation that lies exactly in the middle between the two extremes. For these situations it is highly practical to construct a box-and-whisker-plot. This plot comprises the median, a box with 25% of the observations lying to the left side of the median (408), a box with 25% of the observations lying to the right side of the median and a 'tail' on both the left and right side, respectively (see figure 1.5 below).

In this way, five parameters are available instead of only one (mean). Improvements may occur in either of these five while in the mean may not change at all! For example, when the farmer culls the cows with longest calving intervals, the right side tail will shorten. Problems in reproductive performance can be detected more easily.

After this exercise, data at the group and individual levels should be analysed.

The report on the analysis or evaluation of data has several parts:

1. Details of farmer and farm inc. address + short farm description (if initial report)
2. The name of the documents analysed (milk recording; milk quality; ect), each of them in a different paragraph.
3. Abstract, discussion of findings (strong and weak areas)
4. Conclusions
5. Intervention plan and advice (both short and mid-long term)

➲ This report should not exceed 5 pages otherwise the farmer will not read it properly!

Remarks

If you are not able to carry out a correct and complete evaluation, or if you cannot find the appropriate answers to your questions, it is best to consult other specialists (e.g. a nutritionist). A logical and sound reasoning is indispensable for proper HHPM execution.

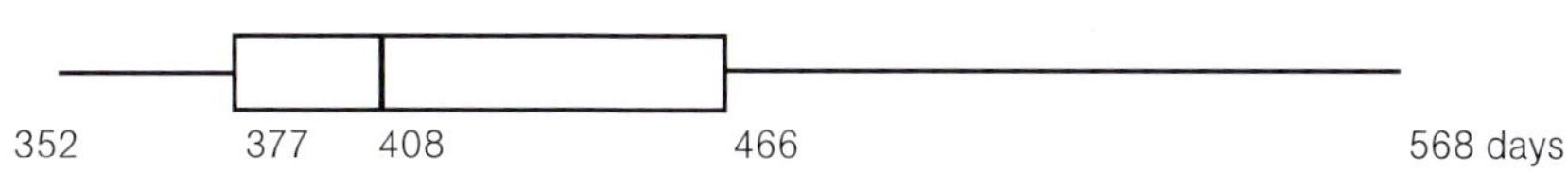

Figure 1.5 Box-plot of calving interval data

Editing a farm visit & analysis report

The editing of a report, irrespective of its contents, is essential. After all, it is the only tangible remainder of the collaborative work with the farmer. The report will allow the farmer to focus on the advice given and allow the veterinarian to follow up the development of the dairy farm.

I. Farm visit report:

During the farm visit, several issues will be discussed with the farmer verbally. Despite this, each visit must be closed with a written report, delivered to the farmer within one week (otherwise it loses its impact).

Indeed, verbal advice is not always adopted by the farmer (memory bias; other priorities). The written report better facilitates adoption of the advice and suggestions by the farmer and maintains his motivation. Moreover, the farmer will be aware that the veterinarian will look for related feedback during the next farm visit.

The farm visit report must be concise and brief (telegram style) and preferably not exceed one page of A4. In order to adhere to this, long sentences must be avoided; rather one should use comprehensible key words.

The report should not comprise general farm data (they are already available on the farm!), nor farm descriptions.

A written report comprises, next to general issues like farm code, veterinarian code and date of visit, several chapters (which can be personalised):

- Findings from the Diagnostic Herd Evaluation on animals (BCS, RF, FC etc.), farm data (milk recording data, etc.) and animal environment & management (hygiene, housing, feeding management, ect.). These findings are to be divided into two categories: strong points and (weak) points for improvement. The highlighting of strong points is just as important highlighting points for improvement as it maintains farmer motivation. Preferably, the findings should be made in an objective manner, for example *"20% of cows with Rumen Fill (RF) score of <2"*; or *"40% of first lactation cows are showing poor leg posture"*. This facilitates the comparison between two subsequent dates (trends).
- Conclusions from findings. This regards the collation of all the information gathered before and during the farm visit; it sets the probability herd diagnosis and prepares the next step. In the report, only the most important conclusions are recorded. Nevertheless, if other issues, which seem minor at that time, remain on the field scoring list, they can, at any time, be made available. Usually, two or three main conclusions are used on which to base the primary actions.
- Action Plan. Based on the conclusions, the interventions and the advice are detailed in the action plan, distinguished by 'short-term' (priority) and 'mid-long term' advice. In general, people cannot manage more than five actions at any one time. Therefore, no more than five priority measures should be noted. If two herd problems were detected, one has to decide which is the more important (influenced by farmer goals and priorities). It can be suggested that they are addressed one by one as opposed to simultaneously to make it more feasible for the farmer to carry out related interventions. The farmer must be kept informed about what is going on, why and how to do it. Mid-long term advice may deal with issues of minor importance; it may also concern more strategic issues, which need ample time for reflection.
- Issues to consider the next farm visit(s). If it is known that there are several issues requiring attention during the next farm visit (for example originating from the farm calendar; see elsewhere in this series), it is best to prepare

Editing a farm visit & analysis report

the farmer in advance for these issues. This does not apply to our activities in the context of the Diagnostic Herd Evaluation, which is more routine. An example is the heart girth measurements of maiden heifers and calves, which are routinely carried out twice a year and may be forgotten all too easily. Moreover, this helps in strengthening the confidence, trust and cooperation between veterinarian and farmer: the farmer sees that the veterinarian aims to take care of his enterprise.

Sometimes it is necessary to add an Annex to the farm visit report (eg. a newly calculated ration; work instruction for using a foot bath properly etc.). This Annex must be on a separate sheet of paper so it does not disturb the routine reporting procedure.

Finally, the report concludes with details of the effective time spent on the farm during the visit (this helps in the adoption by the farmer of all the time spent in the context of a HHM programme and the way the bill is composed).

II. Analysis Report:

The analysis report from the farm data differs from the routine farm visit report. This report is more occasional and may appear, for example, twice a year or according to the specific problem.

Hence, it is more complete, sometimes more complex and more detailed. Photos may even be included to highlight an issue. This report should not exceed 5 pages of A4 and comprises several chapters.

1 **Description of the issue**
It is necessary to have a brief description of the farm (e.g. if becomes necessary to involve other specialists later on). This description provides the background to the problem or technical report. Long sentences must be avoided. When a problem persists, this section contains the arguments for defining that problem.

2 **Analysis of documents**
Each document must be analysed separately and part-conclusions given under the heading of the area the document analysed relates to. For example, the analysis of the yearly reproductive performance will appear under the heading of 'Reproduction'. Each analysis should be distinguished on three levels: herd level; group level; individual level.

3 **Abstract, synthesis**
Like in the farm visit report, the analysed data are synthesised into strong points and (weak) points for improvement.

4 **Conclusions**
As in the farm visit report, conclusions are drawn with an emphasis on the most important. The number of conclusions must be limited in order to avoid confusion for farmer. Remaining points from the analysis must be detailed separately for veterinarian's use in order to action them when needed.

5 **Action Plan**
This is fully comparable with the farm visit reports. For each intervention presented or advice given, identify means of evaluation, as well as a period of time during which the evaluation should take place. Additionally, a prognosis should be given for each management activity intervention. Evaluations are always carried out during subsequent visits; hence these visits have to be planned in advance. Usually, an evaluation criterion is the execution of the routine Diagnostic Herd Evaluation and subsequent interpretation of the results.

The report is not to be left with the farmer with no means of follow-up. The farmer must be aware that there is a follow-up and that he has access to the veterinarian's experience and assistance.

Assessing herd & farm performance

Table 1.19 Reference values for assessing herd & farm performance

	At calving	Early lactation	Mid-lactation	End of lactation	Dry off
Body Condition Score	3.5 to 3.0	2.5 to 2.0	2.5 to 3.0	3.0 to 3.5	3.0 to 3.5
Rumen Fill Score	2	3	3.5	4	4 to 5
Faecal Consistency Score	2 to 3	2 to 3	3	3 to 4	4
Faeces Fibrosity Score	1 to 2	1 to 2	1 to 2	1 to 2	1 to 2

Mobility/ locomotion score	> 85%	Scores 1+2
	< 10%	Scores 3
	< 3%	Scores 4
	< 2%	Scores 5
Hock lesions	< 15%	
Tripling cows	< 15%	
Animal hygiene scores	> 85% (clean)	Score 1
	< 10%	Score 3
	< 5% (dirty)	Score 5
Ketosis	< 5%	
Rumen acidosis	< 2%	
SARA	< 5%	
Milk fever	< 4%	
Digestive disorders	< 5%	
Rumination frequency of the herd	> 85% (good) <65 (poor)	

PLAN OF ACTION
For monitoring actual herd performance

For the short term (priorities; max 5 items)

For the mid-long term

Reference values

Table 1.10 Reference values for HHPM

Parameter	Value
Calving interval:	< 400 days (around 385 days)
Interval calving to 1st oestrus	< 35 days
Interval calving to 1st AI	50 to 70 days
Interval calving to last AI	< 100 days
Interval calving to conception	< 85 days
Interval 1st AI to conception	< 12 days
% of heats detected between 50 and 70 days after calving:	> 85%
% pregnant after 1st AI	> 50%
% pregnant after all AI	> 60%
Age of heifers at 1st calving	24–25 month
% of abortions:	< 4%
% retained placenta:	< 12%
% of cows in suboestrus	< 15%
% of ovarian cysts disease before 70 days after calving	< 5%
% of endometritis cases before 70 days	< 15%
Drops in milk production between 2 recording dates	< 10%
Average milk fat %:	> 3.5
Average milk protein %:	> 3.0
Ratio milk fat to milk protein:	> 1.2 and <1.5
High ratio → ketosis	> 1.5
Low ratio → acidosis	< 1.2 (or 1.0)
Somatic cell count of bulk tank milk:	< 200.000/ml
Total bacteria count of bulk tank milk:	< 10.000/ml
Milk urea nomal level	250 - 350 mg/L
High level means excess of nitrogen (protein) or lack of energy in the ration	> 350
Low level means lack of nitrogen (protein) in the ration	< 250

KEY PARAMETERS

Stress causes an increase in non-esterified fatty acids (NEFA > 0.4 mMol/L) that contributes to abomasal displacement. Serum betahydroxybutyrate (> 1200 ùMol/L) is also implicated.

Indoor temperature in the barn should be between 5 and 15°C as is the thermoneutral zone.

Relative Humidity in the barn must be < 80%.

Luminosity in the barn must be > 200 Lux (you should be able to easily read a newspaper).

Drinking troughs must be clean at all times. Total length must be 600 cm for 100 cows in winter; in summer time this total length must be 900 cm for 100 cows.

Manure scraping on concrete floors and slatted floors: 4 to 6 times per day.

Exercise areas must be as clean and as dry as possible.

Barn walls and equipment must show no signs of fungi or condensation.

When applying smoke powder to assess ventilation in the barn, it should disperse within three minutes during winter and within four minutes during summer. Repeat the test over several farm visits.

Herd claw trimming must be done at least twice yearly by a well-trained claw trimmer. An alternative is to claw trim all cows at drying off but this must be done by a professional.

If a footbath is used, cows should be run through it on three consecutive days per month. Correct footbath size is 300 cm in length, 100 to 150 cm wide and 15 cm high. A pre-cleaning water footbath of the same size can be placed in front of the treatment footbath. Do not use formalin in a footbath prior to animals entering the milking parlour!

Biosecurity

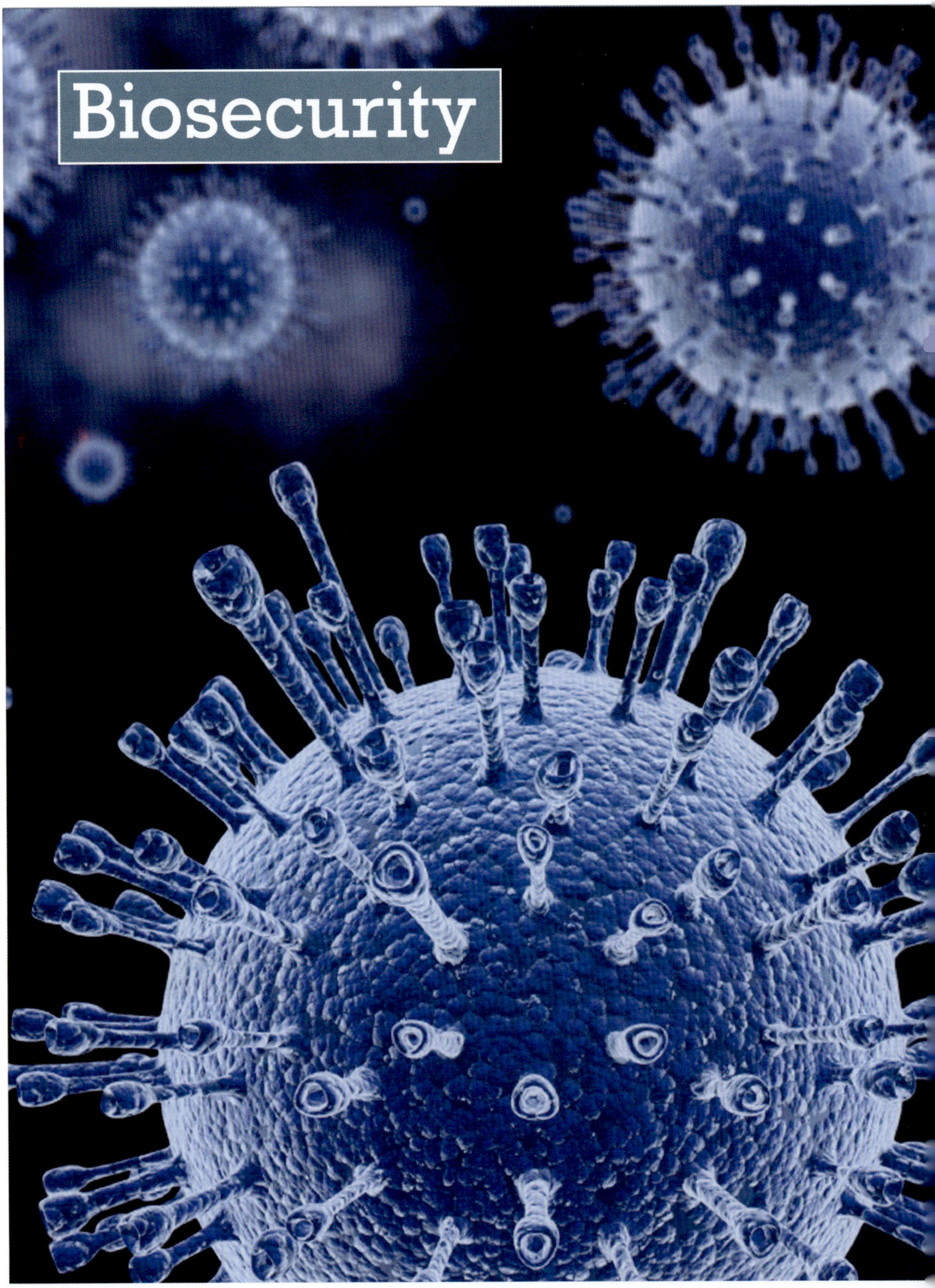

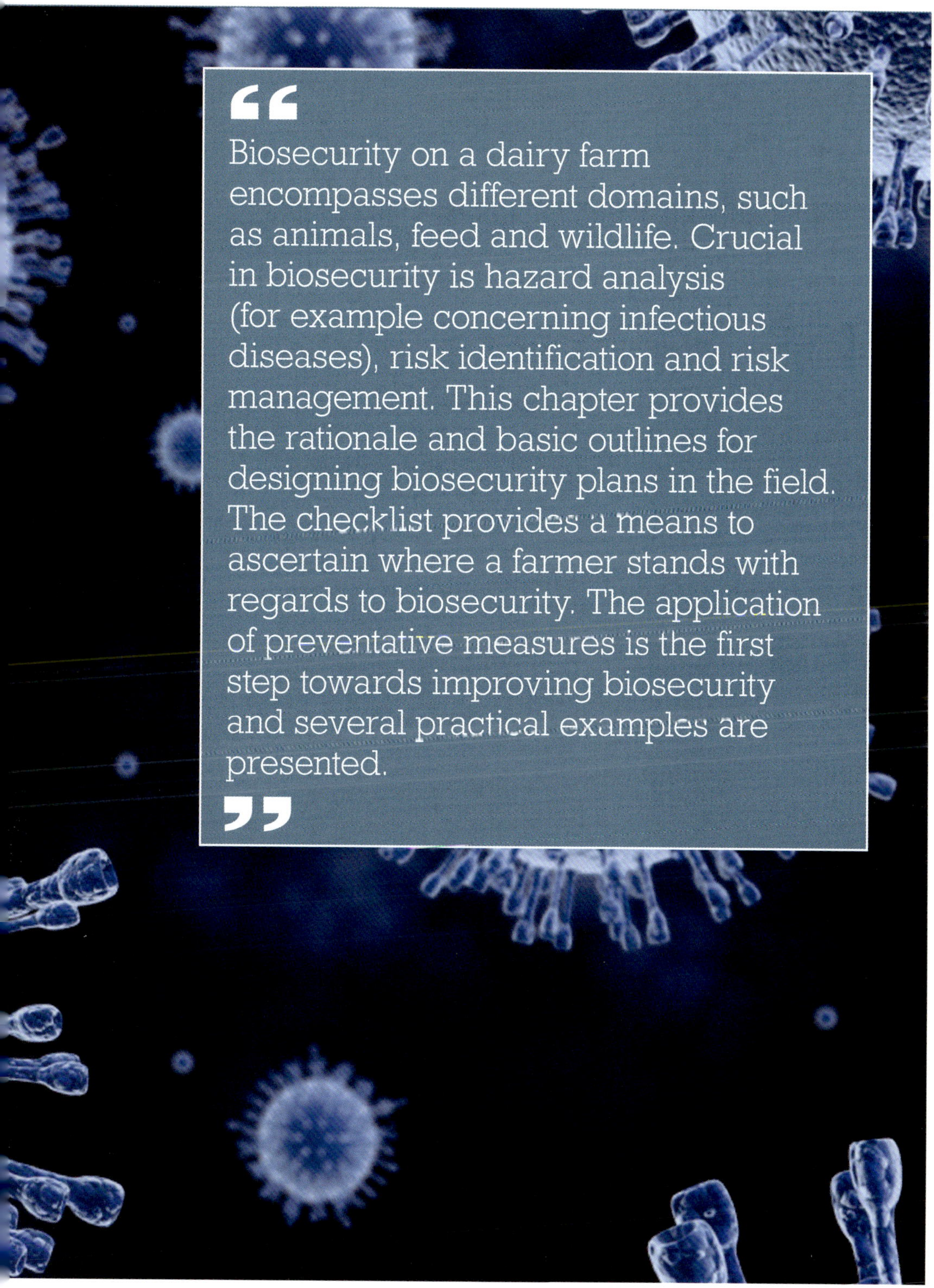

"Biosecurity on a dairy farm encompasses different domains, such as animals, feed and wildlife. Crucial in biosecurity is hazard analysis (for example concerning infectious diseases), risk identification and risk management. This chapter provides the rationale and basic outlines for designing biosecurity plans in the field. The checklist provides a means to ascertain where a farmer stands with regards to biosecurity. The application of preventative measures is the first step towards improving biosecurity and several practical examples are presented."

Biosecurity on farms

Biosecurity = a programme to reduce or prevent the introduction from external sources of infectious diseases, as well as the spread of such diseases once they have entered the farm.

Introduction

Infectious diseases can have a great impact on the economic performance of a farm and may also induce stress in the farmer. Examples of such diseases are: salmonellosis, BVD, IBR, *brucellosis, tuberculosis, paratuberculosis* (Johne's disease). As well as animal and farmer welfare, these diseases have wider implications for the overall standing and image of the dairy industry, including damage to public image, loss of market position and decreased slaughter value. Moreover, in the EU the farmers are held responsible for the products (milk; meat) they put on the market, including the safety of those products. Hence, they are held responsible for any sort contamination (microbiological; chemical etc).

The production process influences the health and welfare of the cattle in the herd, as well as public health. After all, the farmers represent the first link in the dairy food chain! The negative effects of infectious diseases can have implications for both the short- and long–term and depend on, for example, the type of disease/pathogen ,the level of commitment by farmers to address these diseases/pathogens and the concerns of consumers.

Given the important consequences of infectious diseases, the development and implementation of biosecurity programmes should be a high priority among farmers.

Vaccines and antibiotics have been the number 1 choice to prevent, control or reduce the incidence of infectious diseases. However, for certain diseases, they are unavailable or are not sufficiently effective; sometimes, if available, they are poorly administered. In other situations there is limited option for using vaccines (e.g. the EU).

For poorly administered vaccines, one can develop technical working instructions (a kind of 'best practice') including the correct procedure for administering medicinal drugs, as well as correct stock handling, dosage, withdrawal time, etc. Such working instructions must be strictly complied with. Even instructions for the correct use of antibiotics must be strictly adhered to in order to prevent contamination of milk or meat and to minimise resistance development and residues. It is important to remember that vaccines and antibiotics do not replace good biosecurity, even when required for reducing the prevalence of a certain disease (eg. BVD; IBR).

Components of a biosecurity programme

The components of a biosecurity programme are management instruments (Good Dairy Farming Code of Practice). Some of the areas of concern and relevance for veterinarians are listed in Table 2.1

How to design a biosecurity programme

There are five steps to take when designing a biosecurity programme for use on farm:

1 Conduct a written, critical inventory of infectious diseases that are relevant to the farm. Take into account the geography, soil type, housing, animals, people and visitors, transportation means.
2 With the farmer, identify the most important infectious diseases already present in the herd.
3 Identify the most important infectious diseases that are not yet present on the farm but present a threat.
4 Conduct a **Diagnostic Herd Evaluation** (animals, environment, data) to determine the level of risk for the transmission of selected disease pathogens. Formulate goals for the biosecurity programme : for example a 5% reduction in the prevailing clinical mastitis incidence per year; or, the elimination of IBR from the dairy herd in 3 years.
5 Illustrate the biosecurity programme and implement it. Work together with the other people working on/for the farm, including professional consultants to try to ensure compliance. The programme must be updated at least annually.

Areas of concern

Adapted after the text issued by AFIA – BAMN publications, Dorann Towery, 1501 Wilson Boulevard, Suite 1100, Arlington, Virginia 22209, USA (2001).

Table 2.1 Areas of concern & relevance to veterinarians involved in a dairy farm biosecurity programme

New additions	**Cattle, semen, embryos** Note that e.g. Neospora and BVD can be introduced by healthy animals. Moreover, healthy animals can introduce diseases with a long incubatuon period (paratuberculosis).	Check all animals. Test for relevant diseases. Sample milk for bacteriological testing. Vaccinate twice before transportation. Quarantine the animal for 3 weeks before mixing with the herd. Buy semen or embryos or sires from certified traders or with active disease control programmes
Forages and concentrates; water	**Concentrates** **Salmonella spp.** can be found in feedstuffs and in pasture. **Forages** **Salmonella spp.** can be found in forages irrigated with contaminated water. Incorrect harvest or feed stocking may introduce clostridium bacteria. **Water** Water sources can be contaminated and introduce **E. coli** or **Salmonella spp.**, as well as **Cryptosporidium parvum**.	Test water for bacteriological, chemical and nutritive contamination. Ask cattle and feed traders to show the quality assurance for their feeding programme, their stock and their delivery. Prevent faecal and urinary contamination of feedstuffs and water.
Contact between animals	Fences, shows and expositions, errant animals, putting sick animals in the barn, calving pen. Contact between groups of different ages.	Minimise contact between different animal groups. Consider cattle returning from shows or markets as new additions (see above). Minimise contact with non-resident cattle.
Wildlife and other vectors	**Squirrels, rats, mice, foxes** **Salmonella, Brucella, Leptospires.** **Insects** Anaplasmose, Blue Tongue **Rats, mice** **Salmonella, E.coli**	Prevent contact with wildlife. Use pesticides and traps close to feed. Exercise control measures for insects and birds, on and around animals.
Animal health management	**Procedures** -Comply to the Good Medicine Application Code of Practice Note that practices like de-horning, vaccination and implanting could cause disease transmission.	Use disposable utensils. Disinfect other utensils between use on different animals. Use vaccines according to prescription.
Noxae	**Vehicles, persons**	Wash trucks and vehicles after use. Dedicate a special site for dead cattle. Provide strict hygiene instructions for visitors.

Biosecurity on farms

A biosecurity programme is a strategic management tool for the prevention and or reduction and/or dispersion of infectious diseases in the herd. After implementation, the economic benefits and income of the farm will likely improve as programme costs are low.

What are the options to limit the entrance or dispersion of infectious diseases?

The most simple measure is to limit the contact between animals, between animals and people and the movements of animals/people. The greatest risk comes from animals bought in. Not buying in replacements means the farm becomes a 'closed farm' (see elsewhere in the series) and disease risk is reduced. However, this is not always feasible or desired. In all cases, one should realise that each new entrant onto the farm represents a risk of disease introduction.

There are two approaches to minimising this risk:

1 Buy semen or embryos from a reliable source (certified and documented)

2 Test new entrants before introduction into the herd, or buy animals from sources that apply this testing method before selling. Biosecurity programmes have no meaning on a farm if one does not control the potential introduction of diseases.

The movement of pathogens between different animal groups, or between different groups of young animals, within a herd, is comparable to the introduction of disease by new entrants or replacements

Contact –direct or indirect–between different animal groups, such as early lactation, mid-lactation, end-lactation, dry cows or calves must be minimised to reduce the risk of transmission.

A biosecurity programme equates to an assurance policy for animal health and productivity: zero-risk does not exist. It is often more expensive to have several disease-affected cows than to implement a biosecurity programme.

In order to develop an effective biosecurity programme , farmers and employees, as well as the veterinarian need to comply with the actions and points within it. If needed, short, instructive and informative courses should be put in place. Implementation of a biosecurity programme requires time, planning and patience.

Analysis of risks in biosecurity

For the development and application of a biosecurity programme there are three stages:

1 Analysis of risks (identification of problems; amplitude of these problems; probability of occurrence of these problems; impact of these problems)

2 Management of risks (the design and application of a plan of prevention)

3 Communication about the risks (explain the prevention plan to all concerned and emphasise their required compliance)

In the following paragraph, emphasis will be placed on the first of the three stages - analysis of risks

Analysis of risks

One of the goals for dairy farmers is to produce milk of good quality that is free from contaminants for reasonable cost in a sustainable way. Various factors can render the milk contaminated (chemicals; antibiotics) and hence reduce both quality and safety. Among these factors are: bought-in cattle, people, feedstuffs, manure, waste, utensils and equipment, transportation methods. The identification of these risk factors and the evaluation of their relevance is necessary for designing a control programme and, hence, justify the costs of treatment and reduce the losses.

Biosecurity programmes

Four step analysis

For dairy farms, the following four steps have to be taken in developing an analysis of risk programme:

1. Identification of products that leave the farm (milk, calves, heifers, culled cows, manure, etc.) and their value
2. Identification of risks for infectious diseases that have been prioritised (see above), as well as an estimation of the impact of these risks.
3. Evaluation of how these diseases are, will be or could be introduced onto the farm.
4. Evaluation of means and methods of application of a biosecurity programme specific to the selected diseases and farm of concern.

Step 1. Identification of products that leave the farm and their value

For example, farms that sell embryos will be more concerned about *leukosis* or Blue Tongue than those who do not. For other farms, diseases that cause production losses or lead to premature culling are of more importance.

Step 2. Identification of the risks and their priority

Table 2.2 gives examples of diseases necessitating different approaches with regard to biosecurity.

Step 3. How do infectious diseases enter the farm?

Table 2.3 shows differences regarding transmission, incubation, shedding and survival of pathogens in the environment for four example diseases

Step 4. How to develop and apply a biosecurity plan specific to one farm

Table 2.4 shows several routes of transmission for the four pathogens in Table 2.3 The routes include those from one farm to another, within one farm and from one animal group to another. A biosecurity programme must comprise the two major issues of spread between farms and spread on a farm.

When you consider biosecurity measures, any costs and anticipated benefits of those measures must be taken into account. In many cases, management practices have to be changed!

The consideration of the health status of bought-in animals and those of farm of origin is a way to reduce the risk of disease introduction. However, in the case of *paratuberculosis*, the prevention of dispersion of this disease will have limited value if one continues to buy in new cattle, or if one mixes herd mates with bought-in cattle.

A biosecurity programme is not a complete solution for all farms; it is not the golden panacea! Each farm must have its own specific and unique programme! This makes sense as the risk analysis is, by nature, unique with unique results.

Hence, the risk management elements are equally unique. The complexity of the programme will depend on the characteristics of the selected pathogens/diseases, as well as the level of risk reduction desired.

Steps effective in the control of risks will include:

1. The definition of critical goals and products of the farm.
2. The identification and prioritisation of diseases relevant to the farm.
3. The evaluation of transmission risks of those disease pathogens .
4. The evaluation of necessary changes to apply the control programme and the whole biosecurity programme.

Disease & transmission

Table 2.2 Examples of relevant disease conditions that necessitate different approaches with regards to biosecurity (+++is more important than +) After Wells, 1998

Disease condition	Economic production losses	Public health hazard	Dangers for international markets
Mammites	+++		+
Salmonellosis	+	++	+
Paratuberculosis	++	?	+
BVD	++		+
Reproductive disorders	++		
Lameness	+++		
Metabolic disorder	+++		
Cryptosporidium	+	++	
Campylobacter jejuni		++	
Mycobacter bovis		+	+
Bovine leukosis	+		+
Brucella abortus		+	+
BSE		?	+
Blue tongue			+

Table 2.3 Caracteristics of transmission for four pathogens in cattle

Transmission	*St. aureus*	Mb paratuberculosis	BVD	*Salmonella*
Faecal-oral		X	X	X
Nasal / saliva	?		X	X
Milk	X	X	X	X
Uterine		X	X	X
Sexual			X	
Incubation	Days-months	Years	5-10 d	1-4 d
Clinical duration of disease	Days-Years	Weeks-months	2 wks	1-7 d
Duration of excretion	Days-years	Months-years	10-14 d	Week-months
Survival in environment	?	Months-years	<14 d	Months
Growth in environment	Yes ?	No	No	Yes

Routes of transmission

Table 2.4 Routes of transmission for four different pathogens

Between farms	*St. aureus*	*Mb paratuberculosis*	**BVD**	*Salmonella*
1. New cattle entries in the herd	X	X	X	X
2. No quarantine applied			X	X
3. No tests done for diseases	X	X	X	X
4. No vaccination administered			X	
5. Return cattle from shows/markets	X		X	X
6. Other contacts with cattle, feed or water		?	X	X
7. Dispersion by people, vehicles, utensils	X	X	X	X
Within one farm	***St. aureus***	***Mb paratuberculosis***	**BVD**	***Salmonella***
1. Several cows allowed in one calving pen		X		X
2. Calving pen also used for sick cattle	X	X		X
3. Sick cows not separated from healthy animals	X	X	X	X
4. Manure equipment also used for feed mix or for feeding		X		X
5. Water contaminated by manure		X		X
6. Supply contaminated feed		X		X
7. Feeding calves non-pasteurised mastitic milk, non-pasteurised normal milk or non-pasteurised colostrum.	X	X	X	X
8. Contact of animals with others, with feed or with water		X		X
9. No vaccinations administered			X	
10. Dispersion by people, vehicles, equipment	X	X	X	X
11. Teat disinfection, treatment at dry off, udder hygiene	X			

Biosecurity checklist

Table 2.5 Biosecurity checklist for farms			
Section 1 General issues FARMER STATEMENTS			
Best management practices checklist	YES	NO	Veterinarian remarks on farmer statement
I routinely meet milk quality standards (SCC, bacteria, etc)			
My milk and dairy beef buyers offer quality premiums			
I believe it is more profitable to keep diseases off my dairy than to fight them on my dairy			
I agree that taking profit by doing things right the first time is a critical part of biosecurity			
Biosecurity requires some method of permanent animal ID. I have such an ID method in place.			
I can readily track and validate to others the quality represented in my animals			
I have been able to consistently produce and sell more milk per cow per year			
Section 2 Keeping infectious diseases off the farm			
General control practices			
I never purchase or keep animals for or from others			
My vet talks to the seller's vet prior to my buying animals			
My cattle do not attend shows or use community pastures and they are not placed in performance evaluation centres			
My cattle rarely share fence lines with neighbouring cattle			
I purchase and never borrow or use loaner bulls from other farms			
I always buy animals from a paratuberculosis-free certified farm			
I always know the health status of animals brought into my herd			
I never bring in animals without knowing their vaccination history			
I do not buy animals from a herd that has mixed origin cattle			
I limit my purchases to open heifers			
I ask for DHIA somatic cell count information from the seller's herd when buying cows			
I transport animals in my own clean vehicle			

Biosecurity checklist

Table 2.5 Contd. Biosecurity checklist for dairy farms			
Section 2 Keeping infectious diseases off the farm			
Best management practices checklist	YES	NO	Veterinarian remarks on farmer statement
I use a heifer rearer and I have a copy of his/her biosecurity, vaccination and testing programme			
Introducing new arrivals to my farm			
I quarantine new animals for 21-30 days before allowing them contact with my animals. My quarantined area is meant to prevent:			
Quarantined animals from sharing air space with healthy animals			
Quarantined animals from touching my cattle			
Sharing of feeders, waterers, etc			
Isolation – my facilities...			
Allow me to check a quarantined animal's temperature each day			
Provide a medicated foot bath			
Provide separate milking equipment or easy sanitation after milking quarantined cattle			
Allow me to milk quarantined cattle last.			
Controlling farm traffic			
I limit people's access to my animal's pens			
I sign-post the area asking visitors to keep out			
I provide clean boots and coveralls for visitors			
I provide a method for bull calves or other animals to leave my farm without allowing dealers or transporters to enter my barns			
I have dead animals or downers picked up without anyone external entering my barns			
I keep a record of visitors to my farm			

Biosecurity checklist

Table 2.5 Contd. Biosecurity checklist for dairy farms			
Section 3 My use of strategic vaccinations			
Best management practices checklist	YES	NO	Veterinarian remarks on farmer statement
I have a written strategic vaccination plan drawn up for my dairy. If yes→			
I have a list of diseases for which I routinely vaccinate			
I have determined which animals benefit from vaccinations			
I know when my animals need the protection of vaccinations			
I protect against abortion by using a pre-breeding vaccination programme (leptospirosis; IBR; BVD)			
I am concerned about respiratory diseases on my farm (BRSV; PI-3)			
I try to remove carriers of BVD as their presence can overcome the protection provided by vaccination			
I have animals which die unexpectedly examined by my herd vet (necropsy)			
I use modified live vaccine when appropriate or necessary and marker vaccine when indicated			
All incoming animals are vaccinated against IBR, BVD, PI-3, BRSV and leptospirosis on two occasions: 6 weeks and 3 weeks prior to arrival at my farm			

Section 4 Calf management as a specific part of biosecurity			
I or the calf rearer never allow replacement heifers on pastures where manure from mature animals has been spread (micro-organisms may live in soil for up to 1 year)			
All calves are fed colostrum from cows that have been tested clean of infectious diseases (colostrum is not mixed; each calf receives milk from a single cow)			
I avoid adding whole milk to extend purchased milk replacer (milk replacer is considered free of infectious pathogens)			
Replacement heifers are kept separate from other animals for at least six months (kept away from manure of adult animals)			
I ensure that there is no stagnant water in the replacement heifers pens			
Replacement heifers have their own separate source of water			

Biosecurity checklist

Table 2.5 Contd. Biosecurity checklist for dairy farms

Section 4 Calf management as a specific part of biosecurity

Best management practices checklist	YES	NO	Veterinarian remarks on farmer statement
Calves are separated from their dams immediately after birth (no sucking, no searching for teats allowed)			
The calving area is always clean and disinfected			
I consult with my veterinarian (at least) annually about calf care			
Calves are permanently identified before any grouping			
All replacement calves are given a magnetic bolus at initial breeding time			
I have a strategic de-worming plan in place			
I routinely dip the entire navel of newborn calves with strong iodine or other disinfectant			

Section 5 Biosecurity includes testing

Paratuberculosis			
The whole herd is screened using an antibody ELISA test for blood samples			
Whole herd testing is done using a faecal culture test (to measure the level of bacteria shedding)			
Animals testing positive are culled			
Replacement heifers or the herd they originated from is tested prior to introduction into the herd			
Calves or cows testing positive are sold(disease can spread via umbilical cord, manure, animal-animal-contact)			
I have implemented a follow-up testing programme and reviewed the results with my vet			
BVD			
I understand persistently infected animals (PI) as they relate to BVD			
I will cull PI calves in my herd			
I do not keep PI animals as replacement heifers			
I am committed to finding BVD PI animals in my herd (PI immune-peroxydase testing)			

Biosecurity checklist

Table 2.5 Contd. Biosecurity checklist for dairy farms			
Section 5 Biosecurity includes testing			
Best management practices checklist	YES	NO	Veterinarian remarks on farmer statement
I have discussed killed versus modified live virus (MLV) vaccines for BVD with my vet			
Contagious mastitis *(S.aureus; Strep agalact.) (Mycoplasma bovis)*			
I use an effective pre- and post-milking teat dip			
I evaluate my milking machine and equipment (at least) twice yearly			
I check for vacuum stability at least annually			
I check for stray voltage annually			
My milking technique involves preparing cows 1-2 minutes before milking			
Cows with clinical mastitis are tested to determine if the mastitis is Gram-positive or Gram-negative			
I infuse all quarters of all cows with a commercial dry-cow product immediately following last milking			
I vaccinate my cows with E.coli vaccine 6 weeks and 3 weeks prior to expected calving date, and 4-7 days after calving			
I have discussed bovine ulcerative mamillitis management with my vet			
I routinely test the bulk milk tank weekly after adding new heifers to the milking cows			
Salmonellosis *(dublin; typhimurium)*			
I am aware that some salmonella organisms cause disease in humans			
I blood test my cows and my replacement heifers for S.dublin and S.typhimurium every year (carrier cows are culled)			
I isolate sick cows in the hospital area; I discuss antibiotic treatment with my vet			
I provide dry, clean, disinfected calf and maternity pens			
I reclycle flush water from the milking parlour only. All other water is discarded			
I test purchased feed for salmonella 1once per year			

Biosecurity checklist

Table 2.5 Contd. Biosecurity checklist for dairy farms			
Section 5 Biosecurity includes testing			
Best management practices checklist	YES	NO	Veterinarian remarks on farmer statement
I restrict birds and rodents from access to my dairy's animal feed			
I limit the number of cats and stray animals on my dairy			
I do not allow rendering trucks to access feed or animal areas			
I realise my family can be infected with salmonella from raw milk			
Bovine Leukosis			
I never use a needle for more than one animal			
Sleeves are discarded between animals			
Cows that provide colostrum for my young animals are tested for BL			
Neospora caninum			
I am aware that abortions may be caused by *Neospora* infection			
I understand how *Neospora* infection is transmitted			
Replacements are tested prior to introduction into the herd			
Embryo recipients are candidates for testing			
Infection with *Neospora* can influence milk production			

Section 6 Bio-containment: controlling enteric, reproductive, respiratory problems in herd groups			
My facilities provide a clean area for restraint, treatment and isolation of cattle			
My facilities prevent cross contamination of water, feed, manure or equipment between groups of animals			
I have a plan to manage group size, age distribution and animal flow to reduce risks of disease outbreaks			
I carry out taskswith the health status of my animals as my first priority (from handling youngest calves to older cattle first and sick animals last)			
I and my family members or employees use strict sanitation practices if we must go "back" to any animal group or animal			
I have all animals that die of unknown causes diagnosed by a veterinarian (necropsy)			

Biosecurity checklist

Table 2.5 Contd. Biosecurity checklist for dairy farms

Best management practices checklist	YES	NO	Veterinarian remarks on farmer statement
My vet collects blood samples from all cows that abort			
I have all visitors observe our strict sanitation procedures, paying special attention to their feet and clothes			
I send all aborted foetuses to a diagnostic laboratory to determine the probable cause of abortion			
Section 7 Equipment			
I never use the same equipment to feed animals and to clean pens or barn			
I never use common shovels or other equipment to feed animals and to clean out manure			
I never walk or drive through manure while doing routine feeding operations			
I never leave manure hauling equipment in pens with different groups of animals			
Animals are not penned near manure handling equipment			
I routinely clean and disinfect feeding equipment			
I routinely clean and disinfect manure handling equipment			
Section 8 Sanitation			
I always keep young animals away from the faeces of older, adult animals (minimum time period is six months)			
I remove the young calf from its dam immediately after birth			
I always prevent cattle from having access to ponds or streams (surface water)			
I have separate buckets for my skid loader for manure and feed			
I use disinfectants that can kills tuberculosis organisms			
There are no opportunities for manure to enter and contaminate the feed supply or stock			
If manure accidently contaminates feed or water, I treat this as a major problem (direct action)			

Biosecurity checklist

Table 2.5 Contd. Evaluation procedure for biosecurity status on dairy farms

	Score items	
Section 1	7	YES answer = 5 points
Section 2	26	
Section 3	10	NO answer = 1 point
Section 4	13	
Section 5	37	
Section 6	9	
Section 7	7	
Section 8	7	

Total # of scoring items = 116

Maximum number of points that can be achieved by all YES answers = 116 x 5 = 580 points

At a 60 % threshold=	348 points threshold === minimum biosecurity level
Modest evaluation	Poor biosecurity status = less than 350 points Moderate biosecurity status = 350 to 450 points Good biosecurity status = more than 450 points
At a 70 % threshold =	406 points threshold === higher biosecurity level
Strict evaluation	Poor biosecurity status = less than 350 points Moderate biosecurity status = 350 to 500 points Good biosecurity status = more than 500 points

Additionally you can evaluate each section individually, to highlight areas for improvement.
Each section should at least achieve a 60 or 70 percent score to become classified as "fair to good".

Biosecurity plans

Practical management programmes to prevent infectious diseases from entering a farm and, if they do enter a farm,to prevent the spread of such diseases.

Infectious diseases can have a great economic impact on farm revenues and profit. Examples of such diseases are Foot and Mouth disease (FMD), tuberculosis, brucellosis, Bovine viral diarrhoea (BVD), Infectious Bovine Rhinitis (IBR), salmonellosis, paratuberculosis.

Soon, farmers will be held responsible for potential chemical and microbiological contamination of their products at the start of the food chain (liability issue). Vaccines and antibiotics have long been the sole means to control infectious diseases. However, they are neither available, nor efficient or allowed for all relevant diseases. Therefore, risk identification and management programmes, together with best practice guidelines and work instructions, are required in order to prevent such diseases.

A **biosecurity plan** comprises the following major elements:

- A Good Dairy Farming Code of Practice (guidelines) and technical working instructions as general risk prevention measures;
- Specific risk analysis schemes;
- Risk management measures;
- Risk communication strategies;
- Training programmes for farm employees.

Designing a biosecurity plan for a individual farms

1 Conduct hazard identification: what are the diseases most relevant to that farm?

2 Identify the diseases prevalent on the farm and those representing an outside hazard. Identify the important risk factors for each disease identified in step 2 and rank them in order of perceived relevance.

3 Design the biosecurity plan to include hazards, general risks, specific risks, monitoring and diagnostics, general prevention measures, -and specific prevention measures. This step comprises the GDF Code of Practice and working instructions.

Table 2.6 Some examples of risk exposure and management points

Exposure area	Example	Control points
New additions to the herd	Cattle–semen–embryos	Inspection of animals; certificates of origin; diagnostic tests; vaccinations; quarantine three weeks
Feed and water	Forages, water, concentrates	Water quality tests; prevention of site contamination; GMP certificate for concentrates
Animal contacts	Neighbouring cattle; wildlife; cattle shows; clinics	All contact to be minimised; cattle returning from show or market must be considered as “new additions” to the herd
Vectors and vehicles	Rats and other rodents Trucks, lorries, tractor	Rodent control programme; minimise contact; cleaning and disinfection of trucks etc; visitor entrance instructions
Health management		Good Drug Application Code of Practice and working instructions

Biosecurity plans

Adapted after BAMN—AFIA, 2000, Arlington VA, USA

Limiting animal movements onto and within the farm, as well as limiting all possible contact between animals and man/vehicles/vectors are major control measures in biosecurity.

Risk management methods

Risk management methods depend on the hazards and risks identified on a particular farm.

For example:

- Purchase semen or embryos from a certified source only; check documents.
- Test the new animals for freedom from certain diseases before they enter the farm and put them in quarantine for three weeks before allowing integration into the rest of the herd.

It is important to realise that animal movements with regards to production groups (e.g. calves, heifers, first parity animals, dry cows) should be regarded as separate (epidemiological and) management unit in terms of biosecurity; contact between groups must be avoided.

Practical implementation of a biosecurity plan

Firstly, consider and follow all issues addressed above. See the documents BIOSEC 4, 5 and 6 for lists of general prevention measures.

Secondly, consider the following issues:

- What are the products that are being sold on the farm (cattle; milk; cheese; beef; semen; embryos etc)?
- What is the level of economic and welfare loss associated with the disease of concern or the risks addressed?
- How are diseases introduced onto the farm?
- How can you best implement a biosecurity plan?

Table 2.6 presents some examples of transmission risks and disease information needed for control.

General risk factors are commonly put into a GDF guideline, while specific risk factors go into technical working instructions. Examples of the latter are the working instruction for 'Good Hygiene in the Milking Parlour', 'Good Use of the Footbath' or 'Visitors Entrance Protocol'.

It is not necessary or practical to control all possible risk factors on a farm. Multi-factorial diseases can be controlled by eliminating some important risk factors and placing less focus on more minor ones.

General preventative measures

Table 2.7 General preventative measures as part of Good Dairy Farming Code of Practice

General preventative measures	Details	Executed	Planned	Remarks
Housing and hygiene	Cubicles are clean and bedding renewed			
	Hygiene conditions are checked			
	Barn climate is regularly checked			
Feeding of lactating cows	At each feed change the ration is reformulated			
	The mineral status of each ration is checked			
	Forage quality is checked regularly			
	Body condition scoring is routinely carried out			
	Rumen fill scoring is routinely carried out			
	Faeces quality scoring is routinely carried out			
Feeding of dry cows	Rations are recalculated at each change			
	Rations are checked for low energy status			
	Body condition scoring is routinely carried out			
Feeding of transition cows	Ration fibre content is checked routinely			
	Energy status is checked routinely			
	Protein status of ration is checked			
Feeding cows after calving	Ration energy and protein status is checked			
	Fibre content in rations is checked			
	BCS, RF, FQ are scored routinely			
	Feed intake is checked			
Feeding of calves and maiden heifers	Rations are recalculated at each feed change			
	Growth is checked twice yearly			
	Mineral status of rations is checked			
Use of medical products	Up-to-date records are present			
	Dry, cool, clean storage			
	Herd Treatment Advisory Plan present and up-to-date			
Disease records	A disease record is kept of all animals treated			
	Laboratory results are kept			

General preventative measures

Table 2.7 General preventative measures as part of Good Dairy Farming Code of Practice				
General preventative measures	Details	Executed	Planned	Remarks
Mastitis	Cows are kept standing after milking			
	Udder, teat and thigh hair is clipped			
	Performance of milkers is checked			
	Barn climate conditions are checked			
	Waiting and resting areas are clean			
	Milking machine checked twice yearly			
	Vacuum level on milking machine checked routinely			
	Regulator routinely cleaned			
	Pulsator cleaned routinely			
	Pulsator function checked			
	Teat liners replaced after 2500 milkings			
	All rubber material checked weekly			
	Teat-end callosity is scored monthly			
	Udder preparation is a dry method			
	First milk is checked routinely			
	Teat disinfection is carried out correctly			
	Mastitic cows are milked last			
	Interval surveys are carried out routinely			
	All mastitic cows before 60 DIM are sampled for bacteriology			
	Mastitis cases are treated following the HTAP			
	Drying-off is done following the HTAP			
	High SCC cows are checked			
	Chronic mastitic cows and high SCC cows are identified for culling			
	Cows leaking milk are identified			

General preventative measures

Table 2.7 General preventative measures as part of Good Dairy Farming Code of Practice

General preventative measures	Details	Executed	Planned	Remarks
Claw and hock lesions	Manure scraping is done 4-6 times/day			
	Broken slats are replaced			
	Floor maintenance is up-to-date			
	Waiting area is dry and clean			
	Access to cubicles and areas is easy			
	Locomotion scoring is carried out routinely			
	Herd claw trimming is done twice yearly			
	Foot bathing is carried out each month for three days			
	Scoring waggling cows and poor posture			
	Primary claw lesions during herd trimming are recorded			
	Clinical lesions are recorded in a diary			
	Claws of young stock are treated with formaldehyde			
	Herd Treatment Advisory Plan present			
	Pathways to/from pasture are checked for stones and gravel			
	Length of pathways to pasture is limited			
	Number and functioning of watering points at pasture are checked			
	Muddy areas are eliminated			
	Chronically lame cows are culled			
	Milk urea level is optimized			
	Rations are checked for potential to cause acidosis			
	Reproductive performance of the herd is checked at least twice yearly			
Reproductive performance	Housing and hygiene: see above			

General preventative measures

Table 2.7 General preventative measures as part of Good Dairy Farming Code of Practice				
General preventative measures	Details	Executed	Planned	Remarks
Reproductive performance	Dry cows remain housed			
	Dry cows are kept in two groups with different rations			
	Feeding of cows: see above			
	Body condition scoring is carried out each month			
	Fresh cows are checked for negative energy balance			
	Heat detection at least twice /day for 20 min			
	Heat detection in early morning and late evening			
	Detected and suspected heats are recorded			
	Recorded heat dates are used to predict subsequent heats			
	Luminosity in the barn must be > 200 Lux (be able to read a newspaper)			
	Just prior to detection, increase light intensity			
	Separate calving pen, not used for sick cows			
	Pens are thoroughly cleaned and disinfected between individual cows			
	DIY-AI practice by the farmer is checked			
	AI semen is checked for quality			
	For cows using more than 3 AI, a sire with high fertility is sought or cows are culled			
	Interval from calving to 1st AI is limited			
	Low-producing cows are bred early after calving and preferably with beef sires			
	Retained afterbirth is treated properly			
	Endometritis is treated properly			

'Closed farm' Major characteristics

Defined as a farm where one has taken preventive measures to avoid, in the best possible manner, that infectious diseases enter the farm.

The most important characteristics are that:

1 Visitors are not allowed on the premises, or appropriate sanitary measures are taken should they enter (e.g. boots; clothes; hygiene barrier)

2 Contact between animals of different ages or different species is avoided

3 Communication between neighbouring farms with regards to health status and hygiene.

4 The farm entrance is organised in such a way that lorries and other temporary vehicles drive on a so-called 'dirty road' (where farm traffic does not pass, including animals, cars, people and equipment)

5 Dogs, cats and birds cannot enter the buildings

6 Surface water does not serve as drinking water for animals

7 Manure from other farms is never used, unless the farm supplying the manure is certified for certain diseases

8 The farms' own machinery is used; if machinery from third parties is used, it must be cleaned and disinfected prior to use

9 Animals due for culling are separated from other animals; dead animals are stocked along the dirty road for ease of disposal and away from the rest of the herd and buildings .

10 The farm does not participate in cattle expositions or shows or Embryo Transfer.

Risk management

Table 2.8 Example protocols for elements in risk management protocol

General management
Do not purchase new cattle, goats or manure from unknown sources or sources with a lower health status than the farm in question.
Check weekly that contact with other cattle or wildlife whilst out grazing is avoided.
Make sure that all non-farm equipment, trucks and materials are cleaned before entering the farm.
Ensure on daily basis that there is a clean and well-organized animal health care box containing medicinal products, several disposable (or at least) disinfected syringes and disposable (or at least new) needles available.
Medicinal products are stored and locked away in a dark and cool place (particularly vaccines) and an up-to-date record is available.
General pen hygiene
After an animal has been removed from a pen/ box: · put the box outside in the designated cleaning area · remove straw, manure and dirt · clean it with high pressure water · disinfect, rinse and let the pen or box dry
Calf health care
During feeding animals must be checked for clinical signs of illness.
Severely affected animals must be separated from herd mates and put in a separate sick pen.
Affected calves should be treated according to the "Young Stock Treatment Advisory Plan"
The check results are registered on the calf-rearing-chart with date/time and name of observer

IBR prevention

After H.van Weering, GD Deventer, The Netherlands, April 2010

Table 2.9 A farm checklist for IBR prevention

YES	NO	N/A	**Measure** (N/A = Not applicable)
			You purchase cattle from farms that are not certified IBR-free
			You voluntary put cattle from farms not certified IBR-free out to pasture or in housing
			You participate in shows where there are cattle not free from IBR
			You graze your calves and maiden heifers next to fields where cattle from other farms graze at a distance less than 3 m
			Your cattle sometimes escape from their pasture and have contact with other cattle from other farms
			Neighbouring cattle sometimes escape and have contact with your cattle
			Do you use semen from organisations that are not EU-certified?
			Professional visitors may enter your farm without complying with the hygiene rules you have set? (they use boots, clothes, gloves used on other farms)
			You use transportation means that are also used to transport non-certified cattle for other farms
			You share utensils and equipment with other farmers who are not certified IBR-free
	All questions answered with **NO → the risk of IBR is minor**		

One question answered with **YES → there is a realistic risk of IBR.**
Two or more answers **YES → there is a high risk of IBR introduction**

For **bulk milk tank examinations** the following decision approach could be used:

1. No antibodies are found in the tank → less than 10% of infections are expected → approach = serology of cows, vaccination, certification
2. Antibodies are found in the tank → more than 10% infections are expected → approach = vaccinate according to the official control programme guidelines.

When one or more risk factors are found preventive measures should be taken

Parasites & pests

Biosecurity checklist for dairy farms - part of the Good Dairy Farm Codes Of Practice

The checklist forms an essential contribution to farm profit by reducing the costs of production by keeping both endo- and ectoparasites, as well as their vectors, under control. (Ref Milk and Dairy Beef Quality Assurance Program, DQA, Stratford, Iowa, USA, 2000) There are five elements to this quality risk management component.

Table 2.10 The quality of production is paramount for profitable farm income (15 scoring items)

	YES	NO	Remarks of veterinarian
I usually pass all official quality examinations of my dairy			
I appreciate to get a bonus for milk and/or beef from my cattle			
I want to do things correctly from the start			
I think it is better to prevent diseases, parasites and pests from entering my farm than to treat them once they have entered			
I sell increasing amounts of milk each year			
I see the relationship between flies and other problems on the farm, as well as the nuisance they represent to workers and visitors			
I have a written an action plan for dealing with parasites and pests. It comprises chemical, biological and cultural control			
I know that each parasite and pest has its own life cycle and multiplication method			
All my cattle are properly identified			
I can validate to third parties the quality of treatment my cattle receive			
Only if capital is lacking would I withdraw my control programme			
I have discussed the relationship between insects and diseases with my veterinarian			
I realise that parasites and pests present more of a problem in at particular times of the year			
I understand that single treatments are not enough to control the problem			
I participate in other biosecurity and quality risk management programmes			

Conclusions and remarks:

It is not if I should treat my cattle but rather when & how

Table 2.11 Control of endoparasites (8 and 12 scoring items)

	YES	NO	Remarks of veterinarian
It is possible that I lose milk due to endoparasites			
I have developed a written Herd Treatment Plan with my veterinarian			
I understand that my cattle become infected when they graze contaminated pasture			
My cattle have access to an exercise area			
My cows are grazed throughout their lactation			
Losses of body weight and body condition at the start of lactation seem related to endoparasite infection			
My veterinarian or the laboratory uses acknowledged methods to detect endoparasites			
I understand that infection causes weight loss and growth retardation and can result in a delay in the first service (AI) for maiden heifers, resulting in them calving about 2 months later than usual			
My treatment plan is based on individual treatment of cows during the first 100 days of lactation			
The treatment level is based on faecal egg counts			
I know the problems caused by each endoparasite species			
For the treatment of maiden heifers at pasture, I apply a strategic programme targetting key times of year			
My treatment programme comprises different formulations and types of anthelmintics (drench, paste, additives, etc)			
I am familiar with the different anthelmintics			
I know that treating cattle after endoparasites have caused damage and losses is not much help			
Climatic changes during a season may help in designing control programmes			
I apply the principle of pasture rotation every year to ensure pasture is free from grazing for part of it			
I read the drug labels and discuss with my veterinarian before administering any drugs			
I avoid injecting young cattle to prevent skin/muscle lesions (except in the neck)			
My goal is to follow a cost–effective endoparasite control programme			

Conclusions and remarks:

Checklists for control of parasites and pests

Table 2.12 Management tools to control flies (17 scoring elements)

	YES	NO	Remarks by veterinarian
I have educated myself on pest control products and good neighbourhood principles			
I have a written insect control programme,which includes evaluation by culturing			
I only apply approved insect control products			
I apply a rotational programme with different insecticides to maintain their efficacy and avoid development of resistance			
The pest control programme includes flies of all species that are relevant to cattle husbandry			
I use other parasites (wasps) to reduce the number of flies			
Insecticides are locked in a locker secured by a key			
I know that insects are resistant against pyrethroids			
I put emphasis on fly control among calves/maiden heifers			
I know that most flies have a cycle of 10 to 21 days (from egg to larva, to pup) and I adapt the barn cleaning frequency to constantly reduce the size of fly populations			
The milking parlour is treated, as well as the housing			
I use traps, sticking tapes and other products to eliminate insects from the milking parlour			
My programme also comprises pour-on, ear-tags and/or powder			
I know that flies can spread mastitis pathogens, causing milk loss, stress, body weight loss and bacterial contamination of milk			
Flies and other insects around the feed bunk hamper optimal feed intake and, hence, milk production			
I know that calf hutches are predilection spots for flies (95% of flies originate from 5% of farms)			
Regularly, I spray chemical products onto walls, ceilings, as well asareas around (not in!!) the feed-bunk			

Conclusions and remarks:

Checklists for control of parasites and pests

Milk and Dairy Beef Quality Assurance Program, DQA, Stratford, Iowa, USA, 2000)

Table 2.13 Control of lice (12 scoring items)

	YES	NO	Remark of veterinarian
I know that lice remain on the animal throughout their lifecycle			
I know that cattle get nervous and hard to handle when infested			
I always take hair samples from healthy cattle			
I quarantine newly-entering cattle and treat when needed			
I conduct regular inspections in the herd to avoid sudden outbreaks occurring			
I know that the period to control lice is between December and March			
I know that young stock that remain housed are more likely to have lice all year around			
My cattle are housed in free-stalls			
Young calves are housed in individual hutches			
Lice treatment includes powder, pour-on and/or spray			
Treatment should be repeated 21 days after the first treatment			
Carrier cattle are housed separately or sold			

Checklists for control of parasites and pests

Table 2.14 Rodent control and bird control (6 scoring items)

	YES	NO	Remark by veterinarian
I have regular evaluation sessions to detect rodents, birds and other pests			
I have put up bird nets to prevent them from entering the houses			
Ever year I evaluate my control programme with my vet			
I handle my dogs and cats in such a way that they do not enter the buildings nor have access to the feed			
Rats and mice may be an intermediate host for parasites. I have a 3-point plan: * create obstacles (fill in any holes > 2 cm) · set traps and lay down toxic substances (high and low locations) · eliminate areas where they are able to hide			
I am sure that my cattle cannot access any poison laid down for pests			

This checklist comprises a possible **total of 70 scoring points.**

To **classify farms**, the following intervention thresholds could be used:

- Poor situation = immediate action needed, when <35 points scored
- Intermediate situation = some improvement is advised: 35 to 45 points
- Good situation = minor items require addressing 45 to 60 points
- Excellent situation = continue with current control programme 60 to 70 points.

Biosecurity protocol

After htpp://nmconline.org/transl/NMCchecklistNA_fr.pdf NA & www.mastitisnetwork.org

Table 2.15 Biosecurity protocol for contagious mastitis pathogens

General prevention measures dealing with transmission	Tick here if applied
Cows with a confirmed infection and cows with a persistently high SCC (> 200.000/ml) are segregated from their herd mates for a few months in order to reduce transmission probabilities	
Infected cows are segregated from herd mates for milking (separate units are used, or infected cows milked last)	
Cows with chronic St. aureus infection and cows with other pathogens that do not respond to treatment (*Pseudomonas, Arcanobacterium, Nocardia, mycoplasma*) must be culled or permanently isolated	
Check udder health status of first lactation cows (CMT; bacteriology)	
Implement a management practice to prevent maiden heifers reared together from suckling one another	
Do not feed infected or potentially infected (high SCC) milk to calves	

Prevention measures for the sale or purchase of cows	Tick here if applied
Check individual cows SCC data and bulk tank SCC data at the seller's office. Check thoroughly (CMT; bacteriology) the individual quarters before purchase.	
Get aseptically collected milk samples from all quarters for bacteriological culture before purchase.	
Quarantine the recently purchased cows upon their arrival and milk them separately until confirmation of absence of udder infection	
When purchasing lactating cows, first conduct a test on antibiotics residues. In the case of dry cows and late gestation heifers, ensure that the seller writes down all treatments given to the cows (product; date; dosage; withdrawal period; etc) and test the milk after calving for presence of antibiotic residues if required (see note).	

Note: testing for antibiotic residues is not necessary when the seller signs a certificate confirming that the animal has not been treated with antibiotics for at least the last 60 days prior to this sale, or is free from antibiotic residues, or, if treated, the appropriate withdrawal period has been observed.

Biosecurity of feedstuffs

The management quality of the farmer is the key to disease prevention and control on the farm

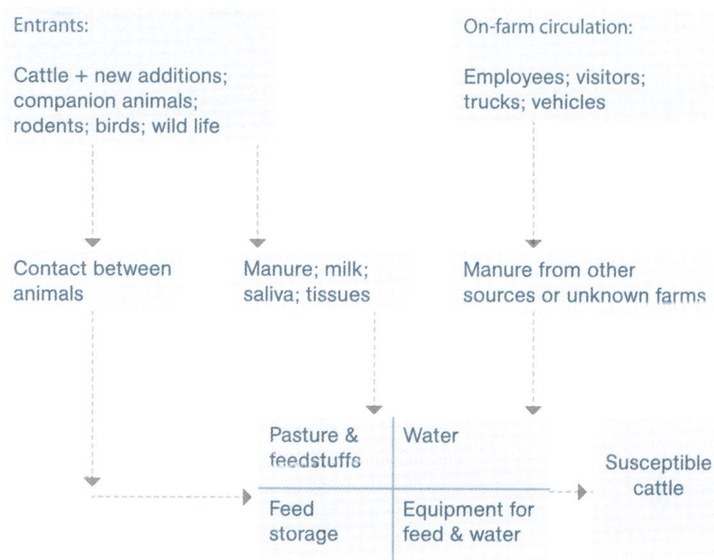

Figure 2.1 Contaminated water and feed may impair the efficacy of biosecurity plans on cattle farms

The contact between animals, the drinking water, the feedstuffs, products of animal origin, manure and environmental sources are the paramount risk factors associated with prevention and control of diseases. The feedstuffs, the ration and the way they are managed should be part of a biosecurity plan on the farm. Knowing the way in which diseases are transmitted is critical for the prevention of such diseases.

Think about issues that are not included in the biosecurity plan

Step 1 is to determine the various sources of contamination on the farm in order to define the routes of transmission (Table 2.16).
Discarded milk is described as non-pasteurised whole cow milk, milk from (treated) sick cows or colostrum. Powdered milk for calves is a sufficient example of alternative option because diseases are not transmitted through powdered milk if correctly stored and prepared. Due to BSE, ruminant proteins cannot be offered to dairy cattle making it important to check the labels of delivered feedstuffs in order to conform to the regulations.

The impact of production method and feeding management

Several feedstuff production methods help to minimise risks of disease introduction and transmission. Feedstuffs, even when harvested correctly, may become contaminated at a later date.

Regarding the method of feed production:

- Ask producers about their quality assurance system and tests for mycotoxins, pesticide control, control of contamination of production lines, personal hygiene and laboratory control tests
- Pelleting may reduce the burden of bacteria if the temperature and duration of treatment are adequate. High temperatures may kill Salmonellae (55°C for 1 hour, or 60°C for 20 min)
- All chemical products and pesticides must be stored separately in a secure, cool, dry location and must be clearly and correctly labelled
- Storage areas must be cleaned between deliveries

Biosecurity of feedstuffs

Table 2.16 Control points for orally transmitted diseases

Disease orpathogen	Manure	Water	Discarded milk	Pasture	Forage/ roughages	Cereals/ concentrates
BVD-DBV	XXXX	C	S			
Coccidiosis	XXXX			C	C	
Cryptosporidium	XXXX	C		C	C	
Giardia	XXXX	C		C	C	
E. coli	XXXX	C	XXXX	C	C	C
Salmonella spp	XXXX	C	XXXX	C	C	C
Campylobacter	XXXX	C	C	C	C	C
Listeria	XXXX		S		XXXX	
Paratuberculosis	XXXX	C	S	C	C	

XXXX= primary source of contamination; S= secondary source; C= other sources (rodents; faeces; tissues or liquids of animal origin)

- All feed equipment must be cleaned between deliveries and use on different farms.
- Mixing of feedstuffs and deliveries to different farms may well contribute to the spread of contamination.

Regarding feed and manure equipment:

- Use high pressure equipment, with or without steam, to clean silos, feed bunks and storage areas. Never use manure equipment for feed and vice versa.
- Check all feedstuffs for the presence of mould, sand, soil, manure and other contaminants
- Discarded milk must be pasteurized before being given to calves; milk from one cow can easily contaminate an entire group of calves.
- The milk powder mixer and other equipment must be cleaned after each meal; buckets must be stored separately.
- There are feed additives available for the preservation of feedstuffs (e.g. organic acids)
- Fermentation acids (silaging) can help to reduce the number of pathogens. Poor coverage, poor enveloping, poor bird control and high relative humidity can all contribute to the degradation of feedstuff quality
- Whenever a silage clamp or silo is opened and/ or re-opened, it must be checked for mould, soil and manure contamination and contaminated areas must not be fed to cattle.
- If moulds or soil are a recurrent problem, the harvest and storage method must be re-visited
- The feed bunks must be thoroughly cleaned at least once per day; feed refusals must not be stored for longer than 24 hours; give those refusals to the eldest maiden heifers
- Feed bunks must be constructed from non-porous material, as pores may harbor pathogens.

Biosecurity of feedstuffs

After BAMN, the Bovine Alliance on Management and Nutrition, AFIA, Arlington, USA, 2001

Regarding a feeding plan, records and labels:

- A label with a guarantee and composition must be present with each delivery.
- Medicated premixes must be stored and utilised dry and correctly; the mixer must be thoroughly cleaned after each medication.
- All labels and bills must be kept on file.
- Prepare a feed plan for each production group (calves; maiden heifers; bulls; dry cows; early lactation; mid lactation; end lactation cows). Dry cows can be kept in two groups: early dry period (far off) and late dry period (transition and close up cows).
- Define the feeding targets for each production group and the way in which performance and health are to be evaluated.
- Assess feed intake for each group and/or each individual in order to better evaluate performance and health.
- Quality test all feedstuffs and record the analysis results. Adjust the ration composition when needed (production level; growth rate; body condition score; health).

Regarding cereals, proteins, roughages and humidity:

Silage and hay can be a source of contamination (bacteria; moulds). Correct silage management can limit these risks.

- Protect the feedstuffs, silages, bales and feed bunk from contamination with manure and dead animals (Clostridium spp; botulism; E. coli; Salmonella spp).
- Do not allow birds, dogs, cats, rodents or wild life access to the feed, feed bunk or cattle housing. Additionally, do not allow them access to dead animals or tissues (afterbirth).
- Applying manure on crops, such as maize or alfalfa may contaminate the subsequent feedstuff. Such application must be carried out sufficiently earlier than harvest time to minimize the risks of paratuberculosis, salmonellosis and colibacillosis.
- Install nets against birds where needed to reduce the risk ofsalmonellosis.
- Ensure that conditions for a good harvest and storage of feed, including silage, are met for all products.

- Some tips and tricks:
  - DM of maize silage must be between 28 and 32%; 35 to 45% for alfalfa haylage.
  - Follow directives or manuals for a good silaging –check proper height, length and density.
  - Use good quality material for baling and for covering silage clamps/silos.
  - Check silage pH (<4) to avoid listeriosis and Clostridium spp to develop.
  - Organic acids could be added to silage: > 3% lactate for maize; 7% lactate for alfalfa silage; < 2 to 3% acetate; < 1% proprionate; < 0.1% butyrate. Too little lactate production indicates poor silage fermentation.
  - The size of the silos must be according to herd size. If feeding rate is too slow, moulds and bacteria will develop.
  - Hay bales must have a DM content of 85 to 90% and must be well packed.

Water quality and distribution systems:

- Water quality must be tested at least once a year (microbiology; chemistry); coliforms and enterobacteria are relevant. Watering points must be well-protected against manure contamination.
- Drinking places must be easy to clean and well-positioned in the barns and pastures. During heat stress conditions, the total length of drinking space must be 900 cm/100 cows.
- Water loss must be avoided as far as possible.
- Poor smell, colour or taste of water can indicate that a laboratory test on the water or its source is needed (nitrate, nitrite, iron).

Biosecurity plan

After BAMN, the Bovine Alliance on Management and Nutrition, AFIA, Arlington, USA, 2001

Nutritional balance

* Several pathogens are opportunistic in nature and only infect weak animals. A good ration composition and good feed intake can contribute to good immunity and general disease resistance (for example, a negative energy balance after calving may lead to ketosis; ketosis may lead to higher mastitis incidence). Minerals and vitamins too may contribute to better immunity. Good feeding contributes to better physical resistance of skin and mouth.

A biosecurity plan in practice

A biosecurity plan must be written and strictly followed on a daily basis. It is the responsibility of all people involved to adhere to it.

- Make it clear to all visitors that a biosecurity plan is operational on the site and they are expected to comply with the regulations laid out in that plan. It is helpful to have written instructions available for them.
  * Routine activities are best planned and organized. Always work in the same order from younger to older animals.
- All traffic to and from the farm must be under control (visit protocols; cattle delivery protocols; cattle reception protocols).
  * Cars and trucks that may carry manure particles on the wheels must be banned from inside the farm premisesand must stay on a parking lot or use the "dirty road".
  * Overalls and boots on the farm must be cleaned and dried daily; visitors are not allowed to use their own boots and overalls.
  * Visitors should only enter the animal houses or feed storage facilities unless absolutely necessary.
  * Visitors' materials and cars must be cleaned before entering and when leaving the farm
  * Rain water must be brought under control and properly stored.
  * All areas of still water must be eliminated from the premises.
  * All traffic of animals to and from the barns must be controlled, especially lorries that transport dead animals and living animals.
  * Control programmes against flies, rodents and birds must be in place around feed stores/clamps, feed bunks, feed machines and barns.
  * A cleaning and disinfection protocol, as well as a maintenance protocol must be developed and strictly complied with.

- People and their training/instruction.
  * The following people are involved in biosecurity: nutritionists, AI technicians, veterinarians, consultants from processing plants or feed industry. If needed, a brief, practical training programme should be developed for certain farming areas for employees. Examples of such areas are: cleaning & disinfection; colostrum management; good farming practice of milking; good TMR handling. The veterinarian and nutritionist are well-positioned to organize such training courses on-site if requested by the farmer.

LELY ASTR
© www.lely.com

Calf rearing

"The rearing of young stock on a dairy farm can be considered as an individual business enterprise. This chapter addresses many issues of rearing. First attention is given to, particular events in specific rearing periods together with specific risks followed by checklists for evaluating rearing management quality. Also included are several protocols for IgG testing, checking calves for respiratory disease, diarrhoea and rehydration, as well as heart girth measurement, ration composition, housing of calves, proper dehorning, hygiene, BVD control, biosecurity and for estimating estimation of losses related to mortality."

Rearing periods - birth to 2 months

From Boersema JSC. 2006. MSc thesis University Utrecht NL

Table 3.1 Potential risk factors in calves	
Disorders	Potential risk factors identified from the literature
Disorders around birth	Bull calf – More muscular breed sire used – Posterior position at birth (milk fever in dam; severe stress; low vitality)– 1st parity
Diarrhoea in 1st week of age	Inappropriate or lack anti-scours vaccination (resistant bacteria; GVP not applied) – Poor hygiene at calving (too many calving cows in the same pen; calves born on slatted floor; lack of attention from farmer; lack of time) – Poor housing hygiene (lack of attention from farmer) – Poor colostrum quality (unhygienic collection; low IgG level; unhygienic feeding; poor storage practices; using colostrum frozen more than 1 year ago; dilution with water; poor thawing practices) – Too long an interval between birth and the calf's 1st meal – calf does not receive first colostrum milked– Insufficient quantity at each meal – Group housed too early (too little space; no individual pen) – Inside housing – No preventive antibiotics administered – No free-choice salt provided – additional teats are removed too early – Large herds – Too little attention from care-taker – Sudden changes in feeding practice – No routine monitoring visits by veterinarian – Heat stress in calves
Diarrhoea in older calves	Use of maternity pen as sick pen – No individual calf hutches – Poor rearing hygiene practice (improper housing and feeding; damp bedding material) – Milk replacer fed without antimicrobials or the equivalent – Sick calves not isolated – Drinking water pH >8 – Roughage from pasture where manure was spread without ploughing in – Sudden changes in feeding or ration – lack of attention from care-taker
Navel disorders	Infected navel (dirty calving pen; dirty calf hutch; no navel disinfection applied; navel suckling by others; calf born on slatted floor) Navel hernia (genetic cause? sex effect? twin birth) – Navel cord too short (poor manipulation during birth eg C-section; posterior position at birth)
Poor weight gain	Infection in calf (navel disorder; diarrhoea; respiratory) – Feeding poor quality colostrum (see under diarrhoea in 1st week of age) or hay – Feeding poor quality milk replacer (inappropriate composition; poor quality water used at mixing; milk powder not stored dry; incorrect preparation or supply (temperature; poor mixing; dirty conditions; inappropriate feeding temperature) – Stressful calf-handling – poor water quality – Too much hay or grass given – Ad libitum milk replacer given (hampers intake of hay = poor rumen development = stress and susceptibility after weaning) – inappropriate concentrate composition – Feeding mastitic or antibiotic-contaminated milk – Stress/infection at dehorning (poor dehorning practice; incorrect timing of dehorning; stressful handling; too small headspace in feed rack) – Lack of concern of care-taker
Respiratory disorders	Poor barn climate (temperature, humidity, draughts; not adapted to climatic changes) – Housing older calves with younger, or too many age groups mixed – Ammonia level in air too high – Lack of attention from care-taker – Season – Poor colostrum management – Poor record keeping – Animal stocking density too high – BVD infection present in herd – Poor bedding material quality – New cattle purchased – Grazing in summer – History of disease (diarrhoea; respiratory)

Rearing periods - birth to 2 months

Disorders	Potential risk factors identified from the literature
M. paratuberculosis infections	Unhygienic conditions around calving (calving pen not cleaned and disinfected) – Unhygienic birth – Calf not removed from dam immediately (no supervision; bull calf; other work) – Calf born on slatted floor – Poor colostrum management (see above) – Contact with older cattle
Incorrect identification of calf	Several cows calve at the same time – Calves born at the same time in the barn or at pasture – Mixing up ear tags – Poor record keeping – Ear tag of dead calf placed in another calf – Computer crash and loss of data
Incorrect extra teat removed	Poor record keeping – Incorrect identification of teat to be removed – Inappropriate teat removal practice

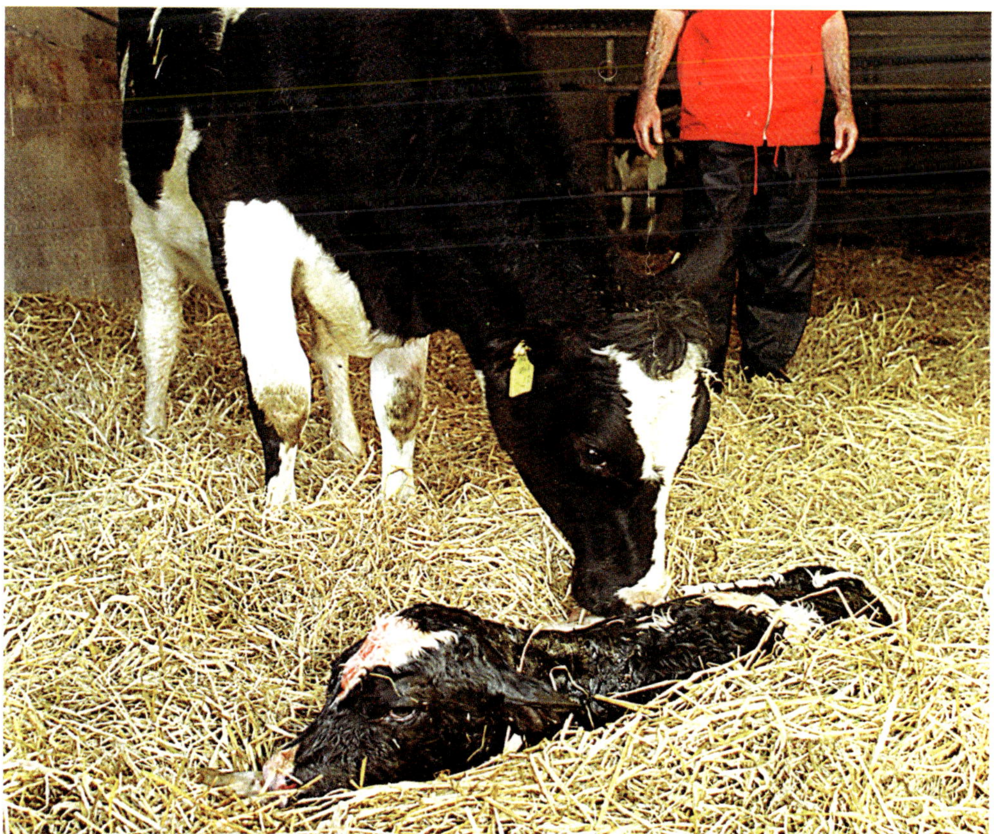

Rearing periods - 2 to 15 months

Disorder	Potential risk factors identified from the literature
Poor weight gain	Infection causing disease in calf (navel infection; ear infection after tagging; diarrhoea; respiratory disease) – Overcrowding – poor quality of feed or water-- Insufficient dietary energy/minerals – Light intensity in barn too low– inappropriate weight at insemination
Respiratory disorders	Herd stocking density too high – Care-taker too busy -- Poor barn climate-- Sudden weather change -- Poor animal management (no hair clipping at the beginning of barn period) -- Poor grazing management -- Poor lungworm infection prevention strategy
Diarrhoea in older calves or maiden heifers	Herd prevalence (eg salmonellosis) -- Farm size -- Fasciola hepatica infection -- Feeding roughage from fields where manure was applied and not ploughed in during the same growing season -- Poor hygiene practices in calf rearing (damp bedding) -- No isolation of sick animals -- Offering water with a pH >8,0 -- Lack of concern of care-taker -- Feed related -- Sudden change of diet
Locomotion problems	Poor claw trimming -- Overcrowding -- Poor hygiene in cubicles -- Too little space in cubicles -- Older animals -- Inside housing / zerograzing -- Large herds -- Previous Mortellaro infections -- Animals in heat
Gastrointestinal parasites	Poor grazing management – Poor worming strategies
Liverfluke infections	Animals grazing on wet land (presence of ditches; high ground water level) -- Presence of snail habitats -- Poor "worming"
Trichphyton verrucosa	Contact with other infected animals -- Poor hygienic practices -- Incorrect treatment of infected animals -- Poor climatic conditions (high humidity; high temperature) – Purchase of infected cattle
Fattening of udder tissue (± 8 mo age)	Feeding rations that are too high in energy between three and 10 months of age
M. paratuberculosis inf.	Contact with adult cattle or with manure before 1 year of age – Insufficient separation of calves from adult cattle
Not pregnant < 15 months of age	Lack of proper heat detection -- Inappropriate bull selection -- Incorrect insemination technique -- Lack of mount detection -- Heifer not in oestrus before 15 months of age -- Light intensity too low (fewer oestrus behaviour signs) – Anoestrus / suboestrus – inappropriate body weight for insemination

Rearing periods - 15 months to 3 weeks before 1st calving

Disorder	Potential risk factors identified from the literature
Abortion	Herd prevalence of several abortion-causing agents -- Large farms -- Buying heifers from other farms -- Grazing heifers on pasture next to infectious animals from neighbouring farm
Poor weight gain	Inappropriate weight at insemination – Infection-causing disease in calf (navel infection, ear infection after tagging, diarrhea, respiratory disease, gastrointestinal parasites) – Overcrowding – poor quality of feed or water -- Insufficient dietary energy and minerals provided
Lameness	Poor claw trimming practices in young stock – No routine screening of claws in heifers -- Overcrowding -- poor hygiene in cubicles -- Too little space in cubicles -- Older animals -- Inside housing / zerograzing -- Large herds -- Feeding management failures
Respiratory disorders	Herd stocking density – Care-taker too busy – Poor barn climate -- Sudden weather change – poor ventilation in housing -- Poor grazing management -- Inappropriate lungworm infection prevention strategy
Gastrointestinal parasites	Poor grazing management -- inappropriate worming policy -- Grazing heifers that have never been outside.
Trichophyton verrucosum	Contact with other infected animals -- Poor hygiene practices -- Inappropriateg treatment of infected animals -- Poor barn climate (high humidity; high temperature)
Liverfluke	Animals grazing on wet land (presence of ditches; high ground water level) -- Presence of snail habitats -- inappropriate "worming" policy
Mastitis	Teat / Udder-sucking -- Flies in cubicles -- Milk leaking before calving – Poor general hygiene – Poor animal hygiene
Over-conditioning	Rations fed that are too high in energy between 3 and 10 months of age
Heifer comfort	Cubicles too small in relation to body size -- Introduction of heifers into the herd (rank fights) – Poor hygiene in calving pen and cubicles -- Health threats like mastitis, lameness and abortion -- Poor adaptation to "lactation-rations" -- Poor barn climate (high humidity; high temperature) -- Feeding management failures – Poor quality of feed and/or water

Rearing periods - around first calving

Disorder	Potential risk factors identified from the literature
Dystocia in heifers	Bull used with low calving ease index -- Heifer body condition < 3 (inappropriate close-up ration composition or diseases) -- 1st parity heifer -- Younger age at calving (<24 month) – Poor social interactions with herd mates – Milk fever / loss of feed intake
Retained placenta	Over-conditioned at calving -- Stress around calving -- Assistance during labour -- Decreased general immune-responsiveness (BVD infection; ketosis) – sub-optimal mineral status
Trauma in newborn calf	Dystocia of dam -- Calving pen too small-- Stressed dam (steps on calf) -- Born on slatted floor -- Born at pasture without supervision too close to a ditch -- Poor management intervention
Incorrect identification of newborn calf	Many cows/heifers calving at the same time -- Calves born without supervision in close-up group or at pasture -- Incorrect recording of events/animal numbers-- Mixing up ear tags -- Tag of dead calf used on another, living calf (allows calf to be removed sooner from farm) -- Computer crash without backup resulting in loss of data
Mastitis	Teat / udder-sucking – Flies -- Poor hygiene in cubicles -- Milk leaking prepartum -- Milk fever -- Blood in milk -- Teat oedema -- High peak milk flow -- Teat canal protrusion – Over-conditioning of heifer
Milk fever (not likely in heifer)	Higher parity -- High milk production level -- High body condition score -- Positive cation-anion balance in diet -- High calcium level in feed -- Season -- Climate -- Housing – Lush pastures -- Lack of exercise -- Length of dry period -- Prepartum milking --Loss of feed intake
Displaced abomasum	Negative energy balance prepartum and subsequent ketosis -- High body condition score -- Suboptimal feed bunk management -- Prepartum diets containing >1,65 Mcal of net energy for lactation/kg of dry matter -- Winter and summer season -- High genetic merit -- Low parity -- Milk fever – Stress – Non-social adaptation of heifers in milking herd
Lameness	Poor claw trimming – Overcrowding -- Large herds -- Poor hygiene in cubicles -- Insufficient space in cubicles -- Inside housing (on slatted floor) / zero-grazing -- Unequal or loose slats -- Floors (slatted or not) without dung remover -- Feeding failures (ration formulation, ration changes)
Udder oedema	Positive cation-anion balance in ration during late gestation -- 1st parity heifer
Respiratory disease	High herd stocking density – Care-taker too busy -- Poor ventilation and barn climate -- Sudden weather change – poor ventilation in housing -- Poor grazing management – Inappropriate lungworm infection prevention strategy

Rearing periods - around first calving

Disorder	Potential risk factors identified from the literature
Liverfluke	Animals grazing on wet land (presence of ditches ; high ground water level) -- Presence of snail habitats -- Inappropriate worming policy
Poor heifer comfort	Cubicles too small in relation to body size – Introduction of heifers into milking herd (rank fights) -- Poor hygiene in calving pen and cubicles -- Health threats like mastitis, lameness and abortion -- Poor adaptation to lactation-rations -- Poor climatic conditions (high humidity; high temperature) -- Feeding management failures – Poor feed and or water quality -- Excessive social interactions
Poor feed intake	Poor quality of feed and water -- Overcrowding/ rank fights -- Discomfort due to disease or heat stress (Thermal humidity index) – Dystocia and postpartum problems -- Negative energy balance and ketosis – Over-conditioning -- Leaving calf with dam.

© www.lely.com

Calf rearing checklist

Adapted after R.Sellers/J.Rydell/BAMN, 2001; G. van Trierum, Internatl Dairy Topics, 2005) Σ= 80 items

Table 3.2 General inventory checklist for calf rearing

1. Dry cow management	Best management	Moderate	Poor management
Average BCS	3.0 – 3.5	-	< 3.0 & > 3.5
Supply of minerals and vitamins	As prescribed	-	Sub-optimal
Hygiene status	Good	Moderate	Poor
Udder/teats are clean	Yes	-	No
2. Calving pen	**Best management**	**Moderate**	**Poor management**
Present	Yes	-	No
Hygiene status	Good	Moderate	Poor
Cleaned/disinfected before each calving	Yes	-	No
Calf separated from the dam	Immediately	Within 1 day	After 1 day
3. Birth management	**Best management**	**Moderate**	**Poor management**
Navel disinfection 7% I	Yes	-	No
Calf is rubbed dry	Yes	-	No
Housing of calf	In hutch or single pen	With its dam	In straw pen
Cleaning/disinfection before arrival of calf	Yes	-	No
4. Colostrum period	**Best management**	**Moderate**	**Poor management**
1st colostrum within 2 h	Yes	No	Left with dam
Quantity 1st colostrum	1 ½ - 2 ½ L	-	< 1 ½ L or > 2 ½ L
Colostrum quality	Serum IgG > 2000 mg/L	Serum IgG 1000-2000	Serum IgG < 1000 mg/L
1st colostrum also for 2nd feeding	Yes	-	No
2nd feeding max 8 h after 1st feeding	Yes	-	No

Calf rearing checklist

4. Colostrum period	Best management	Moderate	Poor management
1st colostrum stock in freezer + clearly labelled	Yes	-	No
Colostrum intake/day	4-6 L	-	< 4 L or > 6 L
Colostrum meals / day	3	2	Unlimited
Colostrum hygiene	Good	Moderate	Poor
5. Milk replacer mixing	**Best management**	**Moderate**	**Poor management**
Water quality checked	Yes	-	No
Mixing temperature	40-65 °C	-	< 40 or > 65°C
Concentration	± 125 gr/L	± 100 gr/L	Unknown
6. Milk replacer feeding	**Best management**	**Moderate**	**Poor management**
Drinking temperature	40-42 °C	lukewarm	< 38 °C
Use of feeding schedule	Yes	No	Intuitively
Check quantity fed	Yes	-	No
Check drinking temp.	Yes	-	No
Litres at 1st feeding	1 ½ - 2.0 L	< 1 L	> 2 L
Drinking system	Automated	Teat-bucket	Bucket
Hygiene status	Good	Moderate	Poor
7. Supplying cows milk	**Best management**	**Moderate**	**Poor management**
Feeding schedule used	Yes	No	Intuitively
Drinking system	Teat-bucket	Trough	Bucket
Dilution	No		Yes
Additions to milk	Yes	-	No
Mixing of cows milk	No		Yes
Antibiotic or high cell count milk supplied	No		Yes

Calf rearing checklist

7. Supplying cows milk	Best management	Moderate	Poor management
Maximum quantity per meal	2-3 L	<2	>3 L
Fermentation applied	Yes	-	No
Pasteurization applied	Yes	-	No
8. Single pen housing	**Best management**	**Moderate**	**Poor management**
Days in single pen	5- 10 days	< 5 days	> 10 days
Extra water supplied	Yes	Yes + electrolytes	No
When is water supplied	Between meals	After the milk	Not applicable
Concentrates supplied	Calf pellets	Cow pellets	No
Roughage supplied	High fibre	Maize	No
Hygiene status	Good	Moderate	Poor
9. Group housing	**Best management**	**Moderate**	**Poor management**
Age difference/group	< 2 weeks	-	> 2 weeks
Water supplied	Unlimited	Limited	No
Freshness of water	Always fresh	Daily	Not fresh
Concentrates supplied	Unlimited	Limited	
Type of concentrates	Calf pellets	Cow pellets	Beef cattle pallets
Frequency of concentrates.	Twice daily	Once daily	Unlimited
Roughage supplied	High fibre	Maize	Silage
Freshness and quality of roughage	Good	Moderate	No attention
10. Health status	**Best management**	**Moderate**	**Poor management**
Diarrhoea	No	0 - 2 weeks	> 3 weeks
Colour of faeces at five weeks age	Brown	Yellow	Grey
Faecal consistency	Solid	Cake	Runny

Calf rearing checklist

10. Health status	Best management	Moderate	Poor management
Coat condition at five weeks age	Shiny	Dull	
Coat condition at 10 weeks age	Shiny	Dull	
Feed digestion	Good	Moderate	Poor
11. Barn climate	**Best management**	**Moderate**	**Poor management**
Air inlet	Yes	-	No
Wind-breaker netting	Yes	-	No
Additional heating avail.	Yes	-	No
Air ventilation	Yes	-	No
Air distribution pattern	Good	Moderate	Poor
Humidity	60-80 %	-	< 60 or > 80%
Air quality – NH3	Good	Stench	
Infections older calves	No	-	Yes
12. Monitoring & evaluation	**Best management**	**Moderate**	**Poor management**
Mortality rate first three months of age	< 5%	> 5%	Unknown
Weight gain first three months of age	800-900 gr/day	-	< 800 gr/day or > 900 gr/day
Concentrate intake at six weeks age	> 1 kg/day	½ - 1 kg/day	< ½ kg/day
Weaning age	6-10 weeks	< 6 weeks	> 10 weeks
Weight at three months	> 110 kg	90 – 110 kg	< 90 kg

Calf rearing checklist

Table 3.3 Calf rearing checklist

		Best practice	Moderate	Poor
Group-wise housing	Frequency of concentrates/day	2x per day	1x per day	Ad libitum
	Forage type	Fibres (hay)	Maize	Silage
	Fresh forages	Yes	Not always	No / unknown
Health evaluation	Diarrhoea	No	0 – 2 weeks	> 3 weeks
	Colour of faeces	Yellowish	Brown	Grey
	Consistency of faeces	Solid	Pudding	Liquid
	Coat condition	Shiny	-	Mat
	Digestion of feed	Good	Average	Bad
Barn climate	Air inlet	Yes	-	No
	Windbreak nets	Yes	-	No
	Heating is possible	Yes	-	No
	Ventilation	Good	-	Bad
	Distribution of air	Good	Average	Bad
	Quality of air	Good	-	Bad
	Infection of older calves is possible	No	-	Yes
Evaluation parameters	Consumption of concentrates at six wks	> 1 kg	½ kg	< ½ kg
	Weaning age	6 – 10 weeks	-	< 6 or > 10 weeks
	Body weight at three months of age	> 110 kg	90 à 110 kg	< 90 kg
	Age at first service	> 15 month	14 – 15 month	< 14 month
	Weight at first service	> 380 kg	-	< 380 kg

NOTE: If a calf is seen going to other calves for navel drinking, following a milk meal, it indicates that this calf drinks its milk too quickly, which could result in poor digestion and retarded growth.

102633

Evaluating calf rearing

Adapted after Interact Agrimanagement BV, Netherlands, 2004

Table 3.4 Quick reference chart for evaluating process on dairy farms

Age period	Target	Implications and factors to address	Scores (1 to 5)
At calving	Clean calving pen. Remove calf immediately from dam	Lower DM intake of dam. Infection transfer via faeces	
From calving to 6 hours	First colostrum fed twice (temperature; IgG quality; method → see documents CALF 3 and 4 for reference values)	Diarrhoea and other disease risks	
From calving to 3 days	Colostrum 3x 1 ½ L per day (check IgG concentration)	Risk of Johne's disease when fed from own dam	
3 days to 4 weeks	Milk replacer (2x 2 ½ L per day) Calf concentrates (pellets 1 kg) Fresh, clean water ad libitum Individual housing on straw	If more than 2 ½ L given = saturation effect. Infection transfer: diarrhea risk. Reduced palatability and intake. Protein% of different sources varies. Chemical & microbiological contamination risks. Contact risks & cross contamination.	
4 to 8 weeks	Weaning at 85 kg body weight 1 ½ - 2 kg concentrates/pellets Good quality hay Fresh, clean water ad libitum Small groups in pens with straw Ventilated barn	If weaned older, saturation effect is possible (see above). Protein levels in rations should be monitored and average daily gain should be checked. Alternative sources of roughage/fibre may be required. Mind feed competition among calves.	
8 weeks to 6 months	At 6 months = 180 kg weight (heart girth measurement) & BCS 2+. Good quality hay. Calf concentrates/pellets (< 3 kg). Crude protein% less than 17%. Fresh, clean water ad libitum. Thick straw and ventilated barn.	Hay must not originate from manure spread parcels because of Johne's risk. If needed, use other type of pellets; otherwise growth delay and fattening may occur. Test on chemical & microbiological contamination of feed	

Evaluating calf rearing

Age period	Target	Implications and factors to address	Scores (1 to 5)
From 6 to 9 months	270 kg weight & BCS 2.5 Ration with 16% crude protein. Fresh, good quality water ad libitum.	Low protein% and poor energy to protein ratio leads to fattening.	
From 9 to 13 month	At 13 months = 370 kg & BCS 3- Ration with 16% crude protein. Fresh, good quality water ad libitum. House on slatted floor, in groups of 6-10 animals.	Heart girth mesured at 1st service = 167 cm. Low protein% and poor energy to protein ratio leads to fattening. When > 10 calves per group = feed competition; variable growth rates. Solid floors often = dirty floors.	
From 13 to 24 months	At 24 months = 570 kg (without calf) & BCS 3+ Good quality grass silage with good energy level Crude protein in ration to be decreased down to 15%. Fresh, good quality water all day. Optimal vitamin and mineral supply	Sub-optimal blood Cu + Se + vit E + Co and pathology may occur. Check blood levels of Cu, Se, Co and Vitamin E	
	At five weeks prior to expected calving date → move to close-up group. Provide more energy-dense, palatable ration	Reduced palatability = reduced DMI Enhanced body condition loss with low energy rations Check anion-cat ion-balance (urine pH; edema)	

NOTE: In this Table various items in column 2 correspond with items in column 3 on the same line.

Testing colostrum (IgG) quality

There are two principal methods for assessing the IgG quality of colostrums on the dairy farm:

1. by colostrometer using fresh colostrum;
2. by refractometer using the blood from calves.

The colostrometer

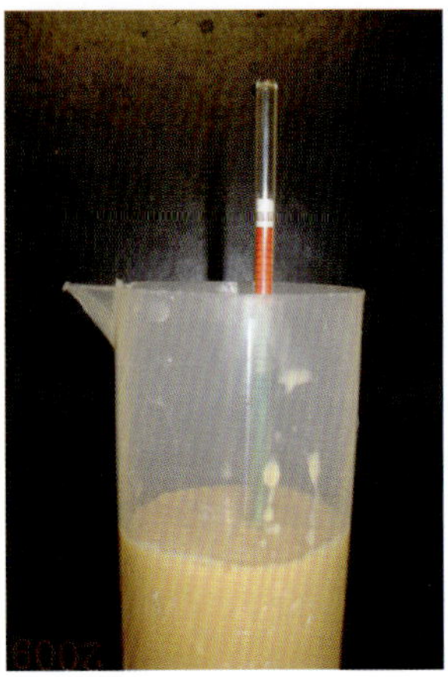

The colostrometer is a practical, simple, cheap (mostly plastic) instrument used to assess the IgG content of fresh colostrum on the farm. It is basically a densimeter and measures the IgG content in a semi-quantitative manner.

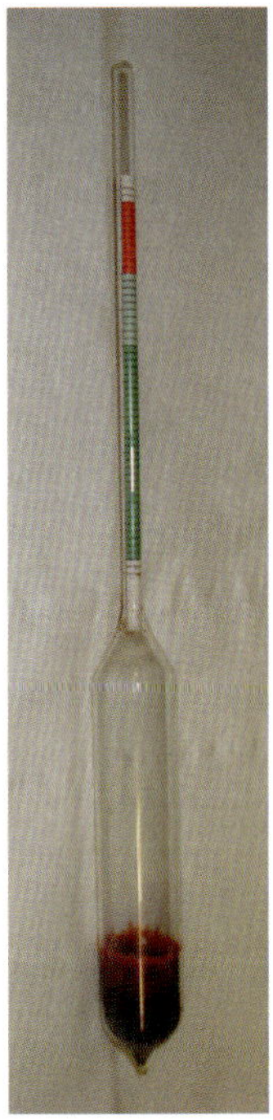

It distinguishes between three classes of IgG (measured at 22°C or 72°F):

< 20 mg/ml colour is **red**
→ insufficient IgG content

20–50 mg/ml colour is **blue**
or sometimes **yellow**
→ doubtful IgG level

> 50 mg/ml colour is **green**
→ adequate IgG content

Colostrum should be tested as a routine management procedure before being fed to calves.

Testing colostrum (IgG) quality

Photos by courtesy of John Mee
reference: Kathy Lee, Michigan State University Extension, 2011

The refractometer

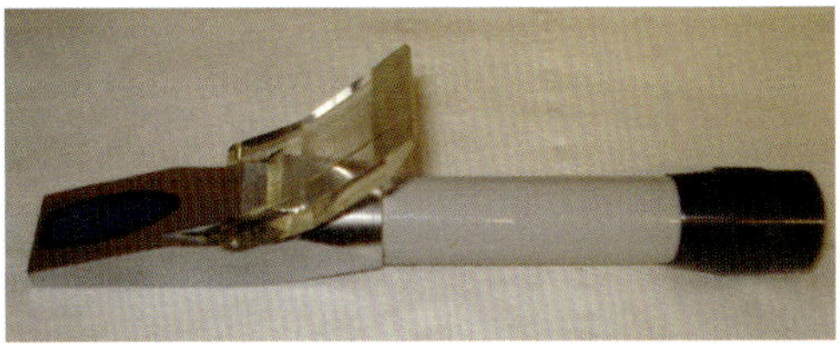

A drop of calf blood is placed into the chamber (on the left in the photograph), the flap is closed and values are reported on a screen behind an optic on the right of the photograph.

5 g/L	**Very poor IgG level**
5-10 g/L	**Insufficient IgG**
10-15 g/L	**Sufficient IgG level**
> 15 g/L	**Good level of IgG**

Another option to test colostrum quality (IgG) regards the **BRIX refractometer**, which is much less costly (about 90 US dollars) than the one previously named which costs about 600 US dollars.

There is an optical and a digital version. The digital version provides a figure, a score. The interpretation is as follows: >

22%	**= good quality**
18-21%	**= fair or moderate quality**
< 17%	**= poor quality**

The BRIX refractometer can also be used to check the adequate mixing of milk replacer powder in water and the consistency (= total solids). See the BRIX website for conversion tables: from BRIX values to gravity values, and *vice versa*.

Respiratory disease in young calves

Amy Stanton, University of Guelph, Canada, www.dairyherd.com/newsletter October 2011

Respiratory diseases in young calves are responsible for great economic losses in production. An early diagnosis and intervention are important issues to reduce the impact of such diseases.

The table 3.5 below comprises some major signs to detect disease in young calves.

Table 3.5 Basic respiratory diseases in young calves

Basic signs	Details
Decreased appetite	Especially in ad libitum milk fed calves, this is an important sign
Decreased energy & lethargy	Affected calves will lie down more frequently
Decreased thirst	Not easy to detect unless by close observation
Apathy	Calves do not care about human interference
Energy conservation	Short lying with head back on the body
Depression	Poor hair coat quality; dull eyes; lowered head
Fever	
Other signs	
Statue standing	With the back curved; neck below the chest
Lethargy*)	Little to no response on stimuli
Decreased willingness to approach	Affected calves not showing curiosity
Growth delay	Affected calves not fitting the standard curve

* For lethargy, a scoring system has been developed:

Score 0	Calf is already standing when person enters
Score 1	Calf stands in response to vocal encouragement
Score 2	Calf stands after vocal call and gate opening
Score 3	Calf stands when person enters pen + 3d call
Score 4	Calf stands only after person makes contact

The more deviating signs are observed in combination, the higher the probability that indeed the calf is suffering from disease. Signs on their own are not (disease-) specific.

IgG concentrations. Milk protein/energy

GD, Deventer, NL, 2008

Table 3.6 IgG concentration and thresholds in blood from newonatal calves	
< 5 g/L	very deficient
5 – 10	insufficient
10 – 15	sufficient
> 15	good

Light in the cattle barns

For each 10 m² of surface, a NEON light of 40 watt must be placed at a height of 3 m. During the day, there should be sufficient light in the barn to easily read a newspaper.

A six-hour period of 'night' per 24 hours should be observed. This practice will promote better feed intakes, reproductive performance and milk production, especially during winter.

Table 3.6 Milk test for energy and protein

	Beta-Hydroxy butyrate (BHB)	Urea	Explications
1	Normal	Normal	No issues
2	Deficient (> 0.1 mMol/L)	Normal	Negative energy balance (NEB)
3	Normal	Deficient (> 6.6 mMol/L or < 3.3 mMol/L)	Insufficient dietary protein
4	Deficient	Deficient	See (2) and (3)

Elevated BHB levels at the start of lactation indicate elevated fatty acids in the liver originating from the dry period (poor transition management).

Elevated BHB levels during mid-lactation indicate insufficient dietary energy or a reduction in feed intake (due to the ration itself or disease). See body condition scoring and rumen fill scoring under HHPM or MONITOR.

Diarrhoea & rehydration

GD Deventer, NL 2008

Diarrhoea in young calves is always a source of worry for the farmer due to the risk of death. A recent study has shown that the annual mortality rate of young calves is around 9%, while 62% of these cases had been affected by diarrhoea (Heinrichs & Kehoe, 2009).

It is highly important to attain a rapid diagnosis and administer treatment quickly to try to avoid **the greatest risk – dehydration. Calves become dehydrated rapidly**, especially in the first weeks of life.

Diarrhoea may cause body weight losses of up to 10 to 15% per day. **Body weight losses beyond 8% warrant IV fluid therapy**, while losses of 14% or more are lethal.

The farmer and his employees must be capable of correctly and rapidly diagnosing this condition by observing the young calves at least once per day. They must then be able to start appropriate treatment.

If the diarrhoea persists in spite of treatment, the farmer must call the vet who should take faecal samples for microbiological analysis and, if necessary, administer antibiotics.

To estimate the level of dehydration in young calves, the following scoring system has been devised (Table 3.7 , Heinrichs & Kehoe, 2009):

Table 3.7 Scoring Chart for scoring dehydration in young calves

% of dehydration	Clinical signs
5 – 6	Diarrhoea; calf still drinks (reflex remains); no other signs
6 – 8	Moderate depression; calf still drinks, skin fold stays up for 2 - 6 sec; the eyes are retracted; calf is weak
8 – 10	Severe depression; calf stays lying; skin fold stays up > 6 sec, eyes are very retracted; the mucosae are pale and dry
10 – 14	Calf no longer stands up; extremities are cold; skin fold stays up; calf is in a coma
> 14	Death

Dairy farms should have a **protocol** for the diagnosis and treatment of young calves. That protocol should comprise actions for answers to questions, such as: "*When is oral rehydration treatment needed?*" and "*How much rehydration product should be administered?*", as well as general prevention & control measures.

For a proper **assessment of the diarrhoea situation, three scoring charts** have been developed that score the:

1. severity of the diarrhoea (Table 3.8)
2. respiration rate of calves (Table 3.9)
3. clinical appearance of affected calves (Table 3.10)

Diarrhoea & rehydration

Table 3.8 Scoring chart for determining the severity of diarrhoea in young calves	
Score 1	Calves show normal faeces, like **pudding**
Score 2	Calves show faeces with less consistency, like **yoghurt**
Score 3	Calves show true diarrhoea; like **syrup**
Score 4	Calves show true diarrhoea, like **fruit juice,** faecal material still present
Score 5	Calves show diarrhoea like **water,** without faecal material; with blood or mucus

Table 3.9 Scoring chart to determine the respiration quality of calves with diarrhoea	
Score 1	Calves show normal respiration; no other signs
Score 2	Calves may cough a little and have nasal discharge; regular respiration
Score 3	Calves show rapid respiration and cough frequently
Score 4	Calves cough frequently; rapid respiration
Score 5	Calves show irregular respiration and chronic coughing

Table 3.10 Scoring chart to determine the clinical appearance of calves with diarrhoea	
Score 1	Calves are alert and active
Score 2	Calves have ears hanging down; are hardly responsive to stimulations
Score 3	Calves are somewhat depressed; head and ears hang down
Score 4	Calves are depressed; head and ears hang down; will not stand up
Score 5	Calves are very dull and lie down all the time

On the basis of these three scoring tables, it is possible to calculate a total, mean score per day for one calf or for a group of caves. When the total mean exceeds a certain threshold (for example score six or eight), then rectal temperature should be taken, and rehydration and/or antibiotic therapy should be initiated.

Example of calculation for observations in a group of 15 calves:
The results are for Table 3.8: 10 calves score 3 and 5 calves score 2;

Table 3.9: 9 calves score 2 and 6 calves score 3;

Table 3.10: 6 calves score 2 and 9 calves score 4.

The total mean will be 124/15= 8.3, which is higher than the threshold of 6 (or 8). These calves must be addressed and rectal temperature taken. Rehydration and antibiotic therapy should also be initiated

Note that a score of two and above, the animals are already considered to have a 'disorder'. Scoring two in each category would result in a threshold score of 6 (= 3 x score 2) per day. Less strict thresholds may be set; however, you are likely to be less rapid at taking action.

➲ The solutions for oral rehydration of a diarrhoea case must not be confused with electrolyte solutions!

The latter are most often used in calves after weaning or in adult cattle during stress. Table 3.11 gives the suggested concentrations of ingredients for oral rehydration solutions (Heinrichs & Kehoe, 2009).

Diarrhoea & rehydration

GD, Deventer, NL, 2008

Table 3.11 List of suggested concentrations of ingredients for oral rehydration solutions

Ingredient	Concentration in mMol/L	Concentration in g/L	Remarks
Natrium(sodium), Na	70 to < 145	< 3.3	Osmotic regulation
Natrium bicarbonate	50 – 80	4.2 – 6.7	Counters metabolic acidosis
Natrium citrate	50 – 80	14.7 – 23.5	Counters metabolic acidosis
Natrium acetate	50 – 80	6.8 – 10.9	Counters metabolic acidosis
Kalium (potassium), K	20 – 30	0.8 – 1.2	Maintains pH
Chloride, Cl	50 – 100	1.8 – 3.5	Maintains pH
Glucose (dextrose)	< 200	< 36	Source of energy
Glycine	< 145	< 10.9	Stimulates absorption

Feeding young calves

Table 3.12 Comparison of the concentration of Fe in whole milk and in colostrum

	Colostrum	Whole milk
Average	1.16	0.31 mg-kg
Difference	0.85	
Min–max	0.55-2.03	0.28-0.35

Table 3.13 Administration of Fe during the first 10 days of life (40 kg birth weight) and the effect on concentrate intake

Control group	42.8 kg concentrates
Study group	52.7 kg concentrates
Difference	9.9 kg ($p < 0.06$)

Are there any differences in the fibre content in the various protein sources for calves?

Table 3.14 Fibre content in various protein sources (reference value : < 0.5% fiber)

	% crude protein	% crude fiber
Whole cow milk	34	--
Concentrates of soya proteins	67	3.6
Isolate of soya protein	86	0.2
Hydrolysate of wheat protein	82	0.5

Body weight: measurement of heart girth

Alain Dequin; April, 2002

Table 3.15 Body weight: measurement of heart girth

Holstein Friesian: STANDARDS

Centimeters	Weight in kg	Corresponding Age
75	45	birth
100	100	3 months
120	150	5 months
128	178	6 months
135	200	8 months
143	243	9 months
148	250	10 months
157	300-315	12 months
165	350	13 months
172	420	15 months
180	475	18 months
185	500	20 months
187	520	21 months
190	550	23 months
194	580	24 months
196	600	25 months
202	650	26 months

Remember that **around 8 months** of age, female calves start to develop functional udder tissue. Over-feeding around this age may lead to fat infiltration of udder tissue and subsequent lower than expected milk production.

It is important to note that the **weight** of the maiden heifer is much more important than its age regarding the timing of **first insemination**. Normally, the combination of age and weight determines the timing of the first insemination. First heats should be detected between 12 and 13 months of age. Inseminations may then be carried out between 13 and 14 months of age.

A maiden heifer needs 0.8 ha of grazing for every month of her age for the first year and 0.7 ha of grazing for every month of her age in the second year. For example, a maiden heifer of 11 months old will need 11 x 0.8= 9 ha in her first year.

Below a girth tape and rope in operation.

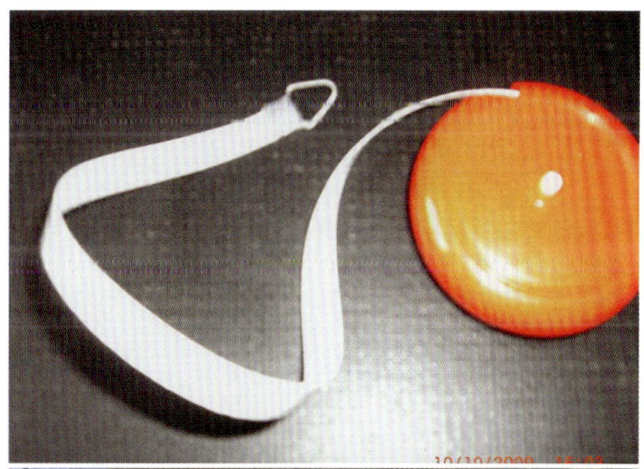

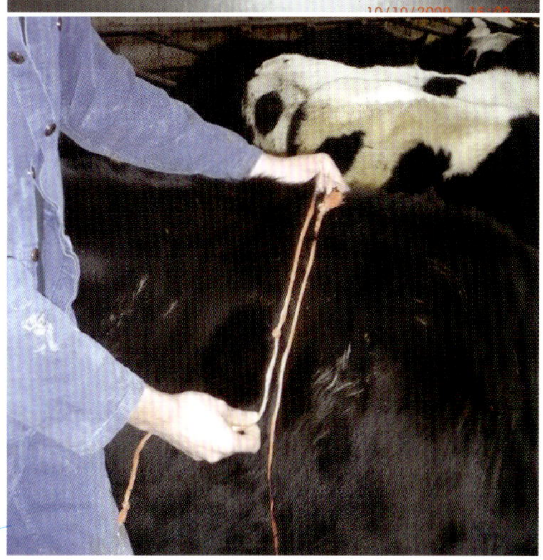

Heifer body weight

Patrick C. Hoffman (University of Wisconsin, USA)
International Dairy Topics (2007) 6 (6) 7-9

Table 3.16 Field chat for the practical assessment of body weight in maiden heifers

Age of maiden heifers (month)	% of mature body weight	Actual body weight (kg)	Your maiden heifers(cm→ kg)
1	6.5		
2	9.7		
3	12.8		
4	16.5		
5	20.2		
6	24.0		
7	27.7		
8	31.4		
9	35.0		
10	42.5		
11	46.3		
12	49.9		
13	53.7		
14	57.4		
15	61.1		
16	64.7		
17	68.5		
18	72.2		
19	76.0		
20	79.6		
21	83.3		
22	87.1		
23	90.8		
24 before calving	94.0		
24 after calving.	85.0		

Using the Field Chart: take the tape measure and determine, for each calf or heifer, the heart girth; put the results in cm in the fourth column. Next step is to convert the cm into kg body weight and put the latter into the third column. By using the reference values (see elsewhere in this series) you can compare the actual results with the target body weights.

Ration composition

Following numerous experiments carried out investigating rearing calves on five different rations over a 10 week period, it was noted that a calf consumes around 38 kg of milk powder (milk replacer).

The experimental rations are given below, together with their respective DMI for the 10 week period (Table 3.17; *International Dairy Topics, 2010*).

- Ration 1 hay ad libitum (a)
- Ration 2 concentrates ad libitum (b) + 5% straw (c)
- Ration 3 concentrates ad libitum (b) + hay ad libitum (a)
- Ration 4 concentrates ad libitum (b) + maize silage ad libitum (d) + 5% straw (c)
- Ration 5 maize silage ad libitum (d) + 5% straw (c)

Table 3.17 Consumption of respective ration feedstuffs (kg DM per week) for five different calf rations

Week#	Ration 1 (intake of feedstuff a)	Ration 2 (intake of feedstuffs b and c)	Ration 3 (intakes of feedstuffs b and a, respectively)	Ration 4 (intakes of feedstuffs b, d and c, respectively)	Ration 5 (intake of feedstuff d)
2	0.03	0.07	0.14, 0.00	0.04, 0.01, 0.01	0.01
3	0.11	0.20	0.30, 0.01	0.17, 0.01, 0.00	0.04
4	0.19	0.40	0.49, 0.02	0.20, 0.02, 0.00	0.08
5	0.37	0.68	0.67, 0.02	0.56, 0.03, 0.00	0.17
6	0.59	1.06	1.00, 0.03	0.88, 0.04, 0.00	0.32
7	0.80	1.29	1.30, 0.05	1.18, 0.03, 0.01	0.44
8	1.03	1.66	1.84, 0.06	1.75, 0.04, 0.01	0.61
9	1.43	2.09	2.35, 0.14	2.08, 0.06, 0.01	0.86
10	1.79	2.56	2.88, 0.12	2.71, 0.11, 0.02	1.10

Ration 3 resulted in the best intakes of all the experimental rations.

Maize silage is not an appropriate feed for calves during the first 6-10 weeks of rearing due to the lower DM and fibre content compared with other forages. Fibre is paramount for young calves in order to avoid rumen acidosis!

During the first and second weeks of rearing, water should be given restrictedly because of the liquid feed; however, from week two onwards, fresh, clean water should be available ad libitum.

Housing of youngstock

After Jeanette FISHER, Heifermax, Tasmania, Australia, in Int Dairy Topics vol 9, nr 6, November 2010

Optimal housing comprises four key elements:

1. Animal comfort (welfare)
2. Good barn climate (ventilation; temperature; relative humidity (RH))
3. Appropriate isolation
4. Economics.

In terms of welfare, calf comfort means 'lack of stress'. Good calf comfort promotes good disease resistance and efficient feed conversion.

Factors that positively impact animal health and welfare, also have a positive impact on farm economics.

Common housing design & management problems:

- **draughts**: an opening is left between the flooring and the walls causing calves to lay down in draughty areas.
- **stagnant air:** good air flow at head height may induce areas or corners where air does not move at all. For example, single calf pens with three sides closed will create stagnant air at the back of the pen. Stagnant air can also be found in the middle of large barns, or in calf pens where the solid barn wall is at the back of the pen. Calves kept in these stagnant air pockets may succumb to respiratory disease earlier than their herd mates.
- **practice of keeping calves warm**. Farmers sometimes keep doors and inlets in naturally ventilated barns closed to 'keep calves warm'. However, this practice induces an increase in Relative Humidity (RH), ammonia and pathogen build-up.

Practical suggestions for improvement:

1. Create good ventilation around the calves while avoiding draughts. Air movement must be above the calves, not at calf level! However, the situation may need to be readdressed during periods of heat stress (see under the heading of CLIM 01, 02 and 03).
2. Minimise draughts by ensuring solid barn walls start from floor level and are at least 1m high.
3. Side-wall curtains must be open at the top (unless there is a tendency for rain or snow to be blown in).
4. Side-wall curtains must be rolled upwards, not downwards.
5. Use a hand-held meter to measure air flow, air speed, wind chill factor, RH, dew point and heat stress potential. These meters are widely available and cost-effective.
6. Ammonia levels in the air can be measured using a small gas analyser. Ammonia levels in the barn of > 5 ppm indicate ventilation is poor and the bedding material in the calf pens poorly managed.
7. Design single calf pens such that the front and back are open, while sidewalls are solid; free-standing wire mesh pens with walkways between them are an alternative.
8. Air flow in a poorly ventilated barn with a solid back wall can be improved by removing the upper parts of that wall and replacing them with 'spacers', positioned about 30 cm back (outside) from the original position, allowing easier ventilation (Figure 3.1).

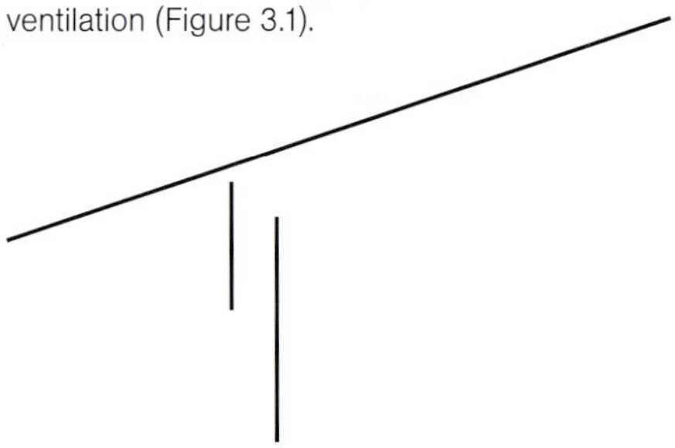

Figure 3.1 Implementing 'spacers' in a solid back barn wall to improve ventilation

Housing of youngstock

Optimal calf housing addresses animal welfare issues, is labour-efficient and promotes good animal health, allowing for optimal growth during the rearing period.

9 Rotating roof ventilators placed in the rear wall of the barn can be a simple and cost-effective method to eliminate stale air from that barn.

10 The thermo-neutral zone for neonate Holstein calves is between 10 °C and 25 °C and for calves aged 1 month is between 0 °C and 22 °C. Obviously, calves prefer lower ambient temperatures than man. Ventilation is more relevant than warmth.

11 If warmth is needed for calves, it is more appropriate to supply solid feeds, while, at the same time, ensuring sufficient straw is available for nesting.

Common design and management problems of isolation provision for calves:

- calves are housed close to cow traffic areas.
- calves are housed in the same barn as adult cows or maiden heifers.
- individual pens allowing calf-to-calf contact and the distribution of faeces between calves.
- bedding material is not completely removed between different calves; pens are not fully cleaned between calves.
- calf-rearers and calf drenches can become vectors for infection transmission.

Practical suggestions for improvement:

- Calves must be separated from maiden heifers and adult cows; cows should not walk too close to calf areas (effluents; aerosol).
- Calf-to-calf contact must be avoided until at least two weeks of age
- Calf hutches in a paddock are an example of good isolation and excellent ventilation (airborne and soil pathogen load is low)
- Calf-rearers must understand the numerous different possibilities for disease transmission and the role good hygiene (including personal hygiene) will play in limiting disease transfer
- If grouped calves are to be reared successfully, additional animal care is needed: group pens are more economic and labour-efficient, but can be detrimental to calf health.

Common design and welfare management problems:

- over-crowding.
- draughty and badly ventilated pens.
- damp and dirty bedding in pens.
- inadequate feeding levels and feeding management.
- limited or no access to drinking water.
- calves kept in large groups.
- large variation in age of calves within one group.

Housing of youngstock

After Jeanette FISHER, Heifermax, Tasmania, Australia, in Int Dairy Topics vol 9, nr 6, November 2010

Practical suggestions for improvement:

1 Provide sufficient area for resting:
Calves < 8 weeks old: 2.2-2.7 m2/calf/ hutch about 2 m2 in group pens Temporary housing is preferable to over-crowding!
2 Calves must be dry and draught free at all times.
3 Calves must be provided with fresh, dry, ammonia-free bedding.
4 Calves must have easy access to clean, fresh water sources and good quality feed.
5 Calves must be kept in single pens or in small groups (< 5 calves) to avoid social stress.
6 Calves must be managed in a set routine and any changes must be made gradually.

Common design and management problems in farm economics:

- failure to renovate or rebuild calf houses in good time.
- limiting the size of a new calf house to limit costs (potentially creating over-crowding and other problems).

Practical suggestions for improvement:

1 Monitor the appropriateness of current buildings and evaluate the ideal solution to any inappropriate housing (e.g. renovating; rebuilding; reorganising).
2 Determine whether current calf morbidity and mortality, poor growth rate and/or poor health are attributable to housing conditions; assess the costs involved.
3 Evaluate the predicted herd size over five years and note how many calves are expected to be reared over that period. Evaluate what the calving pattern is expected to look like (seasonal; spring calving; etc).
4 Assess and compare the costs of renovating/ rebuilding with that of building a new facility.

Once the decision has been taken to make changes to the calf housing:

1 Determine the housing style most appropriate for the individual farm.
2 Define the lay-out of the new barn with respect to the efficiency of labour, feeding practices, animal handling.
3 Determine how calves will be delivered to the pens (develop a protocol).
4 Determine whether calves stay in one pen or are moved between groups with increasing age (develop a protocol).
5 Ensure inclusion of specific 'sick calf pens' isolated from other healthy calves, as well as 'reception pens' for new arrivals, if any. Ensure inclusion of 'heat care pens for warming up hypothermic calves.
6 Determine how milk (or milk replacer) will be delivered, stored, prepared and fed to calves (develop a protocol).
7 Determine how grains (concentrates) will be brought to the calf house, stored and fed (develop a protocol).
8 Define how bedding material will be supplied (top-dressing) to calves in the barn.
9 Define how drinking water is supplied to all calves, allowing free access.
10 Identify an area where equipment can be cleaned and disinfected, where hot water is available, as well as a freezer and refrigerator and where a computer and white board can be present.
11 Identify an area where calves can be measured (heart girth measurement), or scale weighed, and also dehorned.
12 Determine how calves are moved to and from the barn (develop a protocol).

Finally, determine:

1 The most appropriate style of housing for the local climate, herd type and geography (hutches versus barns) of the farm.
2 The housing that is most efficient with regard to cleaning (labour- and cost-efficient); where cleaning out, rebidding and feeding is simple.
3 Whether group pens for calves would save time while retaining good health and growth performance.

Mortality in calves

Source : LAV-LLG-LKV-TSK (Sachsen-Anhalt, Germany, 2006)

Table 3.18 Economic losses associated with calf mortality

Element	Loss	Cost/element (€)	Total cost on 100 effective calvings (€)	Total cost on 50 effective calvings (€)
Mortality (peri-natal)	9%	115/d	1035	520
Mortality (pre-weaning)	9%	150	1350	675
Probability of culling cows following calf mortality	11%	950	940	470
Loss of milk production from culled cows	3.5%	0.28/litre	810	405
Increased # of inseminations	0.14	25/ insemination	30	15
Interval between delayed calvings	13 days	1.10/d	130	65
Total losses			4295	2148

Reducing calf mortality means limiting farm income losses´

Calf mortality is an expensive business. Often, however, the real economic loss is not put into perspective and dairy producers fail to appreciate the financial impact of poor calf management. Table 3.18 above gives an idea of the potential losses from poor calf rearing management and highlight that these calves are the future of the dairy herd.

Bovine Viral Diarrhoea

(BVD) strategies for cow—calf operations

The following strategies are categorized according to whether BVD is suspected or not.

Any strategy should be focused on:

- The search for Persiste Infected (PI) animals as they represent the greatest threat.
- Complementing any biosecurity programmes.

In the USA, about 10% of beef cow herds have at least one PI animal and about 0.25 - 1% of calves born are PI calves. The latter form the source of infection spread.

Modified Live Vaccine (MLV) vaccination will, in general, not induce morbidity or mortality in PI calves, however, there are case reports that indicate that PI calves may become moribund or die.

About 50% of PI calves are clinically 'normal' and may enter, unnoticed into the breeding herd in good condition.

About 7% of PI calves' dams turn out to be PI; the other 93% of PI calves usually had dams with a normal immune response to BVD and were not PI.

The greatest economic losses associated with BVD are related to early embryonic mortality, increased incidences of abortion and high calf mortality. Additional costs include treatment costs, calf morbidity, treatment failure and reduced weight gain.

Endemic BVD may occur in vaccinated herds, due only to the presence of a PI animal.

Vaccination alone will not solve the BVD herd problem

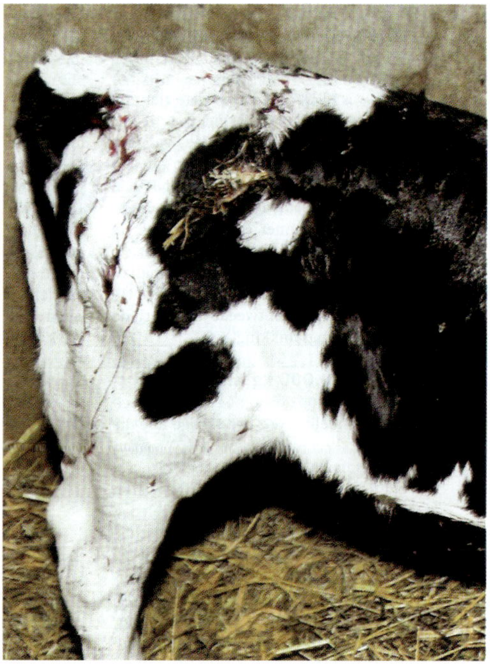

Above: calf showing signs of BVD

Bovine Viral Diarrhoea

Academy of Veterinary Consultants, 2004.
The AABP The Bovine Practitioner,
February 2004, 38(1)105-112

History: Poor reproductive performance in spite of good nutrition and bull fertility. High calf morbidity and mortality in spite of good hygiene and nutrition. There is laboratory confirmation of transient (acute) infection or PI animals.

Conduct appropriate testing in the herd to find the PI animals before the start of the breeding season:

1. All calves with IHC test;
2. All cows without calves (open; dead calf) with IHC, Ag-Capture ELISA, Virus Isolation, PCR;
3. All replacement bulls and heifers (raised or purchased) with IHC, Ag-Capture ELISA, VI, PCR

Test negative

Keep animal in the herd; the PV- is high

Test positive

Heifers, bulls, cows:

Sell PI animals to slaughter

Calves:

Remove calf and dam from breeding herd; if diagnosed by VI or PCR do a virus confirmation at retest in three weeks;
euthanize the calf; test the dam

Test the dams

Test negative:

Return dam to the herd

Test positive:

Sell dam to slaughter

Notes

A all cows still pregnant at time of testing must be removed from the breeding herd because the PI-status of the foetus is still unknown;

B absence of confirmed PI-calves does not guarantee absence of a BVD problem; if suspicion remains, conduct retesting for the next round of calves;

C use IHC, pooled PCR, ELISA of ear-skin samples, or virus isolation;

D implement a complete vaccination programme prior to breeding in the replacement animals and give appropriate boosters in adult cows

Figure 3.1 Protocol for high risk herds (when BVD is suspected

BVD low risk herds

Table 3.19 Protocol for low risk herds (when BVD is not suspected; healthy herd)

History	Good reproductive performance. High percentage of cows that are exposed to a bull. No laboratory evidence of BVD infections or PI animals.
Surveillance Strategy I – Monitor production and health	
Low cost / low sensitivity strategy; slow diagnostic response to PI introduction because milk yield must decline first;	
Monitor overall pregnancy proportion and percent pregnant in first 21 days;	
Monitor stillbirths, neonatal morbidity/mortality and weaning percentage;	
Necropsy and submit tissues (thymus; Peyer patches; spleen; skin; blood) for laboratory analysis on high rate of abortions, stillbirths and mortalities;	
If unexplained suckling calf losses occur (pneumonia; diarrhoea) then send the appropriate samples to laboratory to identify (acute) infection and PI animals;	
Positive tests must be confirmed with other supporting evidence.	
Surveillance Strategy II – Serology type I and II of a herd subset	
Low cost / low sensitivity strategy; serology of non-vaccinated, sentinel animals has been used to identify PI animals in dairies;	
Differentiation of titres due to vaccination or field virus exposition (high titre levels) is difficult and subjective and must include consultation with laboratory diagnosticians for interpretation.	
Surveillance Strategy III – Pooled PCR of blood (all calves)	
High costs / high sensitivity strategy; identifies the PI animals before the breeding season = before bull turn out;	
Delayed response to PI introduction if done after the breeding season;	
Pool samples of 20 to 30 with re-pooling and re-running of positive pools;	
Positive PCR does not differentiate between infections and PI, so other confirmation tests are needed (IHC).	
Surveillance Strategy IV – IHC test on ear-skin samples (all calves)	
High cost / high sensitivity strategy; identifies PI animals before breeding season = before bull turn out;	
Must confirm positive tests if BVD is not suspected because of poor PV+ in herds without prior evidence of PI presence.	

Biosecurity issues

Biosecurity Issues related to the BVD strategies:

The purchase of open female animals

- Heifers and cows must be PI-test negative (IHC; PCR; VI; other tests) prior to introduction into the herd.
- Quarantine for 30 days prior to introduction into the herd.

The purchase of bred animals

- Heifers and cows must be PI-test negative (IHC; PCR; VI) and quarantined until after calving and calf is proven non-PI because PI-status of foetus is unknown;
- Introduce purchased pair to the herd after the calf is born and proven non-PI.

Bulls

- Persistently and transiently infected bulls will shed BVBD virus in semen and other body secretions; transmission of virus may occur after insemination with raw, extended or cryopreserved semen from viraemic bulls; semen used for IA must be collected according to formal guidelines;
- BVD infected semen will not directly cause PI-calves but contact with BVD infected bulls by pregnant cows or heifers can cause foetal infection and PI calves;
- Purchased bulls must be quarantined for 30 days and test PI-negative before turned out with cows.

Fomites

- Virus can survive in faecal material and other body secretions in the environment for hours to days depending on temperature, humidity and sunlight exposure. BVD has been experimentally transmitted from PI-animals to susceptible animals via nose, tongue, injection needles, palpation sleeves.

Embryo Transfer

- Donor and recipients should be tested PI-negative a priori; recipients must be quarantined for 30 days prior to ET; all laboratory fluids used for ET must be free of BVD

Wildlife?

- BVD has been isolated from or serologically identified to infect buffalo, pigs, sheep, deer, elk; deer and elk may shed virus for several days; it is unknown whether PI status can be induced in deer or elk.

Screening incoming cattle for PI animals

- Low prevalence of PI animals (< 0.5%) makes single test strategies as compared with test/confirm test-positive strategies rather expensive for each true positive identified;
- Low prevalence causes even a test with high specificity to have more false-positive results than true positives (the "test to confirm positives" strategy has a high Positive Predictive Value)
- More information about high prevalence populations, such as weight, age, geographic origin may provide guidance for screening only the higher prevalence populations;
- Commingling and transportation of PI animals prior to arrival in the feedlot provokes virus transmission and negative effects of BVD infection prior to screening at arrival.

Purchasing PI-free certified cattle

- All cattle in the group must be tested negative to IHC of skin sample or pooled PCR;
- Economic benefit is determined by multiplying the cost of having a PI animal present (increased pen morbidity; mortality; treatment failure; poor performance) by the expected prevalence for similar cattle (e.g. € 1500 x 0.5% = € 7.50 per head of cattle)

Purchasing PI-low risk cattle

- All cattle in the group originating from farms with complete vaccination programme and a BVD PI surveillance protocol

Biosecurity issues

Purchasing cattle of unknown PI risk

- Cost of unknown status is determined by multiplying the cost of having a PI calf present by the expected prevalence for similar cattle;
- Cost of unknown PI risk is added to other costs for break-even calculation

Communication and feedback for cattle of unknown origin

- When cattle of unknown origin are identified as PI at a feedlot, the consulting vet should notify the manager, herd owner and herd vet and should forward this document.

See also : Slenning B.D. (2003) Bovine Virus Diarrhea Testing Options. The AABP proceedings vol. 36, 45-51

Intoxication by acorns and oak leaves in autumn

In early autumn, non-mature acorns and leaves are often seen falling from oak trees following a storm or heavy wind. These leaves and acorns contain tannins, which can cause lesions in the kidneys and gastrointestinal tract if consumed in high quantities. Maiden heifers grazing at this time of year may be particularly at risk.

Affected animals show a reduced feed intake (sometimes complete lack of appetite), nasal discharge containing blood clots, discharge from the eyes, colic, dark-coloured diarrhoea and foamy urine.

They also tend to have breath that smells of ammonia. Signs appear three to five days following consumption and death usually follows within 24 hours after the onset of symptoms.

Calf feeder hygiene

JSC Boersema, MSc Master's thesis
University Utrecht, 2006, The Netherlands

Table 3.20 Working instructions - hygiene of the automatic calf feeder

	Who?	When?

Every day:
· Before filling the milk automate, clean the milk reservoir
· Clean the teats and milk tubes thoroughly
· Check the cleanness of the mixer
  · If unclean, check cleaning products used, their quantity, their distribution in the feeder (leakage, obstructions)
  → Adopt a fixed day of the week for cleaning the mixer automates.
 	
Each week:
· Conduct a calibration of water and milk powder proportions.
 	
If proportions deviate:
· check the point of distribution of the powder in the mixer (dirt? rests of powder)
· calibrate the automate in case too much powder is being delivered (> 500 ml)
· repeat cleaning and circular rinsing in case too little powder is being delivered (< 500 ml).
· Check exact temperatures of the milk mix (45°C) and distribution (40 °C).
· Ensure the use of good quantities of cleaning products
 	
Every six months:
· Total cleaning and service of the whole automatic feeder (interior and exterior)		

Left: An automatic feeder in situ

Detecting respiratory disease in young calves

Amy Stanton, University of Guelph, Canada, on www.dairyherd.com/newsletter October 2011

Table 3.21 Major signs to detect disease in young calves.

Respiratory diseases in young calves are responsible for great economic losses in production. An early diagnosis and intervention are important issues to reduce the impact of such diseases.

Basic signs	Details
Decreased appetite	Especially in ad libitum milk fed calves this is an important sign
Decreased energy & lethargy	Affected calves will lie down more frequently
Decreased thirst	Not easy to detect unless by close observation
Apathy	Calves do not care about human interference
Energy conservation	Short lying with head back on the body
Depression	Poor hair coat quality; dull eyes; lowered head
Fever	
Other signs	
Statue standing	With the back curved; neck below the chest
Lethargy*	Little to no response on stimuli
Decreased willingness to approach	Affected calves not showing curiosity
Growth delay	Affected calves not fitting the standard curve
* For lethargy, a scoring system has been developed:	
Score 0	Calf is already standing when person enters
Score 1	Calf stands in response to vocal encouragement
Score 2	Calf stands after vocal call and gate opening
Score 3	Calf stands when person enters pen + 3d call
Score 4	Calf stands only after person makes contact

The more deviating signs are observed in combination, the higher the probability that indeed the calf is suffering from disease. Signs on their own are not (disease-)specific.

Dehorn calves with paste

A. Villarroel 2011 Oregon State University

It's a good idea to dehorn cattle that live in confined areas to prevent injuries to humans and other animals. Of the various dehorning methods, dehorning with paste is easy, effective, and economical as well as low-stressto the animal.

Here are the main points to consider when using dehorning paste.

- Apply dehorning paste before calves are 2 days old. After 2 days, calves have figured out how to scratch their heads against something to rub the paste off, and they can also stand on three legs to scratch with the other.
- Using too much paste is the most common mistake beginners make. The result is a big bald spot around the horn area (although the hair will grow back in time). The amount of paste to apply on each horn is about the size of a dime, as indicated in the package insert.
- Don't let the calves get wet for 24 hours after applying the paste. If rain falls on active dehorning paste, it can run off into the eyes and blind the calf. The paste dries in 1 day, after which it is no longer necessary to keep calves dry.
- Apply paste just before feeding the calves with a bottle. It takes a couple of minutes for the paste to start burning, so if you apply it immediately before feeding, calves are kept busy working on the bottle, and they forget about their discomfort. By the time they are done with the bottle, the paste is almost done with the dehorning process, and they will not notice it that much.
- Additionally, research performed with human babies shows that giving breast milk, glucose, or sucrose before a single painful procedure significantly reduces heart rate and crying time compared to using distilled water, a pacifier, or swaddling. So, applying the paste immediately before feeding milk with the bottle may help in two ways: the calves are so busy working on the bottle that they forget their discomfort, and the sugar in the milk may help reduce the pain.
- Producers who have switched to using paste to dehorn calves at birth report great success with no complications, and they like that calves are "done" without showing obvious signs of pain. Only minor head shaking was reported.
- Dehorn newborn or 1 day old calves. Apply the dehorning paste right before feeding. Make sure the paste doesn't get wet for 24 hours.

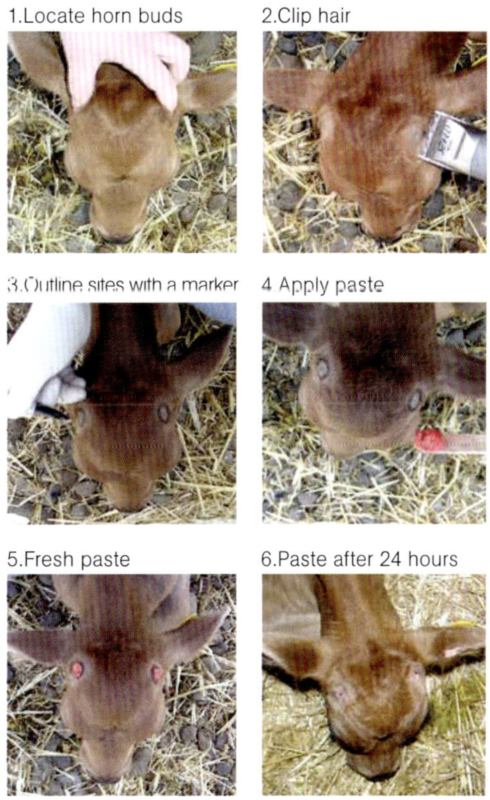

Figure 3.2 Dehorning calves with paste.

Scoring tables for calf diarrhoea

Adapted after Heinrichs & Kehoe, 2009

For a fair assessment of the diarrhoea situation among young stock, three scoring sheets are handled: (1) on the extent of diarrhoea, (2) on respiration of the calves, and (3) on appearance of the calves.

Table 3.22 provides a scoring sheet for assessing diarrhoea cases in young calves; Table 3.23 provides a scoring sheet for assessing the quality and its deviations of respiration; Table 3.24 gives a scoring sheet for assessing clinical appearance of the calves (Heinrichs & Kehoe, 2009).

Table 3.22 Scoring system for distinguishing different degrees of diarrhoea in young calves

Score 1	Calves show normal faeces, with "**pudding**" consistency
Score 2	Calves show slightly less firm faeces which are "**yoghurt-like**"
Score 3	Calves show diarrhoea, with "**syrup-like**" faeces, which are loose to watery; odour
Score 4	Calves show diarrhoea with "**fruit juice**" aspect; faecal material still visible
Score 5	Calves show watery diarrhoea without faecal material, with mucous or blood

Table 3.23 Scoring system for distinguishing different degrees of respiration/deviations

Score 1	Calves show normal breathing, no clinical signs what-so-ever
Score 2	Calves may show a slight cough or runny nose; but respiration is regular
Score 3	Calves show rapid breathing and moderate coughing
Score 4	Calves show severe frequent cough, and rapid breathing
Score 5	Calves show irregular breathing, with severe chronic cough

Table 3.24 Scoring sheet for assessing the clinical appearance of young calves

Score 1	Calves are alert and active
Score 2	Calves have droopy ears, and are slightly unresponsive to stimuli
Score 3	Calves are moderately depressed, with head and ears drooping
Score 4	Calves are depressed, with drooping head and ears; show no interest in standing up
Score 5	Calves are down all the time (side position)

On the basis of the outcome of all scoring Tables, a **daily total** is calculated. When the daily total exceeds a certain threshold (for example an overall average of 8), rectal temperature must be taken, and treatment started (antibiotics and or oral rehydration).

Note that from a score of 2 onwards (for each Table) one can already speak of a disorder. The summing up of 3 scores at value 2 is 6. So, a very strict threshold would be an overall average daily score of 6.

Working instructions

JSC Boersema, MSc Master's thesis University Utrecht, 2006, The Netherlands

Example of calculation:

15 calves were scored, using the three Tables 3.22, 3.23, 3.24

Results of scoring are A (10 x 3 + 5x2) + B (9 x 2 + 6 x 3) + C (6 x 2 + 9 x 4) = 124

The overall average is 124 / 15 calves = 8.3

If the threshold was set at 8, rectal temperature must be taken and treatment of calves started.

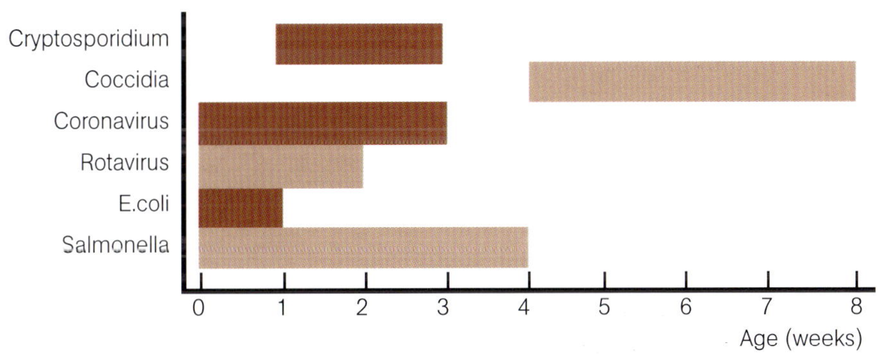

Figure 3.3 Highest risk period for respective infections as related to age

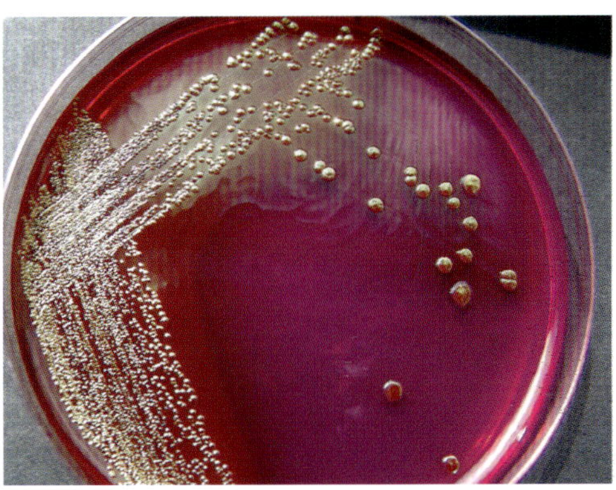

Figure 3.4 Escherichia coli on selective agar

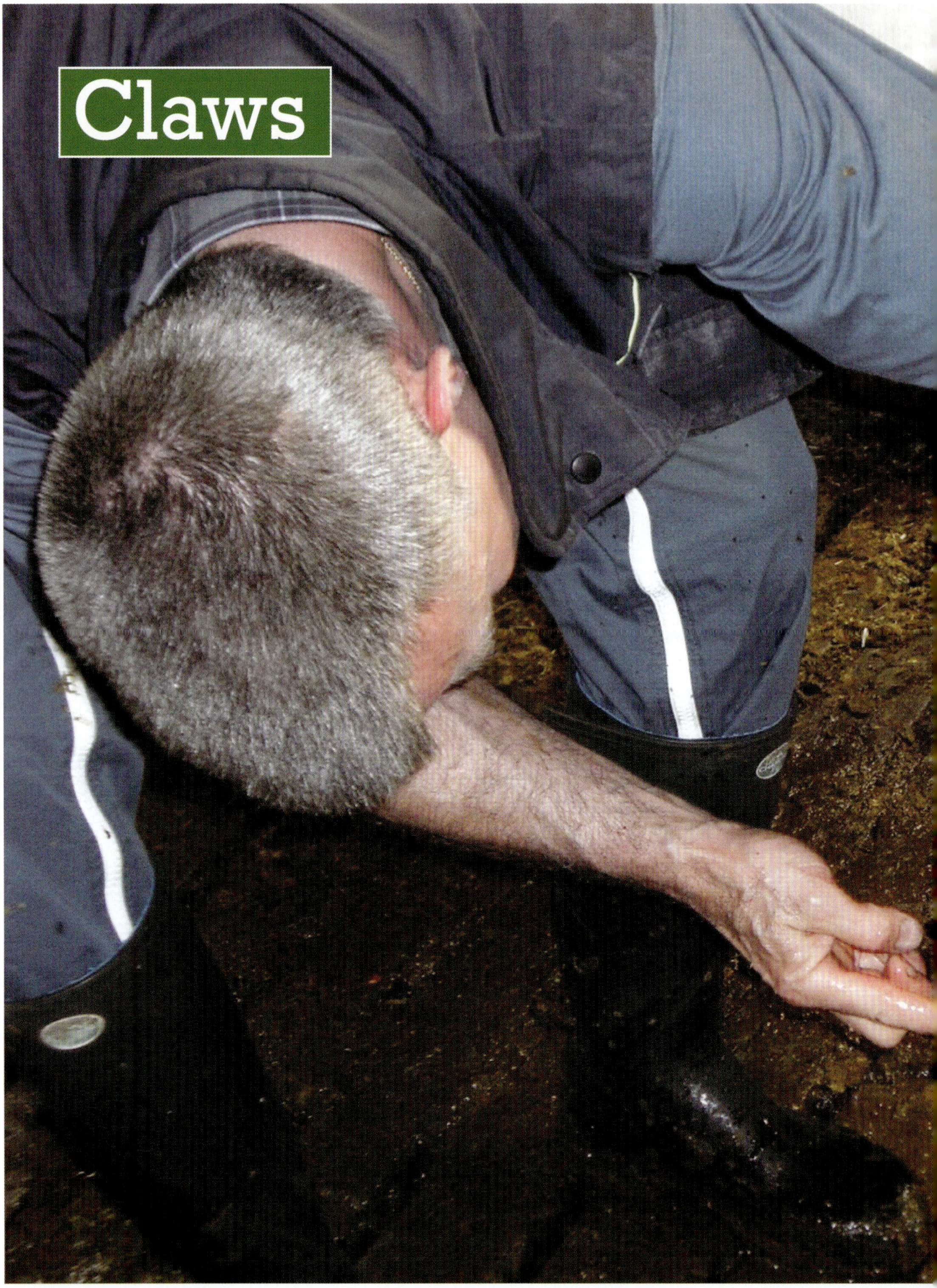
Claws

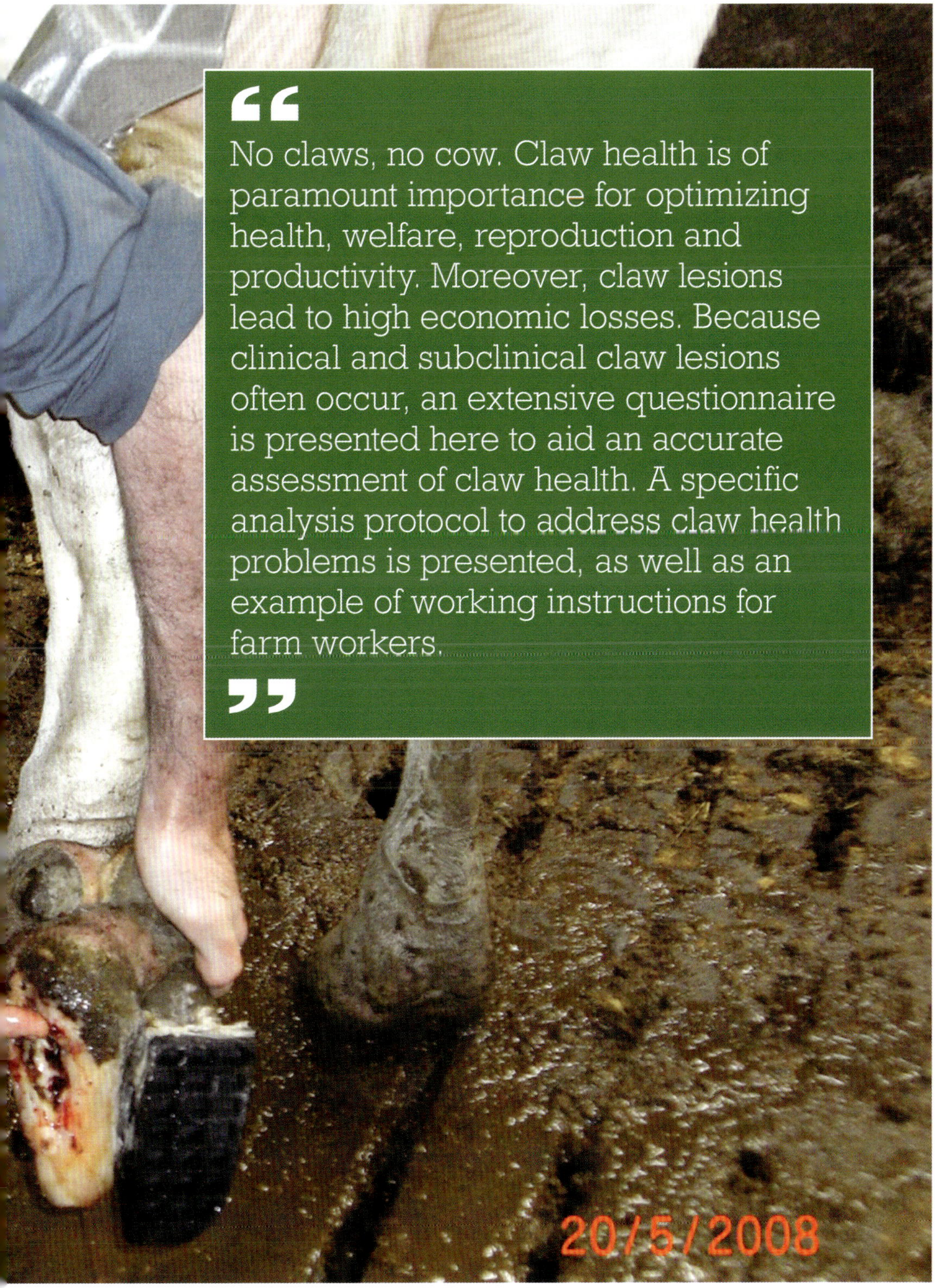

> No claws, no cow. Claw health is of paramount importance for optimizing health, welfare, reproduction and productivity. Moreover, claw lesions lead to high economic losses. Because clinical and subclinical claw lesions often occur, an extensive questionnaire is presented here to aid an accurate assessment of claw health. A specific analysis protocol to address claw health problems is presented, as well as an example of working instructions for farm workers.

Claw trimming

Functional (preventive) claw trimming

Trimming should never be carried out in the month prior to calving.

However, cows may be trimmed on the day of drying off and then again at two to three months after calving (when they are through the high risk period and will more easily recover from any claw lesions).

A second option is to trim all cows twice a year (or more if needed), taking care not forget groups of animals, such as the pregnant maiden heifers

With this option, ensure all cows are trimmed in one day or, at the most, over two consecutive days.

A third option would be to trim cows on an individual basis, for example, cows with highly sensitive claws.

Curative claw trimming

It is essential to trim claws of cows that show poor hind leg conformation, lameness or hoof deformities. Delaying trimming, in this situation, will exacerbate the situation and could result in problems with remaining claws, either by infection or by mechanical overload.

Claw trimming should only be carried out using the appropriate tools and facilities. There exist several simple claw trimming facility options:

- Mobile - these may be placed behind your car
- Installation in a cubicle
- Fully equipped, stand-alone devices that can be placed anywhere on farm.

The following gives an example of a list of animals selected for claw trimming, together with the reason for selection, as part of the routine Herd Diagnostic Evaluation during a farm visit:

Table 4.1 Example of cow selection for claw trimming

Cow ID	Reason for trimming
5671	front claw deformation
5670	hind claw too long
3145	front claw show some deformity
1352	all 4 feet too high and too long
8330	overgrowing soles
1806	lesion on hind claws?
8340	lesions on hind claws?
8329	all 4 feet too high and too long
5648	front claws too long
5693	front claws too long

This list indicates a lack of good claw management, demonstrating the need for routine claw trimming and inspection i.e. a functional (preventative) trimming plan.

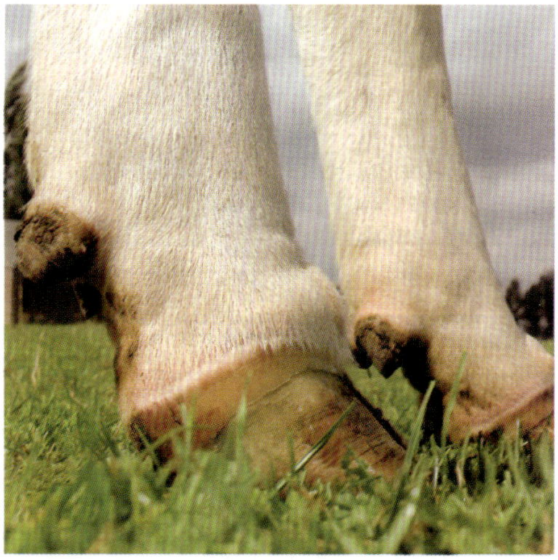

Recording claw lesions

Farm...

Code...

Date...

Veterinarian...

Last Date..

Head of Cow

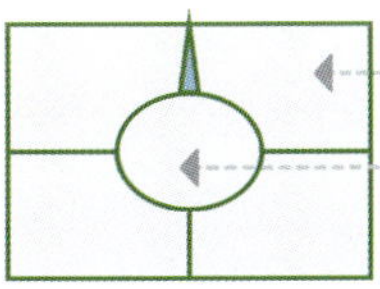

Lesion code

Identification of the cow
Code encircled = Treated

The four feet of the cow

CODES for LESIONS

ID= interdigital dermatitis
DD= digital dermatitis
PL= interdigital phlegmona
LS= sub-acute laminitis
US= ulceration of the sole
WL= white line disorders
TY= tyloma
CC= clinical lameness case

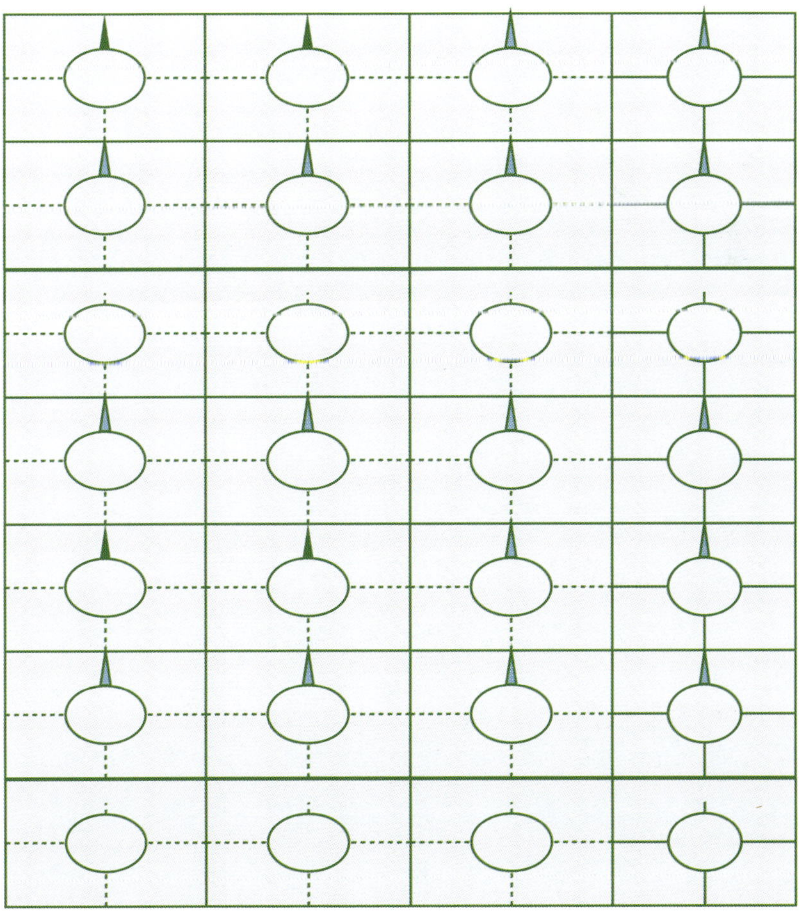

Figure 4.1 Recording claw lesions

Claw lesions questionnaire

National Veterinary School of Nantes, France

1 – Summarising the problem

Client name..

Address ....

1.1 – Presentation of the farm structure of farm, size (cows numbers; quota (if applicable), FTE workers), breed, production level, reason for calling the vet

1.2 – How lameness diagnoses are made by the farmer and how cases are recorded

1.3 –Monthly distribution of lameness cases

	Mar	Apr	May	June	July	Aug	Sept	Oct	Nov	Dec	Jan	Feb
Results from last 12 months												
Total number of cases												
Number of new cases in lactation												

Number of cows in group												
Number of heifers milked												
Number of adult cows milked												

Claw lesions questionnaire

1.4 – Clinical description of the different types of lameness

Localisation of lesions, duration of a case, stage of lactation of affected cows, etc …

1.5 – Treatment of lameness cases

Basic initial treatment	☐ by farmer ☐ by veterinarian ☐ other	treatment, delay to intervention, results
Persistent cases	☐ by farmer ☐ by veterinarian ☐ other ☐ no such cases	

1.6 – Prophylactic measures

Trimming
☐ routinely all cows
☐ cows with claw deformities
☐ in case of lameness

Frequency (times/yr) [ ]

☐ by the farmer
☐ by professional
☐ by others :

Footbath
☐ pass through
☐ stationary
☐ no foot bath

Utilisation
☐ in case of lameness
☐ routinely

Other mode of disinfection of claws [ ]

Frequency (times/yr) & duration (days) [ ]

Product (concentration) [ ]

Bicarbonate/other rumen stabiliser added to the ration
☐ never
☐ in winter ration
☐ year around

Quantity per day supplied (g / cow / day) [ ]

Claw lesions questionnaire

2 – Inspection of the lactating cow herd

At the time of this visit cows were: ☐ housed
☐ at pasture

Calculation of number of cows per lameness assessment category
(must include at least 50 % of the herd)

Signs of lameness		TOTAL
No lameness		
Fore leg lameness		
Hind leg lameness (high on leg)		
Hind leg lameness (claw level)		

Hock condition		
No hock lesions		
Depilation of the hock		
Recent wound on hock		
Old wound or abscess on hocks; thickened tissue		

Hind claw condition		
No anomalies in horn growth		
Long claw walls		

Hind leg posture		
No deviation in posture		
Poor hind leg posture		
Claws are directly under the cow		

Cleanness of cows		
Dirty thighs and/or udder		
< 2/3 dirty and < ¼ crusts		
> 2/3 dirty and > ¼ crusts		

Claw lesions questionnaire

3 – Housing of lactating cows

3.1 – Housing calendar

	Mar	Apr	May	June	July	Aug	Sept	Oct	Nov	Dec	Jan	Feb
Housed day and night												
Housed only at night												
Grazed day and night												
Access to exercise area												

3.2 – Herd size

Average number of cows on zero-grazing in summer season

Maximum number of cows on zero-grazing in summer season

3.3 – Farm design

Housing plan and other observations of the straw yard or cubicle area and exercise area
(indicate the north with an arrow; design; mention length and width of houses and inside areas; indicate the position of the drinking troughs and the feed bunks)

Claw lesions questionnaire

3.4 – Risk factors for inadequate ventilation

Barn is ☐ closed ☐ open

Openings located:

Sidewalls opened ☐ yes ☐ no ☐ monoroof

Traces of humidity / or mould on the wood on sidewalls and/or roof ☐ no ☐ possibly ☐ yes

Other risks of inadequate ventilation

3.5 – Cubicles or tied stall

☐ cubicles

☐ tied stall

Number of places for lactating cows

Number of very dirty cows not sleeping in cubicle

Bedding material

Frequency of addition of bedding material

Quantity of bedding material in kg mean/ cubicle/day

Frequency of manure scraping from end of cubicles

Cleanness on day of visit; presence of manure in >10% cubicles?...

Draw the outline of cubicles and indicate the dimensions in cm (see example below)

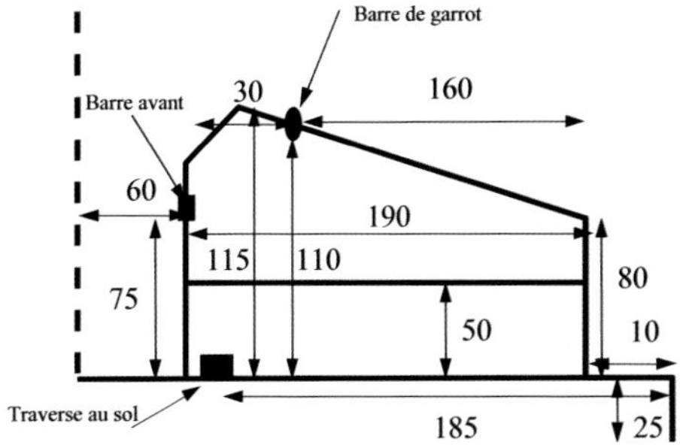

Claw lesions questionnaire

3.6 – Straw yard

Type of surface available for resting

Type of surface of areas not in use by cows

Presence of wet spots in the straw yard
- ☐ no humid zones
- ☐ under the roof-opening
- ☐ at side walls
- ☐ at drinking trough
- ☐ other (please state):

Frequency of addition of straw when cows are housed for 24 hours

Mean daily quantity of straw added

Frequency of cleaning out straw yard

3.7 Internal exercise area

Exercise area is closed
- ☐ yes
- ☐ no
- ☐ in places

Type of surface of exercise area

Frequency of manure scraping of this area

Grooving and manure scraping of exercise area

Presence of a scraper rail on the floor ☐ yes ☐ no

Presence of an automatic scraper
- ☐ no
- ☐ yes + actions watched
- ☐ yes without being watched

Grooved lines in the concrete ☐ yes ☐ no

Slatted floor in passageways between cubicles ☐ yes ☐ no

Presence of standing water in exercise area

☐ no

☐ yes origin of this water:
- ☐ from the roof
- ☐ from milking parlour and waiting area
- ☐ other origin(please state): _______

Characteristics of the floor

Claw lesions questionnaire

Type of floor in the exercise area and feeding area	

Concrete floor re-laid	☐ < 2005 ☐ > 2005 ☐ no	Other (please state)		Washing out nitrite in concrete is done	☐ yes ☐ no

Presence of holes	☐ yes ☐ no	Slippery floor	☐ yes ☐ no	Abrasive floor	☐ yes ☐ no	Other comments:

Presence of steps

☐ no steps
☐ steps at feed rack or feed bunk or drinking troughs (number of steps + height in cm) : ______
☐ other steps:

Describe their position

3.8 – External exercise area

Access to this area during the housing period	☐ yes ☐ no	Type of surface		Frequency of rotation per day or week of this area	

Claw lesions questionnaire

Reasons for using this area

3.9 – Waiting area and milking parlour

Characteristics of the floor

Presence of holes	☐ yes ☐ no	Slippery floor	☐ yes ☐ no	Abrasive floor	☐ yes ☐ no	Other (please state what):	☐ yes ☐ no

Plan of any steps in the waiting area and milking parlour

Indicate whether cows must step up or down; give exact location of steps

Claw lesions questionnaire

4 – Nutrition of lactating cows

4.1 – Ration common to all cows in an individual group

	RATION	From (date)	To (date)		CALCULATION	Adapted by farmer	
☐	TMR			☐	Milk recording	☐ yes	☐ no
☐	Semi-complete			☐	coop technician	☐ yes	☐ no
☐	Individual			☐	feedmill technician	☐ yes	☐ no
				☐	farmer		
				☐	Other (please state)→		
						☐ yes	☐ no

COMPOSITION OF THE BASIC RATION Expressed as ☐ kg milk ☐ Litres milk

	Mar	April	May	Jun	July	Aug	Sept	Oct	Nov	Dec	Jan	Feb	Mar
Production level													

		Quantity in kg/d – state whether DM or as fed*												
FORAGES	DM content (g/kg)	Mar	April	May	Jun	July	Aug	Sept	Oct	Nov	Dec	Jan	Feb	Mar
Grazing														
Maize silage														
Grass silage														
Hay														

*percentage contribution to ration to be given in parentheses

Claw lesions questionnaire

		Quantity in kg/d – state whether DM or as fed												
CONC. / MINE.	DM content (g/kg)	Mar	April	May	Jun	Jul	Aug	Sept	Oct	Nov	Dec	Jan	Feb	Mar

4.2 – Quantity of concentrates (individual basis)

☐ no concentrates offered individually

Brand of concentrates or primary products		
Frequency of distribution based on milk production:	1 kg concentrates for……. kg milk; if more milk then……….	
Growth requirements of 1st lactation heifers taken into account?	☐ no ☐ yes:	
☐ quantity distributed is limited to a maximum of		kg concentrates / day
☐ no maximum		

4.3 –Transition from dry period to peak lactation - multiparous cow with an average milk yield

		Quantity in kg/d – state whether DM or as fed												
FORAGES/ CONCEN-TRATES	DM content (g/kg)	Dry period		close-up period			Beginning of lactation (calving + 1 week = C+1w etc)							
		D+ 1w	D> 1w	C- 3w	C- 2w	C- 1w	C+ 1w	C+ 2w	C+ 3w	C+ 4w	C+ 5w	C+ 6w	C+ 7w	C+ 8w

Claw lesions questionnaire

4.4 – Type and quality of forages available for lactating cows

Check forage analysis results.

If unavailable, conduct a qualitative assessment:

For maize silage:

DM content: ☐ High ☐ Average ☐ Low
Grain content: ☐ High ☐ Average ☐ Low

For grass silage:

Grass species: ……………………………

For hay:

Grass species: ……………………………

4.5 – Type of concentrates available for lactating cows

Concentrates or primary product	Presentation (meal, crushed, pelleted)	Tanned	Specific content*	DM content*	Energy content*	Protein content*	Label or notice
		☐ yes ☐ no					☐ yes ☐ no
		☐ yes ☐ no					☐ yes ☐ no
		☐ yes ☐ no					☐ yes ☐ no

* = exact parameter depends on the national feeding system

4.6 – Distribution method for forages and concentrates lactating cows

Silage

Permanent access to forage	☐ yes ☐ no	Poor silage quality parts are removed before feeding	☐ yes ☐ no	Daily removal of feed refusals	☐ yes ☐ no
Number of cows		Total quantity of feed refusals per day (kg product)		=> Quantity of feed refusals per cow per day	

Claw lesions questionnaire

Number of distributions per day ☐

Number of push-ups per day ☐

Hay

Frequency of distribution (in d) ☐

Number of bales used at each distribution ☐

Weight of the bales ☐

Straw

Distribution of straw in the TMR wagon ☐ yes ☐ no

Other mode of distribution for feed intake ☐ yes ☐ no

Quantity distributed (kg / day) ☐

Concentrates

CB: same quantity given to all cows (basic ration part)
CI: quantity distributed on an individual animal basis

CB	CI	
☐	☐	Food rack
☐	☐	Feed bunk no feed rack
☐	☐	Milking parlour
☐	☐	Automatic feeder

	CB	CI
Number of distributions per day		

Mixed with forage

☐ TMR
☐ TMR + silage harvester
☐ Manual mixing
☐ Top-dressed on forage

4.7 – Weighing of forages and concentrates

How is the weight of distributed silage determined? ☐

How is the weight of distributed concentrates determined? ☐

Selecting cows for a clinical inspection of claws

Which cows should be selected?

Certain groups of cows are more at risk than others. These cows will be the first to show signs of a prevailing problem. These indicator cows should be noticed!

Certain periods in time can also represent an increased risk, such as age, lactation stage and season. Therefore, inspection of claws must be carried out taking these risk periods into account. Late autumn or early winter is a good moment for routine claw inspection.

Young cattle

Healthy and well-formed claws go a long way to ensuring strong, resistant claws when the animal is older. Note that, in some situations, interdigital dermatitis (foot rot) and also laminitis can often be detected in calves from as young as six months of age.

Heifers

The primipares are a group at risk for incorrect feeding and laminitis but also for claw infections. The maiden heifers and first lactation heifers are often the first indicators for Mortellaro disease as older cows may have developed a certain (weak) immunity against the disease.

Cows in lactation

Around calving and during the first months of lactation (< 90 days), the claws are subjected to huge physiological pressures, including feed changes, hormonal changes, changes in housing, social changes and infections, all of which cause stress and potential damage to the delicate structures of the claw. Thus, the cow is more susceptible to disease, such as interdigital dermatitis and laminitis, both of which are two important, primary claw lesion diagnoses. Cows in early lactation require specific attention. If claw lesions are suspected earlier, e.g. the close-up period, they must be addressed immediately.

Which feet to check?

Hind claw lesions are most common - hind claws must be checked. However, the front claws must not be neglected; it is recommended to check these for deformities or other defects.

How many cows should be checked?

At least 20 % of the total number of claws in the herd should be inspected (with a minimum of 15 cows in the herd). Ensure cows are selected in a representative manner and not simply by taking the first 15 cows at the feed rack (remember that lame cows will be slow and probably among the last in the feed rack). All lame cows should be checked. It is best to select a range of animals across the production cycle e.g. early lactation cows, heifers, mid- and end-lactation cows; for example, one cow out of every three to five at the feed rack.

Risk periods

Winter is particularly favourable for the development of claw lesions.

Prevalence of interdigital dermatitis tends to decrease in the summer and Mortellaro disease is more often observed during the autumn and winter periods.

Barn humidity, as well as the presence of steps, stones/gravel in walkways to pastures may contribute to a high prevalence of claw lesions, such as phlegmona.

Putting cows out to pasture can also initiate laminitis, given the high protein contents in spring grass.

General observations on claws in the herd

Checking claw health on a farm is carried out on a herd basis and must be done at different locations to allow for a complete evaluation.

Observation of animals at the feed rack:

A cow with healthy hind feet will stand with her legs directly underneath her and the legs will be parallel.

When an animal experiences pain in a limb or claw, they will bear more weight on the pain-free side. In the case of the hind feet, the outer claws are most commonly affected. This means that that cow will turn its foot to the outside and the hock to the inside to try to relieve the discomfort (Plate 1). Because of this, a claw in a 'normal' position is not necessarily healthy.

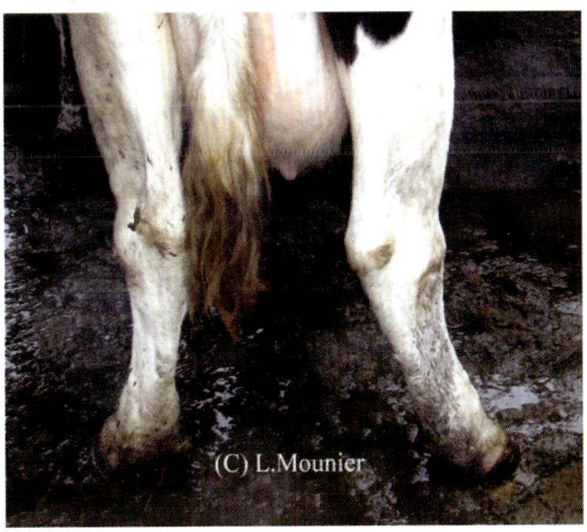

Figure 4.2 Example of poor hind leg posture. Note the out-turned claws and favouring of the left leg.

During the inspection of the claws/legs, the cow should remain standing equally on both hind feet. In case of a painful claw, the cow will alternate between the two hind legs, shifting her weight from one to the other.

Other observations may be carried out with the cows at the feed rack:

- excessive horn growth of the claw can indicate a claw lesion or poor trimming practice.
- other claw deformities: length, angle with floor, ...
- hock lesions may lead to lameness (abscess). The severity of the lesion can be assessed through the surface of the lesion and its depth. Lesions start with a depilation and may result in an abscess with varying amounts of fluid.

Observation of cows while moving: Locomotion score

This must be carried out on a level, non-slip surface. At pasture, locomotion scores will generally be lower than when housed. Refer to the section on **Locomotion scoring** in MONITOR of this series.

Observation during milking:

In the milking parlour, the claws and feet can be visibly assessed at eye level making milking an ideal time to check the status of claws and hocks.

Questioning the farmer about his claw health practice, trimming frequency, number of cows trimmed during each session, manure scraping frequency and use of foot baths can significantly aid in the analysis and evaluation of claw health.

Note that the observation of claws as proposed above is should always be accompanied by a **Diagnostic Herd Evaluation** *(including BCS, RF, FC, UFF, etc, as well as the assessment of farm data) hence facilitating the identification of herd problems and the farming areas potentially involved.*

Farm visit protocol to address claw problems

1 On arrival, get a general impression regarding the farm and its installations

2 Follow the rules of Good Hygiene Practice (change boots and clothes; wash hands)

3 Report to the farmer (or manager, employee, family)

4 Ask the farmer whether there are lameness problems in his herd of cows or calves

5 **If no**, execute the Diagnostic Herd Evaluation routines (→ continue with **step 8)**

6 **If yes**, ask for the details about the lameness cases (clinical diagnoses; predominant diagnosis; current treatment practices; whether preventative functional claw trimming is carried out and, if so, what the protocol is e.g. twice a year by a professional trimmer) and whether the farmer carries out trimming; if yes, when, why and how. Once you have sufficient detail, then proceed with the rest of the anamnesis: breed of cow; average age of cows or calves; milk production level; reproductive problems (heats; pregnancy repeat breeders) calving problems; other diseases.

7 Ask for details regarding: use of a foot bath (dimensions; frequency; products; contents; renewal); presence of a water bath to clean claws first.

8 Clinical inspection of the animals (Notation= GOOD–AVERAGE–POOR)

9 Body condition scores at early, mid-, late lactation and dry cows; rumen fill scores; faecal consistency scores; undigested fibre in the faeces scores; locomotion scores; specific behaviour; poor posture of hind legs; social interactions, including agonistic and antagonistic behaviour; hock lesions; other signs.

Inspection of housing, climate, hygiene and rations (Notation= GOOD–AVERAGE–POOR).

Characteristics of the barn (temperature/ventilation; humidity; draughts; open sidewalls; dirt; moulds).

Characteristics of the waiting area, exercise area; milking parlour (presence of a working manure scraper [type of system; frequency of operation; efficacy]).

Characteristics of the rations (type of ration [complete or not]; feedstuffs; origin and quality; formulation of rations [by whom; ration limited or ad libitum]; risk of acidosis; conditions in front of the feed rack; position of feed rack).

General hygiene in the barns, the cubicles, the passage ways, waiting area, milking parlour. Cleanness of the cows (scoring thighs and udders).

10 Synthesis of the inspection = summarising strong points and points for improvement → work hypothesis and herd probability diagnosis. Or =summarise lameness problems, including prevailing risk factors.

11 Create a plan of action with advice and/or interventions for the short-term (maximum of five recommendations) and for the mid-to-long-term. Ensure you discuss the draft version of the report with the farmer before producing a final version. Deliver the final version (maximum of one page of A4) to the farmer within five days.

12 Produce a calendar indicating when the next farm visit is required to evaluate progress following the advice given in the preceding report.

Points for improvement

FIELD SHEET

	Strong points	Points for improvement
Animal inspection		
Inspection of barns, hygiene and rations		
Synthesis & conclusions (formulate your reasoning and hypothesis here)		
Advice & proposed Interventions		

Hint! The titles given in the above field sheet can be used as headings in the farm visit report!

Protocol No 1 for problem analysis of claw lesions

Concern(s) or question(s) from the farmer

ANAMNESIS = which cows, when, what, how, which conditions, where?

Confirmation of concern(s) or question(s): By farm data
By claw trimmer
By vet during farm visit

Concern(s) not justified

Exit → **END**

Look for other signs → **None?**

Concern(s) justified

Inspection of herd (healthy cows =reference)

Inspection of affected cows
Inspection of barn, climate, etc
Check farm data

Set working hypothesis

Additional activities:

- Nutrition
- Barn hygiene
- Barn climate
- Traumas
- Details of barns and climate

Diagnoses
At individual level
At herd level

Plan of action

Short-term (maximum of five priority recommendations)
Mid-to-long-term (secondary issues and including a prevention plan)

Evaluations = subsequent visits to evaluate effects of recommendations, and efficacy of treatments. Formulation of other recommendations when deemed necessary. Written farm visit report of a maximum of three to five pages (Tables & Figures excluded).

Figure 4.3 Protocol 1 for analyzing herd claw problems

Protocol No 2 for problem analysis of claw lesions

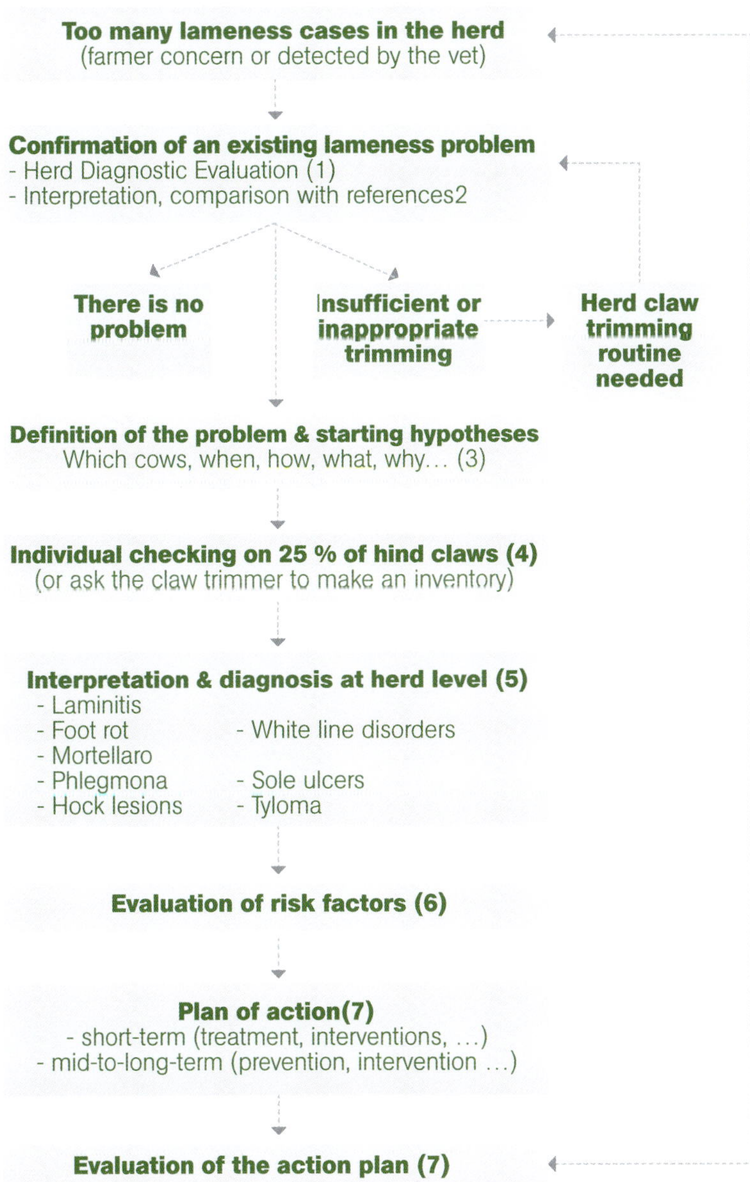

Figure 4.4 Protocol 2 for analyzing herd claw problems

Interpretation & diagnosis of lameness at herd level

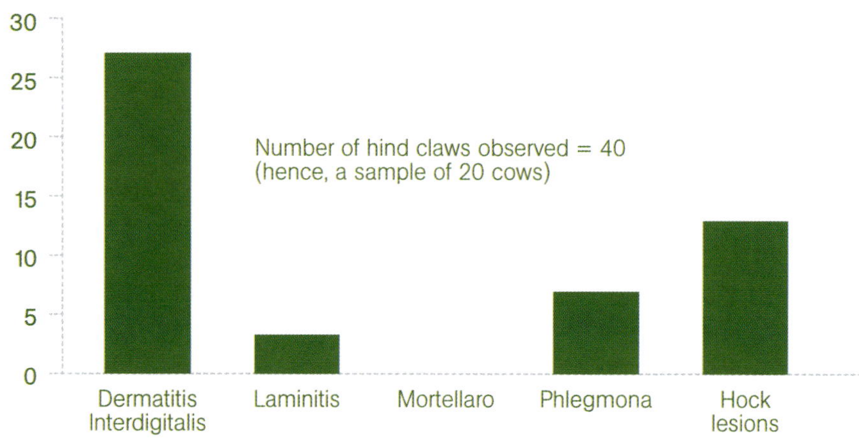

Figure 4.5 Histogram of prevalence of claw lesions in a dairy herd

The diagnostic approach to individual claws (from a sample of cows in the herd) must be interpreted in the context of the herd. Indeed, different lesions may be present on one claw or foot, as well as different claw disorders in the same herd at the same time.

Design a bar-chart (eg Figure 4.5) detailing the important claw lesions detected. Next, set a focus for only the most relevant claw lesions.

➲ Check the animals showing claw disorders and the period during which these disorders are present, within the context of the herd.

Which cows are showing which kind of claw lesions, for how long, how severely, with what characteristics and under what conditions?

➲ Compare the results with reference values at the national, regional or local level

For certain pathologies or in certain regions, the reference values are non-existent or differ largely from other regions. In this case, it is a matter of collecting these data from a group of farms and use the data to create a reference value.

Finally, make a synthesis of all observations, differentiating between strong points and weak points; then, build a conclusion on the basis of a probability hypothesis or working hypothesis. The latter should include the most important claw lesions detected, as well as the associated risk factors on the individual farm.

Table 4.2 An example comparison of claw disorders vs the national, regional and local levels

	Mortellaro	Dermatitis interdigitalis	Laminitis	Phlegmona	Hock lesions
Reference	0	<10%	<5%	0	< 5%
Observed	0	22%	3%	9%	23%

Walk-through foot bath instructions

www.vetvice.nl

Instructions for the use of a walk-through foot bath in herd claw health control

1 The solution to be used in the foot bath must be prepared with 3 to 5 L of formaldehyde 40% in 100 L of water.

2 The foot bath should measure 300 – 500 cm length, 100 cm width with a depth of 15 cm. The foot bath should be placed preferably in the passageway leading away from the milking parlour or at a site where cows frequently pass.

3 The foot bath must be filled with the solution to a depth of at least 10 cm.

4 A small amount of manure may be put into the foot bath to prevent cows from drinking the contents.

5 Renew the contents of the foot bath once 250 cows have passed through it.

6 Run the cows through the foot bath every three to four weeks on three consecutive days. Where there is a high prevalence of claw disorders, such as **interdigital dermatitis**, the frequency must be increased to every two weeks.

7 Cows with sole ulcers or open lesions (e.g. **Mortellaro** disease) should not be run through a foot bath containing formaldehyde.

8 Make sure that cows pass through the foot bath quietly in order to prevent splashing that could lead to teat and skin lesions (teats!).

9 While cows are in the waiting area for milking, make sure the manure scraper is working to clean the passageways and exercise areas. This ensures that the cows come back to a cleaner, dry area.

10 When the ambient temperature decreases below 15°C, add warm water to the foot bath as formalin loses its activity at temperatures below 15°C. If the environment is very dirty, one should install another foot bath, containing only water, before the formalin foot bath to clean the claws from manure and dirt to minimize faecal contamination of the formalin footbath.

Formalin is a toxic chemical; cows must be prevented from drinking it at all times. Moreover, it is also toxic to humans (eyes and skin are highly sensitive and must be protected).

Replacement of formaldehyde

Adapted from A.Schwoerer, AROMSA, 2008

Formaldehyde is a toxic substance with carcinogenic effects on man and animals; moreover, it is an environmental pollutant. It was used in foot baths on dairy farms in order to improve claw health and control infectious claw lesions, such as interdigital dermatitis. Formaldehyde should ideally be replaced by another product with similar or enhanced effects on claw health but without the negative toxic and/or environmental effects.

In a recent field study, three products were tested by the CFPPA in Rheu and IdE, France, for their efficacy as a replacement for formaldehyde:

1. **Product C -** based on sulfamic acid, K-mono-persulfate, and dodecylbenzene sulfonate;
2. **Product F -** based on glutaraldehyde, chloro-alkyl-dimethyl-benzylammonium, and chloro-4-methyl-3-phenol
3. **Product G –** based on Na-tetrahydrate-perborate, and chloro-N-alkyl-(C12-14)-N benzyl-N-N-dimethylammonium

Product F was seen to work against *Fusobacterium spp* and its disinfecting action *in vivo* proved better than formaldehyde (probably due to fewer disturbances by organic matter). It is less hazardous for human health but has still polluting features with respect to the environment. Its undesirable smell is also a disadvantage. Overall, Product F could be an adequate replacement for formaldehyde.

Product C is also reasonably active against *Fusobacterium spp* and demonstrated a more effective disinfection than formaldehyde. Its smell is acceptable and it is relatively easy to use. Hence, Product C could also be a suitable replacement for formaldehyde.

Although Product G was effective against *Fusobacterium spp*, it was not as effective at disinfection compared with formaldehyde, although it still remained more active in presence of organic matter compared with formaldehyde. This product was the safest for humans and the environment of all three tested and was also the cheapest of the products tested, although all three products proved more expensive than formaldehyde. Product G could replace formaldehyde, however, it would be less effective with regards to disinfection. Note that clinical efficacy studies have not been reported.

Formaldehyde solutions

www.vetvice.nl

Formaldehyde Solutions

Considerable care must be taken when using formaldehyde solutions that are available on the market because these products do not always have the same concentration; the concentrations vary between 24% and 40% (these solutions are called formalin).

For a foot-bath, a solution of **4 litres of formalin 37% in 100 litres of water is required.**

If other concentrations are used, the ratio of formalin to water must be adapted. In order to correctly calculate the concentration required, the following formula can be used:

$$\frac{37}{\text{(concentration of product)}} \times 4 = \text{litres of formalin in 100 litres of water.}$$

For example, a product of 24% concentration, would need: (37 / 24) x 4 = 6,2 litres in 100 litres of water.

The total volume of the solution must then be adjusted depending on the size of the foot bath. Therefore, it is also necessary to calculate the volume of the foot-bath in litres.

Remember that, in wintertime, the minimum temperature of the solution in the foot-bath **must be > 12°C**, otherwise the formalin loses its activity. If needed, add lukewarm or hot water to the foot-bath but be sure to take the concentration of the solution into account.

Claw treatment facility

Claw disorders-economic losses

Adapted from Marielle Bruijns, Henk Hogeveen & Elsbeth Stassen Veeteelt August 1 & 2, 2010

About 80% of cows experience a clinical or sub-clinical claw lesion at some point during their lactation. About 25% of the cows show a clinical claw lesion annually. In a herd of 65 cows, these claw lesions cause economic losses of between 2300 and 5000 Euro, with an average loss of 3500 Euro. This means that, at cow level, the average loss is around 53 Euro per cow per year. The losses are caused by several factors, which are presented in Table 4.4 together with their normative values.

Clinical cases cause 66% of the losses, while Mortellaro cases are responsible for the highest losses, followed by sole ulcers, *laminitis* and *interdigital dermatitis* in decreasing responsibility. The milk yield losses are the highest of all sources of loss. In the case of sub-clinical claw lesions, only loss of milk yield and increased calving interval are considered (Table 4.3).

In spite of the fact that the average loss due to claw lesions is 3500 Euro/farm/year, the variation is large (2300 to 5000 Euro) due to variable elements, such as management, feed changes, weather, etc.

Economic losses (Euro/cow/year):

- Milk yield loss (23,70)
- Culling (11,33)
- Increased calving intervals (6,56)
- Additional labour (6,31)
- Discarded milk (2,02)
- Professional claw trimmer (1,68)
- Veterinarian (0,84)
- Treatments (0,73)

Note that these losses may vary greatly between farms and between individual cows.

Both Tables indicate that prevention is better than cure when dealing with claw lesions!

Or, at the very least, to obtain a proper diagnosis as soon as possible, for example by using scoring systems, such as monthly locomotion scoring (see sections on MONITOR) and the additional weekly observation and recording of cows at the feed rack for wiggling on their feet and poor posture of hind legs in order to detect any increasing problematic

Table 4.3. Factors contributing to economic losses due to claw lesions

Source of loss	Value or cost (€)	Remarks
Loss of production	0.12 /l milk	8% per clinical lesion; 3% per sub-clinical lesion
Premature culling	240-913 /cow	Culling caused by lameness
Increased calving intervals	0.70 /day increase	Probability in case of clinical lesion is 60%; in case of sub-clinical lesions 20%
Extra labour of farmer	20/hr	Additional labour for treatments
Discarded milk	0.17 /l milk	Caused by antibiotics in the milk; is higher than production loss because this milk has already been produced
Professional claw trimmer	26/visit/cow	Not included in the calculation are routine herd claw trimming sessions
Veterinarian	30-85 /cow	Visit alone (30) or visit + treatment (85)
Treatments	0.40 to 6/treatment	Antibiotics; claw blocks; curative trimming; etc

Economic losses

Scoring lesions

Table 4.4 Average economic losses (Euro) per case diagnosis per year

	Phlegmona	Interdigital Dermatitis	Mortellaro	Laminitis	White line disease	Sole ulcer	Tyloma	Total
Sub-clinical	---	313	191	473	83	---	48	1108
Clinical	339	279	886	237	113	455	66	2368
Total	339	585	1076	710	196	455	114	3476

trends. Moreover, any cows detected as having a problem must be treated as soon as possible.

A **herd prevention plan for claw disorders** should be designed, comprising, at the very least, routine herd trimming by a professional claw trimmer once or twice yearly, a proper, daily manure scraping scheme (four to six times per day), the provision of correctly balanced rations (energy, protein, fibre) and a foot-bath used as prescribed (refer to instructions on the use of a foot-bath elsewhere in this publication).

Scoring lesions -Digital dermatitis in dairy cattle

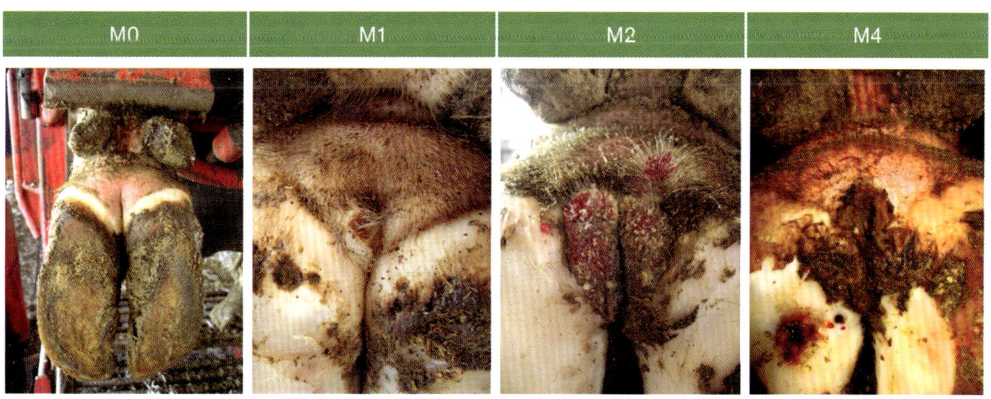

- **M0:** no circumscribed lesion;
- **M1:** early lesion, small and circumscribed, red to grey epithelial defects, less than 2 cm in diameter;
- **M2:** classic ulcerative (bright red) or granulomatous (red-grey) stage with diameter > 2 cm;
- **M4:** late chronic stage characterized by a dyskeratotic lesion (mostly thickened epithelium), surface proliferation or both.

Ref A. Relun, A. Lehebel, N. Bareille & R. Guatteo (Nantes, France, 2010-2011)

Figure 4.6 Illustration of the M-stages used for the scoring of digital dermatitis lesions (adapted from Döpfer *et al.*, 1997).

Climate in barns

> “Housed cattle have environmental requirements to promote optimal physiological function. The barn climate is one of the most important factors influencing that function. It comprises air speed, relative air humidity, ventilation and temperature. Heat stress is one of the most critical problems in housed cattle, problematic, not only during hot summers but in certain housing situations as well.”

Heat stress

Hyperthermic stress mainly causes problems in milk production and reproduction but can also affect other organ systems.

In the event of heat stress, the main actions to be taken are:

1. improve access to cool drinking water
2. provide shady areas in and around the barn and in pasture plots
3. reduce the walking distance from pasture to barn and vice versa
4. reduce the time that cows spend in waiting areas inside or outside
5. improve the ventilation in the waiting area
6. provide cooling in waiting area and in passageways
7. improve ventilation in the animal houses (exercise areas; behind feed rack)
8. provide cooling for close-up cows (from three weeks before to calving)
9. provide cooling for cows at all stages of production
10. provide cooling for young stock.

Methods used to cool the cows differ from region to region (e.g. humid areas vs. arid areas). Evaporative cooling is ineffective under humid conditions.

Suggestion 1.

Recommendations vary widely and there must be some degree of flexibility regarding the method chosen to address heat stress.. Normally 60 cm of trough space per cow is sufficient, resulting in 1 trough for 20 cows. In cubicle housing, troughs are normally located where passageways cross. In case of overcrowding and where access to water is reduced dramatically, additional water troughs should be installed to allow a total width up to 900 cm per 100 cows.

Suggestion 2.

Cows at pasture or in outside exercise areas must have shade provision. Cows exposed to sun at a temperature-humidity-index THI of > 80 during the day produce in general 2 to 4 kg milk less than cows in normal conditions. Shade from trees is adequate but aluminium or other metals or wood can also be used. Porous materials are not effective (e.g. wind break curtains). The **efficacy of materials for providing shade** is shown in Table 5.1.

A minimum of 4.5 m^2 of shade per cow is required and materials used to provide shade must be at a height of 4.50 m and in a north–south orientation to eliminate condensation.

Orientation of the barn should be south-west. A north-south orientation will increase the risk of **direct sun radiation** throughout the day. The afternoon radiation is more intense than morning radiation (note: risk of photo-sensibilisation). In this situation, the cubicles on that side will be used to a lesser extent or avoided completely. In the late afternoon, the sun is low and may still enter the barn. In a north-south orientation, curtains may be used on the west side from 1300 H to 2000 H (if used continually curtains may hamper ventilation).

Heat stress

After AABP proceedings 2002; Smith, Brouk & Harner. And Bonnefoy JM & Noordhuizen JPTM Maîtriser le stress thermique chez la vache laitière. Bulletin des GTV, June/July 2011: 109-116.

Table 5.1 Efficacy of materials used to provide shade

Hay	15 cm thick	1.2	GOOD
Wood	not painted	1.1	
Galvanised metal	white	1.1	
Aluminium	white	1.1	
Nylon with a layer of neoprene	two-sided white	1.0	
Aluminium	standard	1.1	REFERENCE
Galvanised metal	standard	0.9	POOR
Asbest	natural colour	0.9	
Tissue	90% solid	0.8	
Tissue	80% solid	0.8	
Wood	5 cm slat + 5 cm slot	0.5	

Suggestion 3.

Where possible, reduce the walking distance of the cows to reduce exposure to the sun. Sometimes it is not possible to reduce the walking distance (e.g. grazing to milking parlour/barn) and some farmers may keep cows inside during daytime and turn them out at night after the evening milking. Daytime housing requires additional management, such as water provision, ventilation and cooling and, in some case, it is not possible to house the cows at all.

Suggestion 4.

A roof angle of 4 to 12 (± 33%) and an open ridge minimises the increase in the cows´ respiration rate. Side walls should be 430 cm high to allow natural ventilation. The wider the barn, the less natural ventilation occurs. For example, a 6-row versus a 4row cubicle barn has 40% less ventilation and the respiration rate of cows is increased. Other buildings or barns should not be in close proximity (at least 50cm apart) to aid movement of air around the outside.

Suggestion 5.

There are two methods of cooling: (1) by ventilators, (2) by sprinklers or hose. The principle of cooling is to first get the cows wet then dry them.

For moderate heat stress conditions - soak the cows for 5 min, and then dry them using a ventilator at 0.33 m^3 per cow.

For severe heat stress conditions - soak the cows for 10 min, and then dry them using a ventilator at 0.33 m^3 per cow.

Soaking the cows has a greater positive impact on respiration rate than the ventilator has. Respiration rate can be reduced by 45% after 90 min of cooling. Cooling can be provided in the exercise areas behind the feed rack, in the waiting area, in the passageways, at the gates to pasture, however, is should not be provided above the feed bunk or cubicles. Cooling can also be provided outside in sheds. Aa special showering area can also be installed in the barn.

Heat stress

Ventilators are generally 76 to 91 cm diameter and provide 0.47 m^3 of air. One ventilator should be installed for every 10 cows or per 13.5 m^2 surface. Ventilator rows should always be parallel to the long side of the barn e.g. every 2m, at a height of 1.50 m above the cows. Sprinklers should proved 1.25 L water per m^2.

In barns with more than four cubicle rows ventilators with greater capacity (91 to 120 cm diameter; capacity 0.40 m^3 air per cow) or large propellers (800 cm wings) are required. These larger devices must be placed at 9m respectively 12 m intervals length. Ventilators must come on in temperatures from 22 °C upwards. If the temperature rises above 24 °C they must be on every 5 to 15 minutes.

Problems with cooling

There are various reasons for ineffective cooling:

- the diameter of water tubes is too small
- there is an insufficient number of tubes
- the outlet of the sprinklers is too small
- there is too much wind blowing the sprinkler water away
- after cooling, the water cannot be properly collected
- ventilators are poorly positioned in the barn
- electrical capacity of the barn is too small
- maintenance of ventilators is poor
- cooling by evaporation is applied despite humid conditions

Temp		Relative Humidity																		
F	C	0	5	10	15	20	25	30	35	40	45	50	55	60	65	70	75	80	85	90
72	22.0	64	65	65	65	66	66	67	67	67	68	68	69	69	69	70	70	70	71	71
73	23.0	65	65	66	66	66	67	67	68	68	68	69	69	70	70	71	71	71	72	72
74	23.5	65	66	66	67	67	67	68	68	69	69	70	70	70	71	71	72	72	73	73
75	24.0	66	66	67	67	68	68	68	69	69	70	70	71	71	72	72	73	73	74	74
76	24.5	66	67	67	68	68	69	69	70	70	71	71	72	72	73	73	74	74	75	75
77	25.0	67	67	68	68	69	69	70	70	71	71	72	72	73	73	74	74	75	75	76
78	25.5	67	68	68	69	69	70	70	71	71	72	73	73	74	74	75	75	76	76	77
79	26.0	67	68	69	69	70	70	71	71	72	73	73	74	74	75	76	76	77	77	78
80	26.5	68	69	69	70	70	71	72	72	73	73	74	75	75	76	76	77	78	78	79
81	27.0	68	69	70	70	71	72	72	73	73	74	75	75	76	77	77	78	78	79	80
82	28.0	69	69	70	71	71	72	73	73	74	75	75	76	77	77	78	79	79	80	81
83	28.5	69	70	71	71	72	73	73	74	75	75	76	77	78	78	79	80	80	81	82
84	29.0	70	70	71	72	73	73	74	75	75	76	77	78	78	79	80	80	81	82	83
85	29.5	70	71	72	72	73	74	75	75	76	77	78	78	79	80	81	81	82	83	84
86	30.0	71	71	72	73	74	74	75	76	77	78	78	79	80	81	81	82	83	84	84
87	30.5	71	72	73	73	74	75	76	77	77	78	79	80	81	81	82	83	84	85	85
88	31.0	72	72	73	74	75	76	76	77	78	79	80	81	81	82	83	84	84	85	86
89	31.5	72	73	74	75	75	76	77	78	79	80	80	82	82	83	84	85	86	86	87
90	32.0	72	73	74	75	76	77	78	79	79	80	81	83	83	84	85	86	86	87	88
91	33.0	73	74	75	76	76	77	78	79	80	81	82	84	84	85	86	86	86	87	89
92	33.5	73	74	75	76	77	78	79	80	81	82	83	85	85	85	86	87	88	89	90
93	34.0	74	75	76	77	78	79	80	80	81	82	83	85	85	86	87	88	89	90	91
94	34.5	74	75	76	77	78	79	80	81	82	83	84	86	86	87	88	89	90	91	92
95	35.0	75	76	77	78	79	80	81	82	83	84	85	86	87	88	89	90	91	82	93
96	35.5	75	76	77	78	79	80	81	82	83	85	86	87	88	89	90	91	92	93	94
97	36.0	76	77	78	79	80	81	82	83	84	85	87	87	89	89	91	92	93	94	95
98	36.5	76	77	78	80	80	82	83	83	85	86	88	88	89	90	91	92	93	95	95
99	37.0	76	78	79	80	81	82	83	84	85	87	88	89	90	91	92	93	94	96	96
100	38.0	77	78	79	81	82	83	84	85	86	87	89	90	91	92	93	94	95	97	96
101	38.5	77	79	80	81	82	83	84	86	87	88	89	90	92	93	94	95	96	98	98
102	39.0	78	79	80	82	83	84	85	86	87	89	90	91	92	94	95	96	97	98	99
103	39.5	78	79	81	82	83	84	85	87	88	89	91	92	93	94	96	97	98	99	100
104	40.0	79	80	81	83	84	85	86	88	89	90	91	93	94	95	96	98	99	100	101
105	40.5	79	80	82	83	84	86	86	88	89	91	92	93	95	96	97	99	100	101	102
106	41.0	80	81	82	84	85	87	88	89	90	91	93	94	95	97	98	99	101	102	103
107	41.5	80	81	83	84	85	87	88	89	91	92	94	95	96	98	99	100	102	103	104

Figure 5.1 Temperature Humidity Index (THI) chart

Prevention & control measures

Table 5.2 Some major prevention and control measures for situations of heat stress in dairy cattle.

	Control measure
Nutrition	Increase the frequency of feeding to 4-6 times a day; give the highest proportion in late evening or at night
	Reduce the effects of a negative energy balance around calving; maintain feed intake at a optimal level
Drinking water	Increase the number of drinking places available. Total width of drinking places must be 900 cm for 100 cows. Provide water of low temperature (e.g. 10°C) and clean troughs every 2 days. Check water quality beforehand (chlorates, sulfates, microbes)
Barn climate	Create shade over feed bunks (at 4 to 5 m^2 per cow)
	Install and use ventilators (see Table 5.3 on application of cooling)
	Install and use showers, mist-producing devices, or sprinklers (see Table 5.3,)
	Increase the number of sprinkling and drying cycles per day (see Table 5.3)
Management of reproduction and general issues	Do not use natural service (bull spermatogenesis disturbed)
	Provide cooling to dry cows from four weeks before calving onwards (sprinkling, drying, shade, ventilation)
	Create shade at pasture
	Reduce the walking distance from pasture to milking parlor, if possible
	Reduce the time the cows spend outside; reduce the waiting time before milking; reduce the number of cows per group for milking.
	Check more frequently the vitamin and mineral status of the feed (rations): K, Na, Cl, P. If needed, increase K, Na, Mg
	Adjust ration composition (2-3% fat in DM; protein level < 18%; rumen degradable protein < 61 % of total protein; salt; Lysine 1% of DM; water addition (4-5 litre per 20 kg DM). Eliminate citrus pulp from the ration (because it increases heat production)
Calves	Check IgG in colostrum (colostrometer) and in serum of all calves born in heat stress periods (refractometer; laboratory)
Other measures	Reduce cattle density in the barn (alternate groups inside and groups outside); reduce group sizes
	Check behavioral preference of cows for certain cubicle rows and resting areas. If deviations occur, take measures.
	Do not handle (process, treat, load) cattle in daytime. If handling is necessary, it must be for no more than 30 min
	Do load cattle for market or slaughterhouse during the day
	Provide shady work areas for employees; 15 min breaks every hour; more water drinking

Spraying devices & ventilators

Table 5.3 Examples of installing spraying devices/sprinklers and ventilators in the barn

Location of the devices	Installation details
Type of device	
Barn	
Ventilators	Behind the feed rack, not above cubicles or resting areas, not above feed bunk. Install ventilators in a line parallel to cubicle rows
	Diameter 76 to 91 cm. Capacity 0.47 m3. One ventilator for every 10 cows, or for every13.5 m2 of surface area or for every 2 m length of the barn. Height should be 250 cm. Install at an angle of 30° on the vertical axis.
	Air speed should not exceed 375 to 425 L/min, or ~12 to 20 km/hr
	Alternative option = large ventilators of 7 m diameter, for every 20 m of barn length, at an angle of 30° and each covering 400 m2 barn surface area.
Sprinklers	1.25 L water per min, at low pressure
Sprayers	At a height of 150 cm above the cows, at every 2 m length of the barn. Capacity of 300 L/hr.
Waiting area	
Ventilators	1 ventilator of 91 cm diameter for every 40 m^2 of surface area
Sprinklers	18 L of water per hour at a pressure of 300-400 kPa
Sprayers	At 150 cm above the cows, at every 2 m length of the barn. Capacity of 300 L of water per hour.
Milking parlor Ventilators Sprinklers Sprayers	Specifications as named under Barn or, for a 2x14 herringbone, 9 ventilators: 3 above the pit, and 3 above each side. For a 2x6 herringbone, 3 ventilators: 1 above the pit and one above each side. A mist-producing device may be added to the ventilators.
Shower room in barn Ventilators Sprinklers Sprayers	Located somewhere in the barn where it is most convenient for cows and man. One sprayer for every 2 m and 2 ranges of ventilators (60 to 90 cm diameter). Do not forget a rubber covering on the floor.
Entrances to the barn, exits to the pasture plots	
Ventilators Sprinklers Sprayers	Specifications as for Barn

Beware: the ventilators positioned nearly vertically above the cattle may provoke circulation of dust!
See also the review paper by Bonnefoy JM & Noordhuizen JPTM(Bulletin GTV, France, June 2011)

Testing ventilation in barns

For testing the quality of the barn climate, especially the ventilation, it can be indicated to conduct tests of smoke powder. These tests are low-costs, easy to perform and very instructive to the farmer too because the pattern of ventilation is clearly visible. Such tests is conducted by putting smoke devices at each 5 m length of the barn at cow level and observe what is happening. Normally the smoke should have gone within 3-4 min.

Below are listed a few addresses and types of smoke device. Yellow is preferred by author.

Table 5.4 Smoking powder for testing the ventilation in barns

www.feux-artifice.fr	Sécurix no. 1	100 gr	#0953	about € 10
	Sécurix no.3	500 gr	#0954	about € 18
lafeste@lafeste.com	Grains	100 gr	#00865	€ 5.00
		500 gr	#00866	€ 17,00
		1 kg	# 0866-C	€ 24,00
	Slow powder	100 gr	# 00850	€ 12,00
	Rapid powder	100 gr	#00853	€ 12,00
Cordia Protection ; ZAC La Villette aux Aulnes ; 2 Rue Gallillée ; 77290 Mitry–Mory, France	Exercise smoking powder	18 gr 60 gr 430 gr		
www.pyro-partner.de ; Rigistrasse 8–10; 12277 Berlin, Germany; (tel +49 30 720 080 0)	Tabletts AX 18		 # K 7051	€ 7 /5 pieces € 20 /5 pieces
www.waffen-ostheimer.de ; Waffen Ostheimer; 67346 Speyer, Germany	Raucherzeuger		# 49–007	€ 15 /kg
www.ASMC.de ;	COMET Raucherzeuger white	1 kg	#913450	#913450
Smoke grenade UK	white	box of 35 pieces		€ 111,00 (€3,20/piece)
EDRUCO ; Kapelsestraat-West; (zone KMO); Starrenhoflaan 6; 2950 Kapellen, Belgium	white, yellow, green, or red smoke		n°. 12003	

Note- prices are only indicative

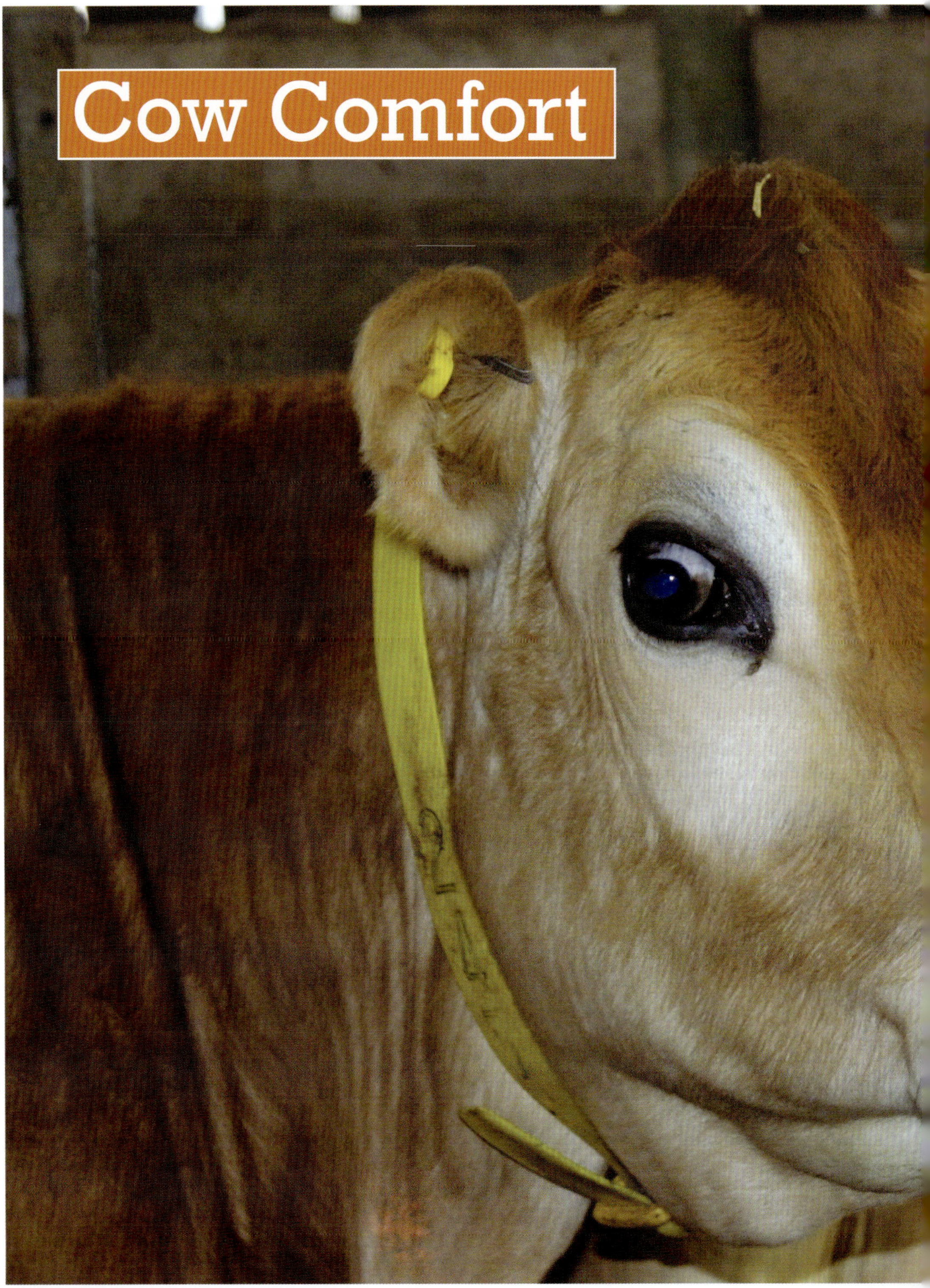
Cow Comfort

"Cow comfort can be considered as the practical translation of various animal welfare concepts. It comprises the sectors of housing & barn climate, feed & water, health (especially claw health) and species behaviour. In this chapter the approach to cow comfort and animal welfare is emphasised from a practical point of view. More specific, technical details of these sectors are addressed elsewhere in this book."

Welfare - frequency of lameness

Eighty dairy farms, with either a loose housing system with straw yards or as tied stalls, were assessed for the prevalence of lameness. There was a large variation in lameness prevalence between the two systems.

Results

1. Farmers do not know the lameness risk factors very well
2. The rate of lameness detection is abnormally low, 8% observed versus 27 to 34% in reality
3. Treatments for lameness are either too late or non-existent
4. The fact that herd claw trimming is rarely executed makes the situation worse
5. The barn surfaces are not very well and there is a lack of cow comfort
6. The steps (cubicles; drinking places) are too high (> 20 cm is bad) and sometimes there are even two steps
7. There are traumatizing areas and or slippery areas in the houses
8. The exercise areas are muddy and or full of gravel
9. The lack of hygiene and the humidity provoke claw lesions
10. The traumas on a hard wearing floor, and gravel are prone to cause phlegmonas.

Primary criteria for cattle welfare

1. The body posture of the cows.
2. The straightness of the spine of the cow, both while standing and while walking (locomotion score).
3. Treatments for lameness are either too late or non-existent
4. Treatments for lameness are either too late or non-existent

To ensure cow welfare better, a herd claw trimming routine must be implemented. Trimming frequency should be every four months. An alternative option is claw trimming twice yearly. Older cows and repeat cases must be trimmed more frequently.(Note that the current Dutch method of claw trimming is to trim the claws twice per year (routine functional trimming) and to trim each cow just before drying off, as well as after peak milk yield (after 100 days lactation). If necessary, older cows and chronic cases must be claw trimmed more often.

Table 6.1 14 key animal needs for cattle welfare*

Feed*	Health*
Drinking water*	Reproduction
Respiration	Grooming*
Excretion	Locomotion*
Resting*	Orientation/ exploration
Safety*	Pain experiences
Thermoregulation	Social interactions*

*= 8 primary issues for assurance & control
Note: these Animal Needs include both positive and negative [risk] indicators.*Metz, 2003, after Dantzer 2002 & Bracke *et al.* 2002)

Cow comfort & cattle welfare

Dr.Jos P.T.M.Noordhuizen, DVM, PhD, Dip EC BHM; Dr. Jan J. Lievaart, DVM VACQA-International, Portugal & France

Animal welfare concepts are various but the most well known concept is that of the Five Freedoms.

In relation to practical application, these freedoms are too vague. Therefore, the biological needs concept has been derived from these Five Freedoms. There are twelve, defined biological needs and the twelve biological needs can be easily translated into practical parameters. Cow comfort represents the combination of parameters in the areas of feed & feeding, barn & barn climate, housing conditions and specific behavioural aspects. Essentially, they represent potential risk areas that contribute negatively or positively to the well being, behaviour, health, production and reproduction of cattle. In this paper the fore mentioned issues will be addressed.

Introduction

Animal welfare has become an important issue in European livestock farming, in EU politics and in society. Concerns of the general public largely originate from the perception of the intensive (industrial) livestock production and from different outbreaks of some highly contagious diseases in recent decades, such as Foot-and-Mouth Disease, Avian Influenza and Classical Swine Fever together with the massive eradication campaigns following these outbreaks. The role of the media in the perception of the general public is a highly influential. Consumer protection is a major issue in EU politics; in several directives and regulations (e.g. 178-2002) this also comprises the issue of animal welfare, for example during animal transportation for trade reasons. Although the physical distance between livestock farming and urbanized areas is narrowing rapidly, the gap in knowledge about farming procedures and animal husbandry between producers and consumers is increasing. The general public has certain opinions about animal husbandry, some of which are largely uninformed. In food chain quality assurance systems comprising livestock production, animal welfare is included as one of four main issues next to food safety, public health and animal health. This implies that the dairy farmer has an important role in providing the animals with the proper environmental and managerial conditions for optimal wellness and, therefore, optimal health and productivity.

This paper addresses dairy cattle welfare from the perspective of cow comfort on the dairy farm. In the context of cow comfort, focus will be on four predominant clusters: barn climate, housing conditions, feeding management and behaviour. The actions the dairy farmer can take to optimize cow comfort and subsequent health and productivity and the role the veterinarian can play to support him in that process are discussed in this paper using several field examples as illustration.

Animal welfare

Animal welfare has been extensively addressed in the literature. Various different approaches have been followed, from basic neurobiological and neuro-edocrine approaches (Dantzer & Mormède, 1983; Dantzer, 2002) to that of the balance between satisfaction status and stress responses, as well as that of adapting the environment to the animal's performance. A synthesis of these approaches is presented in Figure 6.1.

Cow comfort & cattle welfare

Dr. Jan J. Lievaart, DVM
VACQA-International, Portugal & France. Dantzer & Mormede, 1983; Dantzer, 2002

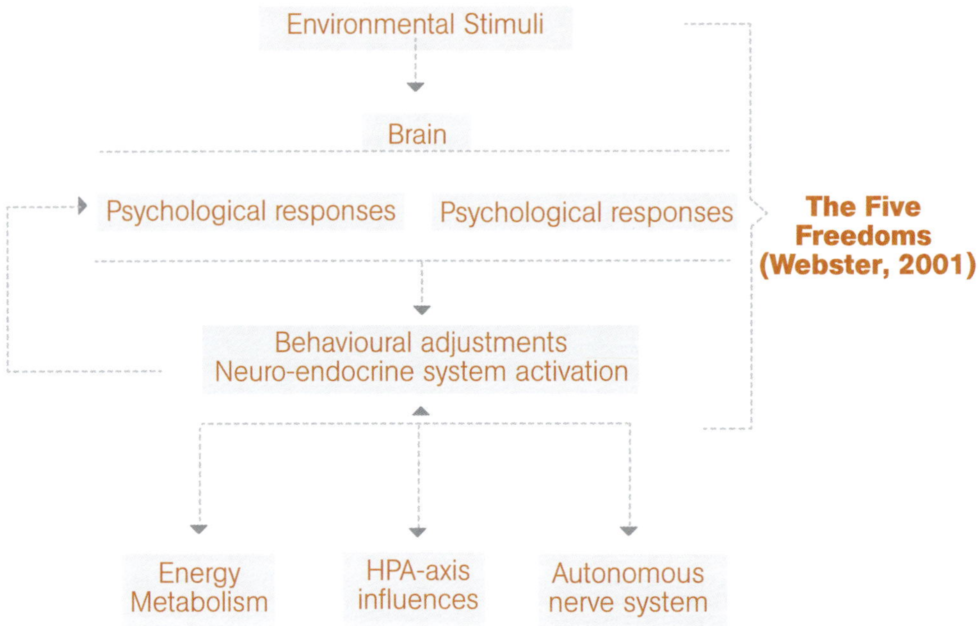

Figure 6.1. Synthesis of animal welfare concepts and pathways

Because these five freedoms are rather vague for facilitating application in the field, they have been translated into biological needs (Bracke *et al.*, 2001). These authors have defined seven primary and five secondary biological needs (Figure 6.1)

The model most frequently referred to, in which the animal is pivotal for defining welfare, is the concept of the Five Freedoms (Webster, 2001). These five freedoms refer to the freedom from hunger and thirst, disease and pain, physical and physiological discomfort, fear, distress and chronic stress, and limitations in conducting species-bound behaviour.

When assigning parameters to these clusters of biological needs, the result is around 100 different parameters describing the seven primary biological needs (Lievaart *et al.*, unpublished). The art is in translating these biological needs into practical parameters, which can be addressed, and preferably scored, in the field. The ultimate example of this exercise is in the concept of cow comfort.

Cow comfort & cattle welfare

Definition
The current status of the biological needs of an animal

Table 6.2 List of primary & secondary biological needs in relation to animal welfare. (Bracke *et al.*, 2001)

Primary biological needs
Feed and feed-related behaviour
Water and drinking-related behaviour
Resting, lying and standing
Locomotion (and claw/leg disorders)
Social comfort (interactions)
Health status
Safety (fear, flight behaviour, aggression)
Secondary biological needs
Excretion (faeces and urine)
Thermoregulation
Exploration and orientation
Grooming, comfort behaviour
Reproduction and rearing

Cow comfort

Cow comfort is the adequate combination of such elements in the immediate cow's environment, which jointly contribute to the cow's wellness, behaviour, health, reproduction and productivity. Cow discomfort is the outcome of the combination of risk conditions associated with the aforementioned elements.

There are four major clusters concerned with cow comfort: [1] Barn and barn climate, [2] Housing conditions, [3] Feed and feeding-associated issues, [4] Specific behaviour.These clusters and their respective elements will be addressed in more detail below.

Barn and barn climate

Firstly, the long barn sidewall should be directed against the predominant south-west wind; the sidewalls should be completely open while the cows should be protected against heavy, cold winds by mobile, light-coloured curtains. When several barns are present, the distance between barns must be more than 30 m to allow adequate air movements. The barn should provide sufficient surface area and volume for the number of cows contained within it. The air-movement at cow-level should be less than 5 m/sec. This is because the thermoneutral zone of cows is between −5°C and +20°C and their thermo-indifference between +5 and +15°C, contrary to that of man, which is much higher (Schrama *et al.*, 1996). Through this improved ventilation and air movement the total bacterial load of barn air should be less than 150.000/m^3 and the moist air originating from cows sweating (25 lt/cow/day) is easily removed (Banhazi *et al.*, 2004). Relative humidity in the barn should not exceed 80%. Moreover, if the farmer cleans the floors from manure/urine regularly, the floors will be much dryer enabling cows to walk unhampered.

Light in the barn should be more than 100 lux (preferably 200) for achieving optimal cow activity and optimal cow observation. This light level can be achieved by using halogen lighting. As a rule of thumb, one should be easily able to read a newspaper in the darkest part of the barn.

Housing conditions

The passageways in a loose housing system should be sufficiently wide to allow for easy and rapid cow traffic e.g. 3 to 4 m. The total walking or waiting area surface for cows should allow an inter-cow distance of 2 to 5 m, comparable to that when out grazing. Ideally, the feed rack should not be a fixed to allow optimal access to feed and rapid in/out movements through the yolk in case of hierarchical confrontations. A sufficient number of feeding places must be provided in relation to the number of cows present. The cubicles, if any, should be

Cow comfort & cattle welfare

the appropriate size for the prevailing breed; this can be observed by the cow's udder and tail lying inside the cubicle. The base of the cubicle should have a downward slope of 5 cm to aid drainage and bedding material must be deep-littered (> 15 cm) or sand, soft and dry to promote cleanness and dryness of cows. The knee-test by the farmer or veterinarian will show whether or not these demands are met. Deep-litter cubicles should be topped-up daily with at least 1 kg straw of 2–3 cm length (plastic, isolation) or 12 kg sand (plastic, cool, with grip) per cubicle. When additional rows of cubicles have been installed in parallel in the barn, then the distance between the cubicle fronts should be at least 1 m to allow cows to easily stand up. Brisket boards should be adjusted to 170 cm from the back of the cubicle, a neck-rail at 165 cm ± 5 cm from the back of the cubicle and at 110 –120 cm height above the bedding.

Water troughs must be easily accessible and protected from freezing. The flooring in exercise areas, such as passageways and waiting area should be clean and dry and possibly coursed but preferably rubber topped for optimal cow activity, oestrus expression, claw health and cow interactions. Floor maintenance means screening for unequal slats, loose slats and ensuring their repair. Optimal locomotion of cows as detectable by the locomotion score, step length, oestrus behaviour, claw health and peri-arthritis cases should be achieved. The majority of risk factors associated with claw lesions relate to floor quality, locomotion, non-hygienic floor conditions and social stress (Somers *et al.*, 2003; Bergsten, 2004).

The milking parlour must be easily and rapidly accessible for cows, fully lit and equipped with additional technology, such as automatic cluster detachment (threshold at 300 or 400 ml/min), teat spraying devices, large milking claw volumes for both rapid milk removal and prevention of teats being washed in (contaminated) milk. In automatic milking systems, specific adaptations must be provided (Ketelaar-de Lauwere *et al.*, 1999). If possible, a dynamically movable floor is installed to improve milkers' labour and well-being. The exit of the milking parlour must equally be easy and round (no 90° corners).

In hot climatic conditions, where heat stress may occur, large ventilators well-positioned in the barn should be installed, while an internal and/or external sprinkler installation is advisable. At pasture, shady places should be available during hot conditions, for example large trees or hedgerows.Separate calving pens and sick cow pens should allow visual contact with the herd. There should be more than sufficient space for each cow. Cleaning and disinfection should be carried out before the arrival of each new animal

Feed and feeding management

High yielding cows show feeding frequencies up to 12 times per day. Therefore, feed should be available throughout the day. The feed bunk must be smooth, clean and dry; daily cleaning must be performed to maximise feed intake. The feed bunk floor must be 10 cm higher than the floor on which the cows are standing to optimize feed intake and prevent front claw lesions. Feed in the feed bunk should be pushed-up to the cows regularly. Feed refusals should be weighted per group and divided by the number of cows to estimate average feed refusals and monitor trends. The feed rack should be easily accessible and preferably not fixed. In the case of a fixed rack, a fastening must be available to facilitate veterinary examination and treatment.

If fed, a TMR should be checked for particle length, as well as homogeneity of mixing, quality of ingredients and sorting by the cow (evident by holes in the feed when offered to the cows). Body condition scores, rumen fill scores and faecal quality scores (consistency; undigested fibres) are suitable management tools to evaluate feeding efficiency.

Specific behaviour

Next to the behavioural issues addressed in the clusters above, there are a few additional features

Cow comfort & cattle welfare

of interest. For example, the ratio between standing time and lying time in a group of cows (e.g. the fresh cows). On a 24 hours basis, the pattern may be as follows: resting with/without ruminating 14 hrs, lying with or without ruminating ruminating 6 hrs, eating at feed rack or while grazing 2-6 hrs. In between, cows will show standing or walking activities up to 5 km per day. Locomotion and step length are indicators for claw health (Somers *et al.*, 2003; Bergsten, 2004). Synchrony within the herd is an indicator of herd social behaviour.

Idle standing may be an indicator of claw health problems, general distress and/or poor cubicle condition. On average, low yielding cows show longer idle standing time than cows with higher yields. The position of cows in a cubicle may also be an indicator of cubicle quality (Dantzer & Mormède, 1983; Dantzer, 2002).

Farm staff/employees should observe the cows on a regular, routine basis in order to, for example, detect disease signs at an early stage and thus be able to deal with them in a timely fashion (Cow Signals, Hulsen 2004).

From the features addressed in the four clusters described above, it is clear that farm management plays a pivotal role in achieving optimal cow comfort. This also includes cow handling in general, cow-man interactions, general health care, including participation in veterinary herd health programmes (Brand *et al.*, 2001), cow observation skills and reproductive efficiency (interval between calving and first oestrus, negative energy balance, cystic ovarian disease, sire evaluation).

Additional welfare-associated management issues refer to Good Dairy Farming Code of Practice invasive treatment, such as asking for veterinary aneasthesia when indicated, timely euthanasia for specific cows, injecting cows in the most appropriate way, timely seeking of veterinary assistance and application of good hygiene practice.

Discussion and conclusions

Optimizing features of cow comfort as addressed in the preceding paragraphs can be considered as the basic elements in providing appropriate conditions for achieving good cow welfare. Features of cow comfort are closely related to the biological needs, which have been previously identified. Moreover, it shows that the farm manager plays a pivotal role in this process and cow observational skills are the key step in evaluating the outcome of adjusting the cows' environment and associated cow behaviour. Several recent publications have been issued all addressing the observation of cows, or Cow Signals (Zaaijer & Noordhuizen, 2003; Hulsen, 2004).

Various aforementioned cow comfort elements may be easily applied to newly designed barns. In existing barns it is often a matter of compromise between existing conditions and adjusting for more ideal conditions. The veterinarian or other professional, may be of assistance in achieving those goals, when he/she is willing to exploit his/her observational diagnostic skills, to invest in developing these skills continuously and to conduct the proper interpretation of results for the appropriate advice.

In order to provide the farmer with some additional time to spend on cow care and cow comfort, various practical tools have been designed that reduce both the required labour and some time-consuming routine practices. Examples of these are: management tools for scoring Rumen Fill, Faeces Consistency, Undigested Fibres in Faeces, Teat end callosity, Udder Hygiene (see MONITOR series).

Useful Reading

Banhazi TM, Seedorf J, Rutley DL, Cargill C, Hartung J. 2004. Effects of airborne pollutants and factors affecting concentrations in livestock buildings. In: Animal Production in Europe, the way forward in a changing world. Proc. of the ISAH conference St Malo (F) 11–13 October, vol 1, 187–199 (Madec F & Clement G, editors)

Bergsten C. 2004. Lameness and claw lesios

Cow comfort & cattle welfare

as influenced by stall environment and cow comfort. In: Proc. of the Europ. Meeting of the French Buiatrics Society, Paris (F) 23–24 Nov., 150–156

Bracke MBM, Metz JHM, Dijkhuizen AA, Spruijt BM. 2001. Development of a decision support system for assessing farm animal welfare in relation to husbandry system: strategy and prototype. J. of Agric. & Environm. Ethics 14: 321–337

Brand, A., Noordhuizen, J.P.T.M., Schukken, Y.H. 2001. Herd health and production management in dairy practice. Wageningen Academic Publishers, Wageningen, The Netherlands, 543 pp, 2nd release

Dantzer R, Mormède J.P. 1983. Stress in farm animals: a need for re-evaluation. J. Anim. Sci. 57: 6–18

Dantzer R. 2002. Le bien-être des animaux d'élevage. In: Conduite à tenir, Proc. Of the GTV meeting in Tours (F) 29-31 May, 265–268

Hulsen, J. 2004. Kuh-signale: Krankheiten und Storungen fruher erkennen. Verlag Roodbont, Zutphen, Holland

Ketelaar-de Lauwere CC, Ipema AH, van Ouwerkerk ENJ, Hendriks MMWB, Metz JHM, Noordhuizen JPTM, Schouten WGP. 1999. Voluntary automatic milking in combination with grazing of dairy cows: milking frequency and effects on behaviour. Appl. Anim. Behav. Sci. 64: 91–109

Lievaart JJ *et al.* 2005. A system to score cattle welfare on dairy farms based on adaptive conjoint analysis studies (unpublished data)

Schrama JW, van der Hel W, Gorssen J, Henken AM, Verstegen MWA, Noordhuizen JPTM. 1996. Required thermal thresholds during transport of animals. The Vet. Quarterly 18: 90–95

Somers JGCJ, Frankena K, Noordhuizen-Stassen EN, Metz JHM. 2003. Prevalence of claw disorders in Dutch dairy cows exposed to several floor systems. J. Dairy Sci. 86: 2082–2093

Webster AFJ. 2001. Farm animal welfare: the five freedoms and the free market. The Vet. J. 161: 229–237

Zaaijer, D., Noordhuizen, J.P.T.M. 2003. A novel scoring system for monitoring the relationship between nutritional efficiency and fertility in dairy cows. The Irish Vet.J. 56: 145–151

Paper originally presented at the 1st Buiatrissima Congress, Bern, CH, 2005

WELFARE

Is a dynamic process

Is biologically measurable (from physiology to clinics)

Refers to 'milieu interne' and 'milieu externe'

Comprises elastic (= freedoms) and non-elastic (= compulsory) needs

Q&A
Comfort & welfare

Q What is the average time-budget of dairy cows during the day?

A Feeding: 3 to 5 hrs per day dispersed over 9 to 14 meals 5 to 9 hrs per day dispersed over 10 to 15 meals

- Social interactions: 2 to 3 hrs
- Resting and lying: 12 to 14 hrs
- Drinking: half an hour
- Ruminating: 7 to 10 hrs
- Walking: 2 to 3 hrs

Q Fear in a cattle herd: what does it mean and what are consequences?

A Increased risk of accidents for man and animal

- Reduced/retarded growth rate in young stock
- Decrease in milk yield
- Reduced performance (health; reproduction; beef)

(Source -X.Boivin, L.Mounier, Le Nouveau Praticien 3-12-2009)

Q How do you calculate the volume of water intake for dairy cows

A Use the following equation
(1.54 x DM) + (1.33 x BW) + 0.89 x DM%) + (0.58 x Tmin) – (0.30 x Pluvio) – 25.65
DMSI is dry matter intake in kg/day
BW is body weight
DM% is dry matter in percentage
Tmin is minimal ambient temperature
Pluvio is mm of rain in that region per year
(Source Cardot V, Leroux Y, Jurjanz S. (2008) Drinking behaviour of lactating dairy cows and prediction of their water intake. J. Dairy Sci. 91: 2257–2264)

Q What are the allowances for water intake of cows

A About 15 percent of the group must be able to drink at the same time

- About 30 to 50 percent of total water consumption must be carried out after leaving the milking parlour.
- Drinking space must be 60cm for 15 cows (4 cm per cow)
- 400 cm per 100 cows; in periods of hyperthermia (heat tress) this must be doubled.
- Drinking troughs should never be in duo setting (because of negative social interactions) nor placed in corners.
- In free stalls with straw yard, the available surface for the cows must be 10 m^2 per cow, including 3 m^2 of exercise area.

Q What rumen parameters should we monitor?

A Ruminal acidosis makes the cows more excited and nervous.

- The rumen pH should preferably be at > 6, when feed particles >2 mm exceed 40 percent of the total fibre content in the ration, as measured in the Penn State sorting boxes and when the ration contains 25 percent NDF in the DM.
- Frequency of rumination in the herd is one of the parameters used to evaluate feed efficiency; on average, > 70 percent of the cows at any time of the day should be ruminating.
- For every hour of rumination lost for whatever reason, the rumen pH will decrease by 0.2 to 0.4 units.

Q What are typical particle lengths in feeds

Particle length (% or product weight)

Length mm	Maize silage	Hay	TMR
>19	3-8	10-20	3-8
19-8	45-65	45-75	30-40
8-3	30-40	20-30	30-40
<3	<5	<5	<20

Q Do brushes improve cow comfort and performance

A Cows using the rotating or other brushes in the barn pass in average five times per day to take advantage of it (De Vries *et al.*, 2007).

Q&A Comfort & welfare

- Moreover, the cows using brushes are much less dirty.
- Good cow comfort increases milk productivity in dairy cows.

A survey conducted in the USA (Schukken *et al.*, 2010) looking at the effect of introducing brushes into groups of cows that had no previous experience of them, showed that the average milk yield significantly increased by 1 kg per day throughout lactation (especially second parity animals) when brushes were introduced, than groups without brushes.

At the same time, cows with access to brushes had significantly fewer mastitis cases than those without brushes. This effect increased with parity.

TOP TIPS

Each cm less of bedding material in the cubicles results in a reduction in lying time of 11 minutes. This means that on a daily basis, the cow rests 2.5 hours less than she should!**

Cows able to rest for up to 14 hrs per day produce 2 kg more milk than cows that rest for only 12 hrs daily.***

**Drissler *et al.*, (2005) J. Dairy Science 88: (7) 2381–2387)
***(Prod.de Lait Québecois, 24 : 18–19)

A CLINICAL WELFARE SCORING MODEL

1.Welfare criteria (in relation to the named 12 biological needs)
2.Target values (both positive and negative)
3.Testing criteria (feasible; reasonable; limited number; performance or management/environment related)
4.Assurance and control (lower thresholds only; cow and means oriented)
5.Recommendations

8
5
6

Disease

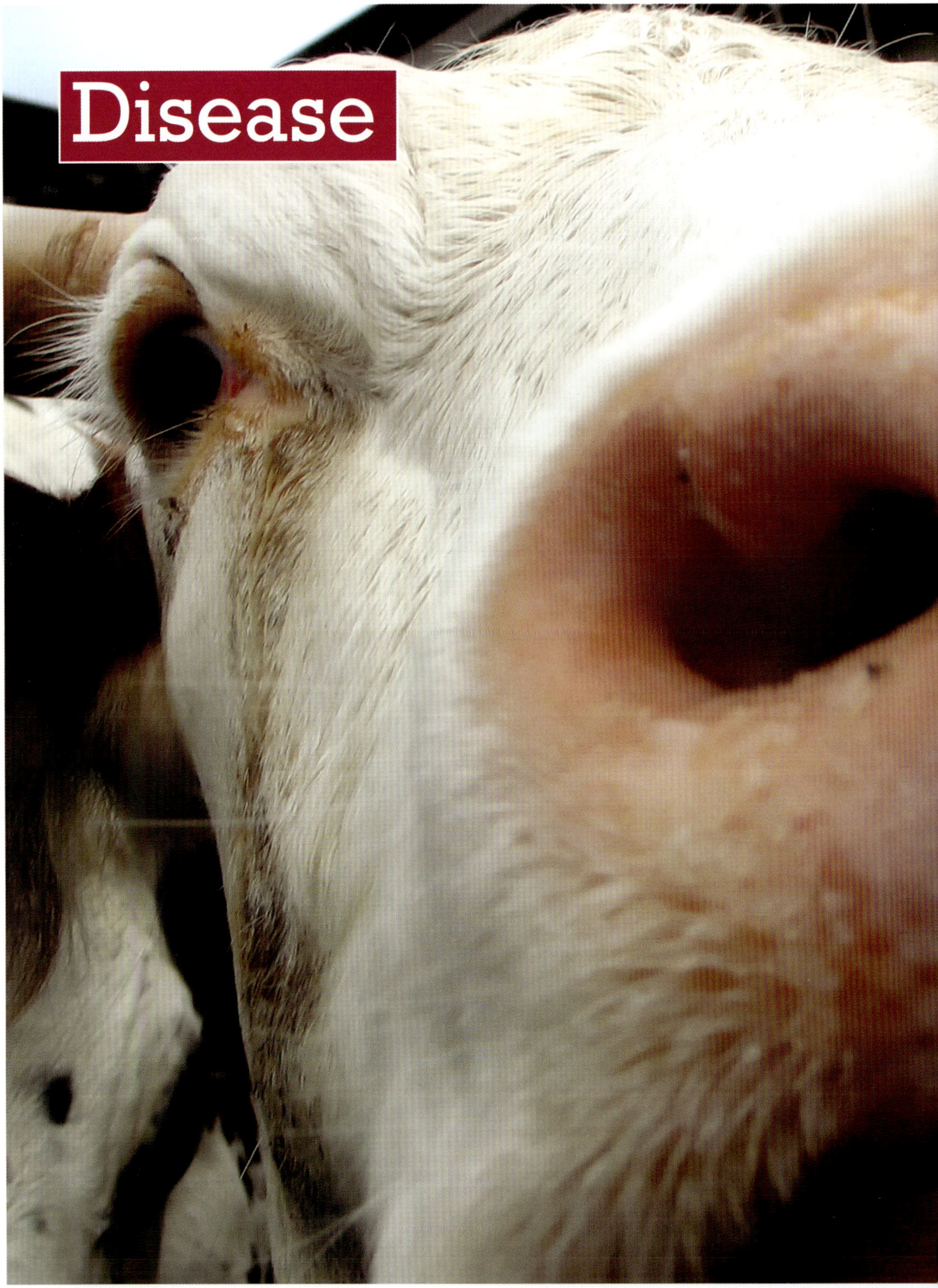

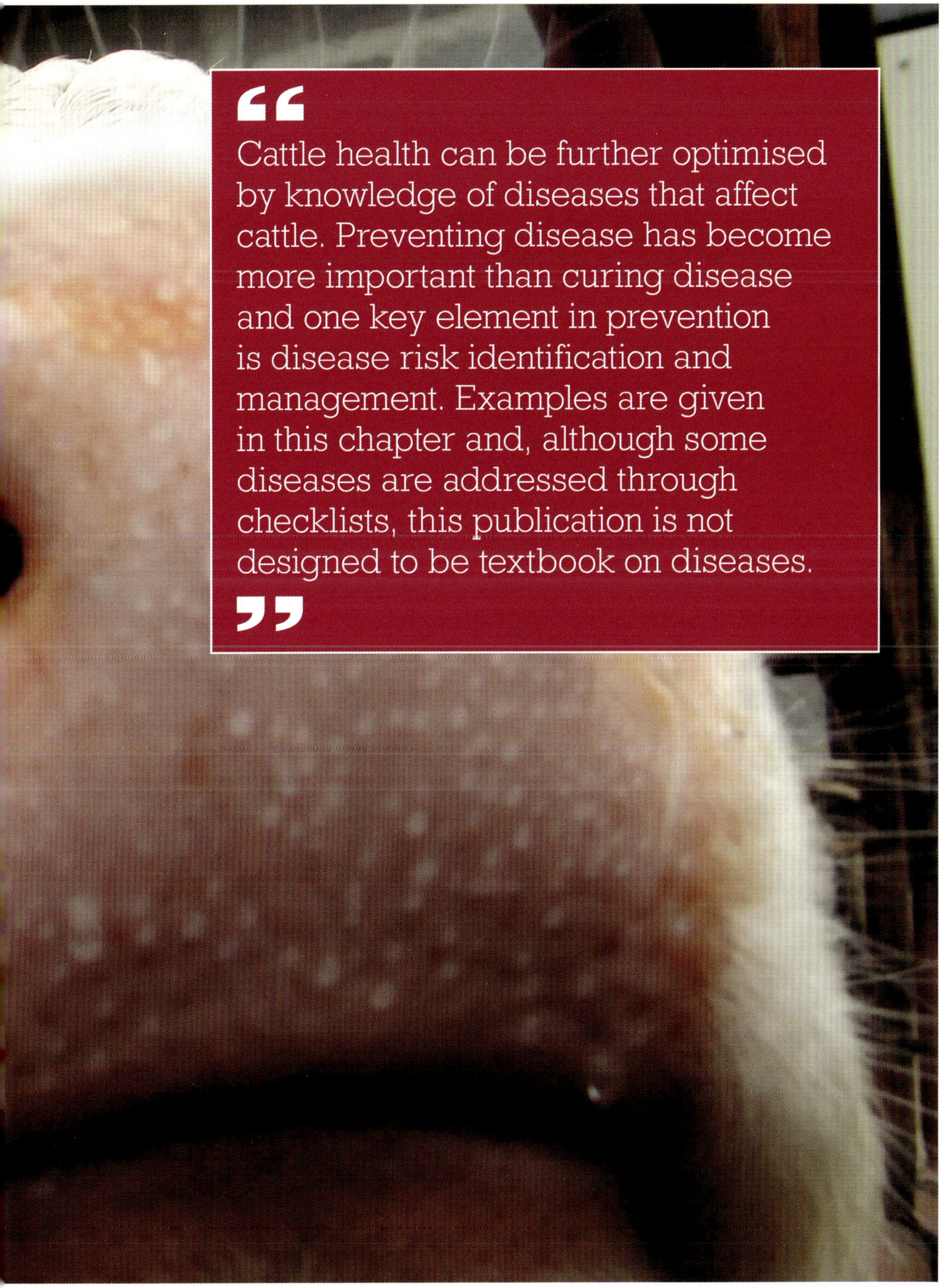

> Cattle health can be further optimised by knowledge of diseases that affect cattle. Preventing disease has become more important than curing disease and one key element in prevention is disease risk identification and management. Examples are given in this chapter and, although some diseases are addressed through checklists, this publication is not designed to be textbook on diseases.

Level of resistance

Adapted from GD Deventer NL, 2008

Table 7.1 Rapid screening test for the level of resistance in cows during the transition period

FARM: **Veterinarian:** **DATE:**

		YES	NO
Feed and feeding	The average BCS of the dry cows is above 3.5		
	Feed intake in dry cows dropped dramatically in late dry period		
Early lactation	Loss of BCS at six weeks after calving is > 1 unit		
	Rumen Fill score after calving is < 3		
Minerals & Vitamins	Dry cows and pregnant heifers receive a standard premix		
		Points scored	
Potential stress factors	Calvings occur in isolation outside the herd (yes = 4 points)		
	At the end of the dry period or on the day of calving, forage quantity differs by more than 50% from that offered in lactation (yes = 2 points)		
	At the end of dry period, the cows are still not adapted to at least two kg of concentrates (yes = 2 points)		
	The calf is not separated from its dam in the first six hrs after birth (yes = 2 points)		
	There are too few cubicles for all cows in the herd (yes = 6 points)		
	There are too few places at the feed rack for all cows to eat (yes = 4 points)		
	There is a lack of cow comfort in the herd (yes = 4 points)		
	Total number of points scored* =		
Herd health	High prevalence of infectious diseases		
	High prevalence of endemic diseases (mastitis, lameness, etc)		

➔ *a total score of points ≥ 6 indicates a lowered disease resistance in the cows.

The general health status of the herd (e.g. IBR, BVD, Salmonellosis) may indicate whether such diseases specifically contribute to this lowered disease resistance.

If the answer to several of the above points has been 'YES' then this is indicative of a situation at relatively high risk for poor disease resistance.

In this situation, a more in-depth analysis is warranted to determine the cause(s).

Risk factors for abomasal displacement

Table 7.2 Risk factors for abomasal displacement	
General information	Number of cows present in the herd
	Number of cows per parity
	Mal-adaptation of cows and heifers in the herd after calving (acifodid; low feed intake)
	Level of milk production (in L, fat, protein)
	Animals of parity 1
Housing	Type of barn
	Type of cubicle and bedding
	Type of exercise area
Nutrition	Quantity of maize silage/cow/day at calving
	Quantity of grains/cow/day around calving
	Quantity of hay (long particles) at calving
	Quantity of fibre in the ration around calving
	Ratio concentrates-to-forages too high
	Proportion of grains in concentrates too high
	Low level of feed intake around calving
Feeding management	Feeding system (conventional or TMR)
	Changes in forages around calving
	Changes in concentrates around calving
	Grazing cows or zero-grazing without or with exercise area
Animal health	Retained afterbirth > 10% of the cows calved
	Milk fever > 4% of cows calved
	Ketosis > 5% of cows calved
	Mastitis, or endometritis > 15% of cows calved

After: Hultren & Pehrson, 1996 ; Grymer *et al.*, 1982 ; Willeberg *et al.*, 1982 ; van Winden,2002 ; AABP #32; Noordhuizen, 2008.

Managing heat stress

Adapted from Smith, Brouk & Harner, 2002 proceedings of the AABP. pp 71-76

Heat stress can have a great impact on –especially- milk yield and reproduction. There are several possibilities for reducing the negative effects of heat stress:

1	Improve access to water
2	Provide shade in/around the houses and at pasture
3	Reduce the distance cows have to walk
4	Reduce the time that cows spend in the waiting area (especially if outside)
5	Improve ventilation in the waiting area
6	Provide cooling in waiting area and exercise areas
7	Improve ventilation in the animal houses
8	Provide cooling for close-up cows (three weeks before calving)
9	Provide cooling for cows after calving and at the start of lactation
10	Provide cooling for cows in mid- and end-lactation, as well as dry cows.

The methods used to cool the cows will vary from region to region. For example, cooling by evaporation will be difficult in humid conditions.

1. Improve access to water.

Recommendations may vary widely, so it is wise to be flexible with regard to values. The usual advice is 60 cm of drinking space per cow, which means one water **point for 10 to 20 cows**. In loose housing with cubicles, one drinking trough is located where exercise areas or passage-ways meet. In situations of high animal density (over-crowding), the access to water is substantially reduced (sometimes by up to 40%). In these conditions, it is advisable to provide additional water troughs.

During periods of heat stress, a total drinking trough width of **900 cm per 100 cows** should be available. It is better to have too much than too little drinking space!

2. Provision of shade

Cows at pasture or in outside exercise areas must have shade. Cows exposed to direct sunlight and with a temperature-humidity index (THI) of > 80 during the day can produce 2 to 4 kg less milk per day.

Shade from trees is adequate but often shade is provided in the form of aluminium or other material. Porous material (curtains for example) is not effective! Table 7.3 highlights the varying efficacies of different materials for providing shade.

About 4.5 m² solid shade per cow is adequate to reduce the impact of the suns radiation. The shade provision must be constructed to a **height of > 430 cm** and with a **north–south** orientation to avoid humid spots developing in the shade.

The orientation of the barn must be southwest. A north–south orientation gives a higher risk of direct and intensive sun radiation during the whole day compared with an east-west orientation. Direct sun radiation during the afternoon is much stronger than during the morning! The cows will not use the cubicles if in a north-south orientation and will look for cooler places or shade elsewhere. This phenomenon can be noted by simply observing the behaviour and preference of the cows.

Late in the afternoon the sun is very low but radiation can still penetrate the barn. In a north–south orientation, it is possible to install curtains on the west side to prevent the sun from penetrating. Ensure the

Managing heat stress

curtains are down from 13.00 hrs until around 20.00 hrs, after which they may be raised again to allow ventilation.

Table 7.3 Efficacy of different materials for providing shade compared with standard aluminium **

Hay	15 cm thick	1.2
Wood	not painted	1.1
Galvanised metal	white	1.1
Aluminium	white	1.1
Nylon avec couche de néoprène	white on 2 sides	1.0
Aluminium **	<---standard --->	1.0
Galvanised metal	standard	0.9
Asbestos	natural colour	0.9
Tissue	90% solid	0.8
Tissue	80% solid	0.8
Wood	5 cm closed ; 5 cm open	0.5

3. Reducing the distance cows have to walk

During grazing and milking, it is not always possible to reduce walking distance for the cows. Some farmers will put the cows inside the barn to avoid direct sun radiation, but this requires that other critical attention points are addressed first! (free water access; ventilation ; cooling cycles etc)!

4. Reduce the time that cows spend outside, when grazed

It is not always possible to have cows inside at all times. At least one could follow the principle of retaining them as much as possible in the barn, provided that the other requirements are met. See also Point 3 above.

5. Improve ventilation

Roofs with an inclination of 4 in 12 (about 33%) and an open top result in more normal respiration rates compared with other systems. A height of the side walls of 430 cm also has a beneficial effect. Both these parameters contribute to effective, natural ventilation.

Ventilation is also affected by the width and volume of the building: the greater the width of the building, the more the natural ventilation diminishes. A barn with six cubicle rows has 40% less ventilation resulting in increased respiration rates in cows compared with a barn with four rows of cubicles.

To maintain optimal natural ventilation, parallel buildings must be separated by a distance of at least 50 m.

6. Provision of cooling : the cooling cycles

During periods of heat stress, cooling should be administered in the form of cooling cycles consisting of spraying water followed by drying off.

One cooling cycle comprises:

- watering by sprinklers or sprayers for 1 to 2 min (1.25 L/m^2/min/sprinkler)
- drying by ventilators for 5 to 6 min (0.33 m^3/sec)
- repeat the above steps for 30 min.

In the event of heat stress, two cooling cycles, each of 30 minutes duration, are should be carried out – one before noon and the second prior to the evening milking.

If temperatures do not drop substantially during the night, an additional cycle can be carried out during in the night. In the worst case scenario, (temperatures remain above 24°C), cooling cycles should be carried out every four hours.

Water provides better cooling than air movement

Managing heat stress

Adapted from Smith, Brouk & Harner (2002) Proceedings of the AABP. pp 71-76)

The most important areas in the building to keep cool are:

- The barn, just behind the feed rack
- The waiting area
- The milking parlour
- Entrances to and exits from the barn.

If the means to install ventilators and sprinklers are unavailable, it is possible to install a **'cow shower spot'** in the barn where cows can enter freely.

For the installation of ventilators and sprinklers or sprayers, it is essential to seek the advice of **specialised engineers**. Too many costly faults can be made when people attempt to install these devices themselves. Examples of errors with do-it-yourself installations are:

- The diameter of the water tubes is too small
- Insufficient numbers of tubes installed
- The outlets of sprinklers are too small
- There is too much air movement resulting in the spray-mist blowing away
- The ventilators are poorly positioned
- The ventilators are poorly maintained
- The air speed of ventilators is too high
- The electric capacity on the farm is too low

Prussic acid poisoning

= hydrogen cyanide (HCN) intoxication

www.dairyherd.com 12th August 2011

HCN accumulation may occur in plants like sorghum, white clover, arrow grass and velvet grass, especially in the (younger) leaves.

Less is found in the stem and hardly any in the seeds. With silage, Prussic acid poisoning is not frequent because the fermentation processes will metabolise most of the Prussic acid.

Intoxication occurs due to **heavy stress on the plants**, such a freezing, extreme drought (heat stress) and other damaging conditions.

Cattle and sheep are particularly susceptible.

Signs occur from 30 to 120 min after ingestion of affected plant leaves.

Signs are:

Excitement	
Rapid pulse rate	The blood is cherry-red in colour
Muscle tremors	
Rapid breathing	Mucosae are bright pink in colour
Staggering	
Collapse & death	

➲ Acceptable levels are < 600 ppm.

From 600 to 1800 ppm signs may occur. Contaminated feed should only be fed in smaller portions mixed with other forage or, preferably, not fed at all.

Above 1800 ppm the forage is extremely toxic.

A potential solution to save affected cows may be 7.4% Na-thiosulphate or 1.2% sodium nitrate intravenously (125 – 250 ml).

Table 7.4 Plants that accumulate prussic acid include (Provin and Pitt, 2011)

Plant	Prussic acid potential
Pearl and foxtail millet	very low
Sudangrass and sudangrass hybrids	low or moderate
Sorghum-sudangrass hybrids	moderate to high
Forage sorghum	moderate to high
Shattercane	high
Johnson grass	high to very high
Grain sorghum	high to very high
Sorghum almum	high to very high
Arrowgrass	low
Velvetgrass	low
White clover	low
Birdsfoot trefoil	low
Chokecherry*	low
Pincherry*	low
Wild black cherry*	low
Apricot*	low
Peach*	low
Apple*	low
Elderberry*	low

*Prussic acid may be found in the leaves and seeds of these tree species but it is not likely that livestock will consume sufficient amounts of these plants. Avoid housing livestock near these trees, especially if traditional forages are lacking/not available.

Nitrate toxicity

Table 7.5 Common crops and weeds that have potential to accumulate nitrate (T.L. Provin and J.L. Pitt; Texas A&M University, Cooperative Extension Service, USA, 2011)

Agricultural crops		Weeds	
Oats	Sorghum	Pigweed	Wild Sunflower
Canola	Pearl and German Millet	Bull Thistle	Canada Thistle
Barley	Corn	Fire Weed	Kochia
Wheat	Sudangrass	Lambs Quarters	Mustards
Rye	Rape	Nightshade	Millet
Sugar Beets	Soy bean	Russian Thistle	Smartweed
Flax	Sudan grass	White Ragweed	Canada Thistle
Alfalfa		Johnson grass	Jimson weed

Harvest as silage - Ensiling tends to reduce the nitrate content of forages. Forages high in nitrate can lose from 40 to 60% of their nitrate content during fermentation.

Harvest forages suitable for silage at the stage of optimal quality and quantity. Excessive nitrate in forages will not always be reduced to safe levels during ensiling. If silage is suspected to be high in nitrate, analyze the feed for nitrates before feeding. The analysis will help in designing rations to prevent livestock losses from nitrate poisoning.

Harvest close to maturity - Crops normally have lower nitrate levels at maturity, so for crops suspected of containing nitrates, silage should be harvested as close to maturity as possible. If the field being harvested as silage has been identified as, or is suspected of, having high nitrate concentrations, raise the cutter head to avoid stalk bases that have the highest nitrate concentration.

In crops, such as oats for hay, it may be wise to harvest the crop at a more mature stage than desirable for maximum yield of digestible nutrients.

Harvest as green chop - Some forages are harvested in an immature stage and fed to cattle in the form of green chop. High nitrate feeds that are piled in mounds and allowed to heat before feeding are especially dangerous to livestock. Heating causes nitrate to convert to nitrite, making the feed about 10 times as deadly. An analysis of nitrate will help in designing feeding programmes to prevent livestock losses from nitrate poisoning.

Nitrate in water - Mature livestock can tolerate higher concentrations of nitrate in their water supply than young livestock. In the case of calves, however, much of their fluid intake is derived from nursing and very little nitrate is secreted in milk.

Livestock drinking well water are not likely to experience nitrate toxicity. Nitrate toxicity from water is more likely to occur when livestock drink water from ponds, road ditches or other surface depressions that collect drainage from feedlots, heavily fertilized fields, silos, septic tanks or manure disposal lagoons.

Nitrate toxicity

As with feed, frequent intake of water containing nitrates appears to increase the overall nitrate tolerance, however, when evaluating possible toxic situations, the nitrate in both the feed and water must be considered because they are cumulative.

Nitrate toxicity is not likely to occur from water containing less than 443 ppm nitrate NO_3, provided that animals are fed a balanced ration that is not high in nitrate and that sound feeding, watering and management practices are followed.

Treatment

Nitrate poisoning can quickly become fatal. When high nitrate is suspected, remove the contaminated feed and provide a high-energy feed, such as barley. A veterinarian should be called immediately to confirm the tentative diagnosis. Because death is a result of oxygen shortage, handle cattle as little and as quietly as possible to minimize their oxygen needs.

Administer a 4% methylene blue solution intravenously at a dosage rate of 4.55 mg. per 0.45 kg. of body weight to treat cattle with nitrate poisoning. Treatment may have to be repeated every six to eight hours because the rumen may be full of forage or feed that contains nitrate and nitrites will continue to move from the rumen into the blood stream. Mineral oil may be given orally to protect the irritated mucous membranes.

(Canada – Government – Agriculture: www.agriculture.gov.sk.ca & aginfo@gov.sk.ca)

Examples of nitrate field tests

A rapid test is the **diphenylamine spot test**. The diphenylamine test is comprised of diphenylamine salt (0.1 grams) dissolved in sulphuric acid (30 ml-36N). A single drop of this acid reagent is placed on a freshly split plant stem. If a dark blue color develops immediately (in the first 5 seconds) nitrate is present. If there is no immediate change in colour, there is no nitrate. The diphenylamine field test indicates only the presence or absence of nitrate. It does not determine the actual nitrate concentration and, as with any field test method, it should be used **only as a screening tool**. Any positive result from the spot test should be followed up with a laboratory analysis for quantification.

Another option may be found at www.nitrate.com/store/ NECi, Lake Linden, MI, USA

1. Dry forage nitrate test (11 US dollars) (e.g. hay):
2. Testing range is from 0 to 9000 ppm nitrate, or 0 to 2000 ppm nitrate-N, or 0.9% N Sample must be ground.
3. Livestock wet feed nitrate test (11 US dollars) (e.g. silage, green chop or standing hay):
4. Testing range as above (1). Use a garlic crusher or equivalent to squeeze out the excess liquid.
5. Soil test (11 US dollars)
6. Water test (30 US dollars)

Salmonella status

Adapted after GD Deventer, the Netherlands, 2008

Table 7.6 General assessment of the Salmonella status of the herd

Risk factors for introduction of Salmonella into the herd	YES	NO	Not applicable
You sell or you graze cattle without a certificate "not suspected of salmonellosis"			
You use manure from other cattle farms			
There are also swine on your farm			
You use surface water to drench the cattle			
Your cattle (sometimes) have contact with other cattle from farms in the neighbourhood			
Visitors (sometimes) use their own boots and clothes on your farm			
Your employees also work on other farms			
Trucks, tractors and equipment belonging to other people enter your farm or grazing without being cleaned or disinfected			
There is no system to control rodents and other animals			
Risk factors for the distribution of salmonellosis on your farm			
Calves are not (always) born in a clean environment			
Calves up to six months age are housed with adult cattle			
Pasture plots/fields are not mown or rested (three months) after application of animal manure before cattle are allowed to graze			
Other risk factors			
It is possible that liver fluke and/or BVD are prevalent in the herd			
Minerals in the ration are possibly insufficient for certain cattle age groups			

General interpretation of the outcome:

Answers	Salmonella risk level
All answers were "no"	very low
One answer was "yes"	moderate
Several answers were "yes"	high

In the second and third situation, one should take action to further limit the risk (information supplied elsewhere in this series)

Risk factors

JF Labbé, SNGTV, Paris, France, 2009

Table 7.7 General inventory of risk factors for certain diseases

Disease entity	Potential risk factors
Environmental mastitis	**Straw yard**: surface/cow; temperature at 10cm deep; frequency of cleaning out; straw addition; straw distribution in the yard
	Cubicles: number of cubicles/number of cows; floor surface condition; cleaning bedding; floor humidity and manure
	Ventilation: air inlets and outlets; ammonia odour; dirt and spin rags; condensation on walls
	Waiting area: surface/cow; inclination of the floor; steps; ventilation; dirt/manure
	Exercise area: accumulation of dirt, mud and manure
Heat detection	Animal density in straw yard and exercise areas, narrow passageways between cubicles; slippery floors; floors yielding lesions; insufficient light; no management planning; no heat detection tools; heat stress
Body condition deviations (feed intake)	Insufficient number of places at feed rack; 1st lactation animals not separated; shoulder lesions at feed rack; poor accessibility to feed bunk; water troughs insufficient and poor quality; poor feed bunk hygiene; stress
Cleanliness of animals	**Straw yard:** surface/cow; straw filling; hygiene; cleaning; **Cubicles:** slat sleepers; surface quality; cleanness in cubicles; diarrhoea cows
Knee and hock lesions (abrasions; abscess)	**Cubicles:** sizes; lay-out; maintenance; access Claw lesions prevalence
Lameness cases	Humid areas like passageways, exercise areas, near troughs; poor manure scraping practice. Traumatising areas: holes in alleys, exercise areas, waiting area; iron parts sticking out; too many or high steps; 90° angles; gravel. Slippery or too hard flooring; scraping practice. Dead zones; herd trimming policy.

Rapid evaluation of BVD prevention

After GD DeventerHolland; March 2008

Rapid evaluation of BVD prevention

This rapid evaluation is used to determine the risk of introduction or spread of BVD infections on farm

	YES	NO	NA	Remarks
1				Calves are separated from any pregnant cows throughout the year
2				The farm is a closed system where cows are only purchased from certified BVD-free herds
3				Cows are transported by the farm's own vehicles
4				Only tools and equipment belonging to the farm are used for cow management
5				Materials brought in by professional visitors are clean (drugs; flasks; needles; syringes; boots; gloves; etc)
6				Only semen or embryos from certified BVD-free bulls are used for serving maiden heifers and cows
7				The embryos and semen currently in stock were purchased after 1998
8				There is negligible risk of heifers and calves making contact with neighbouring cattle. If necessary, preventative measures can be taken.
9				Only certified BVD-free cattle shows or markets are attended.
10				Preventative measures have been taken to ensure cows cannot have contact with neighbouring cattle at pasture
11				There is sufficient distance to buildings and land/grazing of other farms.
12				Neighbouring farms are closed farms or are (certified) BVD-free

(NA= not applicable)

Point 1 is the most important for BVD prevention.
Points 2 – 7 limit the risk of (re-) introduction of BVD.
Points 8 – 12 are relevant but their effect on BVD prevention is limited.

Table 7.8 BVD evaluation results

Results	Risk of BVD	Risk profile
Answered YES to all points	Hardly any risk	--
answered YES to all points with the exception of one	Real risk	NO for point 1 means that the virus can spread on the farm. NO for points 2 – 12 means that the virus can be introduced onto the farm.
Answered NO to point 1 plus answered NO to at least one other point.	High risk	NO for question 1 plus a NO for at least other point means that the virus can be introduced and spread on the farm.

Salmonellosis status

Source : GD – Deventer, Holland, 2008

General evaluation of the Salmonellosis status of the herd

Tick the box that is applicable to your farm

YES	NO	N.A.	
			Risk factors for introduction of disease:
			You sell, or you put out to pasture, cows that have not been tested for salmonellosis
			You purchase or use manure or slurry from other farms
			You have swine on the farm
			You use surface water as drinking water for cattle
			Your cattle sometimes have contact with neighbouring cattle
			Professional visitors sometimes use their own boots and clothes
			Your employees also work on other (dairy or swine) farms
			Trucks, tractors, vehicles and other equipment enter your farm or pastures only if they have been cleaned and disinfected.
			Rodents and other pests are not strictly controlled
			Risk factors for the spread of disease within the farm:
			Calves are not (always) born in clean pens
			Calves up to 6 months of age are housed in the same barn as adult cattle
			Pasture is not (always) mown after manure or slurry spreading before cattle are allowed to graze or there is no three month interval between manure application and grazing
			Other risk factors :
			It is possible that liver fluke and/or BVD are present on your farm
			The levels of certain minerals in ration may be insufficient for certain cow groups

Table 7.9 Analysis of risks for Salmonellis status

The answers:	The relative risk of salmonellosis:
All answers = NO	Very low
One answer = YES	Probable – take action
More than one answer = YES	High – take action

Clinical signs

Source : GD, Deventer, Holland, 2009

Table 7.10 Clinical signs most frequently observed in dairy herds		
	% of farms with signs in one or several animals	% of farms with signs in one or several animals
	Salmonella dublin	Salmonella typhimurium
Diarrhoea	64	81
Abortions	52	28
Pneumonia	40	0
Arthritis	20	0
Necrosis of the tail or ear	4	0
Dead or euthanasia	80	36

Note that about 80% of the cases of infection are subclinical infections!

Parasitology pepsinogen

After GD Deventer, Holland; March 2008

Examination of fresh faeces (mixture of faeces from five calves or individual samples) at 6–10 weeks after going out to pasture.

Table 7.11 Interpretation of pepsinogen values in calf blood (in international units per litre).

Pepsinogen (IU/L)	Interpretation	Treatment advice
< 1.5	Degree of infection is low.	Do not treat
	Animals have enhanced immunity (from a very short period at pasture period or an earlier infection without treatment)	
1.5–3.5	Good degree of infection.	Do not treat
	Immune resistance is adequate.	
> 3.5	Degree of infection is too high.	Treat the animals
	This may cause a loss of growth	

(GD, Deventer, The Netherlands; 2008 // jono)

Herd vaccination scheme

Before designing a Herd Vaccination Scheme, the veterinarian should determine with the farmer the most relevant diseases they intend to vaccinate against. This not only includes prevalent diseases but also potential external disease threats (see also the section Biosecurity).

The most appropriate timing **of vaccination** depends on (1) the disease one is vaccinating against, (2) the age of the animals at vaccination, (3) the season during which one will vaccinate and (4) the individual farm situation.

Furthermore, it is important to address **risk management measures** (e.g. in respiratory disease situations, the barn climate must be checked and improved where necessary).

Table 7.12 shows an example of a herd vaccination scheme using Intervet S.P. products.

After determining the most important diseases to be included in the vaccination scheme, the veterinarian and the farmer could create a **calendar-like plan** to mark dates of vaccination and the respective animal groups to be vaccinated, as well as the products to be used. This calendar can facilitate the planning and actions to be taken.

Table 7.12 Example of a herd vaccination scheme using Intervet S.P. products

Disease	Vaccine	1st vaccination in young stock	2nd vaccination	Booster vaccination
Trichophyton verrucosum	Bovilis Ringvac	> 1 week of age (preferably at age of dehorning)	2 weeks later	No
Lungworm	Bovilis Lungworm	6 weeks prior to grazing	2 weeks prior to grazing	No
Pneumonia	Bovipast	> 6 weeks of age	4 weeks later	No
IBR	Bovilis IBR Bovilis Rhinobovin	> 3 mo. age	No	Every 6 month
BVD	Bovilis BVD	13 mo. of age	14 mo. age, before AI	1 to 2 mo.
Diarrhoea at birth	Bovilis Lactovac	7 weeks before calving	3 weeks before calving	3 weeks before calving

Johne's Disease-free young stock

Johne's Disease is also known as Paratuberculosis

Actions around calving/birth:

- Ensure the dam is given an individual, clean and disinfected calving pen
- Remove the new-born calf immediately after birth

Initial rearing phase - colostrum and milk feeding:

- Ensure the calf receives colostrum from its own (disease-free) dam only
- Following the colostrum-feeding period, only offer the calf milk replacer
- House the calf individually from two weeks of age
- Keep calves separated from older cattle groups
- Provide clean, fresh drinking water at all times
- Supply good quality, clean roughage and good quality concentrates

After weaning:

- Calves must be housed and grazed away from older cattle
- Calves should only be grazed on pastures that are free from spread manure or slurry
- Clean drinking water must be available at all times while calves are grazing
- Avoid housing or grazing with goats

The main risk factors regarding the introduction of Johne's Disease into dairy farms are:

- The purchase of cattle from a farm with an unknown status regarding Johne's Disease
- Young stock having access to pastures previously grazed by other animals
- Using manure from a farm with grazing young stock
- Cattle of all ages having contact with feral ruminants e.g. deer
- Cattle returning from markets, shows or collection sites
- Young stock drinking from natural surface water sources
- Farm's own cattle transported together with cattle from other farms
- Visitors who do not comply with hygiene rules or any biosecurity plan (see section on Biosecurity)

The main risk factors regarding the spread of Johne's Disease between dairy farms are:

- Adult cattle and young stock housed together
- Lack of routine culling
- Young stock grazing with sick cows
- Calves fed with fresh cow milk or antibiotic-contaminated milk
- Calving taking place without appropriate hygiene measures
- Clinical disease cases that are not properly diagnosed, treated nor culled
- When the calving pen is also used for sick animals
- Young stock grazing with sheep and/or goats
- No available calving pen
- Cows housed in an over-crowded loose barn with a manure scraper and/or slatted floor
- Calves fed colostrum from their own (not tested infection-free) mother
- Faecal contamination of drinking troughs
- Young stock grazing on other farms with an unknown health status
- Cattle movement between pens
- Housing of calves in groups without adequate separation
- Cows in a loose housing system
- Using the same tools and equipment for all areas of the farm

Tuberculosis - farm visit report sheets

Dublin, Ireland, Veterinary Document ER 76 B

Farm ID **Date of visit**

Address **Veterinarian**

	YES	NO
Is this a fulltime farmer?		
Is this a livestock transporter		
Is this an agricultural contractor?		
Is this a feed mill agent?		
Is this a registered dealer?		

Herd inventory	Cows	Heifers > 1 yr	Males > 1 yr	Stock < 1 yr
#	Cow-calf	Breeding	Beef	Females
#	Fattener	Beef	Store	Males
#	Dairy cattle	Store		
Other livestock operations?	Deer #	Goats #	Sheep #	Pigs/Poultry #

Farm area = _____ hectares (1 ha = 2.5 acres) Grassland =_____ ha; other=

Number of animals purchased in the year prior to the herd outbreak = ________________

Is there any confirmation that reactors have been introduced into the herd? ________________

If yes, list their ID, date of purchase and farm of origin:

____

____

____

____

Is there any evidence of earlier herd outbreaks? ________________

If there is evidence, when were the last identified reactors culled? ________________

Were reactors housed during the earlier outbreak? ________________

Tuberculosis - farm visit report sheets

Dublin, Ireland, Veterinary Document ER 76 B

Date of last disinfection of the housing? ____________

Date the current reactors were housed? ____________

Was slurry, dung or manure from earlier herd outbreaks spread on grassland? ____________

If yes, give the date of application: ____________

Date the grassland was first grazed by current reactors: ____________

Conclusion:
Are residual problems from an earlier outbreak to be expected as the current source?
If yes, what are the contributing factors: Slurry ____ Manure ____ Houses ____ Other ___

Cattle groups and areas of land established in the nine months prior to herd outbreak

Number of cattle groups in the herd: ____________ Number of clear groups: ____________

Total number of land areas: ____________

Number of land areas related to clear groups: ____________

Land areas not grazed, or grazed by clear groups only: ____________

(see farm geographical map for identifying the land area numbers)

Land area (= pasture plot) where outbreak is likely to have started:

Herds or groups adjacent to or neighbouring the suspected pasture plots

Herd/group ID: ____________

Distance from suspected plot: ____________

High or low contact risk?

Tuberculosis - farm visit report sheets

Dublin, Ireland, Veterinary Document ER 76 B

Badgers

Complete the following table with either 'yes', 'no' or 'unknown'.

Badger data	Affected plots	Adjacent land, forest, etc	Unaffected plots	Adjacent land, forest, etc
Active badger sets				
Other activity (passes, latrines)				
Sightings				
Verified by veterinary inspection				
Any recent disturbances of badgers				

Conclusion:

Have badgers been trapped in the area? ____________________

If yes, by whom? ______________________________ Is the trapper licensed? __________

Were TB-positive badgers found in the area? ________________________

Non-bovine animal species

	Yes	No		Yes	No
Wild deer present?			Examined?		
Farmed deer present?			Examined?		
Goats present?			Examined?		
Other species (please state)?					

Water supply

Streams present in the area? __________ Stagnant water ponds present? ______________

Farm has own well? ____________________ Farm uses mains supply? ____________________

Free-standing water troughs used? ____________________

Water supply could be involved as a source of infection for the herd? ____________________

Tuberculosis - farm visit report sheets

Dublin, Ireland, Veterinary Document ER 76 B

Mechanical sources

One or more of the following modes of infection transfer may be active on this farm.
Please complete the following table by ticking the relevant boxes.

Personnel	Full time	Part time	
Slurry	Own slurry spreader	Contractor	Aerosol observed
Dung	Own dung spreader	Contractor	
Haulage	Own transport means	Cattle transporter	
Shared facilities	Shared cattle crush	Rented house	Shared housing
Associated herds	Shared management	Shared equipment	Cattle movement between herds
Contact with infected cattle	Cattle break-in	Contact at roads	Contact across fences

Conclusion:

Is a possible mechanical source established?

Humans

Was there evidence that humans are involved as the infection source? ________________

Cattle owner (or manager) risk profile

Advice as discussed with and given to the cattle owner (or manager)

Statement of the cattle owner (or manager)
The undersigned hereby states that he/she has provided the information as accurately as possible, to the best of his/her knowledge. Moreover, he/she has taken notice of the advice given.

(A signature is not compulsory; if not desired then they can simply initial the document)

Prediction of transition disorders

Steven Le Blanc *et al.* (2002) Proceedings of the AABP USA

Prediction of transition disorders by metabolic parameters in dairy cows

A study conducted by Le Blanc *et al.* (2002), using data from 19 herds (total of 1076 cows), investigated the retained placenta and left-displaced abomasum incidences (Table 7.13).

The study revealed a positive correlation between pre-calving serum levels of non-esterified fatty acids (NEFA) and the incidence of left displaced abomasum (LDA). Animals with serum NEFA levels >0.5mEq/L are at a 3.5 times higher risk of LDA. Differences in beta-hydroxybutyrate (BHB) were only evident post-calving. Additionally, animals that tested positive for the presence of ketone bodies in urine were 11.8 times more likely to suffer an LDA. Serum BHB concentrations >200μmol/L were positively correlated with a higher risk of LDA of up to 2.6 times. In contrast, serum calcium concentrations of >2.2mmol/L were actually associated with a 50% reduction in the probability of an LDA.

Table 7.13 Incidences of retained placenta and left-displaced abomasum (LDA) across 19 dairy herds (1076 animals in total)

	Incidence (%)
Retained placenta	15
LDA	5

When, from 6 days before calving to calving date, the serum NEFA is > 0.5 mEq/L → then there is a 3.5 x higher risk of LDA (test sensitivity 64% and test specificity 66%)

When urine test on ketone bodies is positive → then there is a 11.8 x higher risk of LDA (test sensitivity 39% and test specificity 96%)

With regards to retained placenta, serum NEFA levels of >0.7mEq/L and cholesterol levels of >1.8mmol/L increased the risk of retained placenta by 1.8 and 1.9 times, respectively. Conversely, neutrophil cell levels of >5 x 10^9/L reduced the probability of the disorder by 40%.

When serum NEFA is > 0.7 mEq/L → then there is a 1.8 x higher risk of retained placenta.

When cholesterol is > 1.8 mmol/L → then there is a 1.9 x higher risk of retained placenta.

When blood neutrophils are > 5 x 10^9/L → 40% less probability of retained placenta.

BHB is a better parameter than NEFA or other parameters to assess the risk of transition disorders in dairy cows

SARA
sub-acute ruminal acidosis

Estimated economic losses Milk Production (milk yield; fat; protein)
Reproduction
Claw lesions (laminitis)
} **€ 3300 / 100cows**

If the pH of the rumen is around 5.8, there will be less NDF digested. (Varga, 2005)

The rumen pH in high yielding cows, to which large volumes of easily fermentable concentrates are given, stays permanently **below the critical threshold of 6.0**! (Broderick, 2006)

The rumen pH in the **case of a TMR** will be :

- **< 5.5** among 17% of cows around 20 days after calving
- **< 5.5** among 28% of cows around 110 days after calving

Problems are most prevalent in parity 1 cattle within the first 125 days after calving and older cows during the latter stages of lactation. (Oetzel, 2003):

The longer the rumen **pH remains below the critical threshold of 6.0**, the greater the concentrations of the following parameters:

- **Acetate**
- Butyrate
- Lactate ±±
- Butanediol
- Ethanol
- Succinate
- Formate

Oetzel GR & Krause KM, (2004). Ruminal pH concentrations and post-feeding pattern of VFA and organic acids in cows experiencing SARA (Proceedings of the 12th ICPD, Michigan USA 2004, in Wageningen Academic Publishers 2006, pp 281)

Mean prevalence of SARA and related other disorders:

- Denmark **22%** (ketosis 26%; abomasal displacement 16% (Enemark *et al.*, 2001);
- New Zealand **20%** (10% abscess in beef cattle; 2-45% SARA in beef cattle (Gibbs *et al.*, 2007);
- Holland **14%** (variation 0-38%); partly during early postpartum period, partly at mid-lactation (Kleen *et al.*, 2004).

FLUOROQUINOLONES AND 4TH GENERATION CEPHALOSPORINES

Remarks regarding fluoroquinolones and 4th generation cephalosporines:

1 Keep these antimicrobial products only for diseases that are difficult to cure and for which antibiotic sensitivity tests have been performed;

2 Do not use these products for treating groups of animals, except in specific cases where group treatment is indicated. The potential benefits should be weighed up against the increased risk of resistance development;

3 Avoid the off-label use of these products as much as possible.

Adapted after www.dier-en-arts.nl

10 criteria for detection of SARA

1 Forage conservation	The pH (< 4.0) and the temperature (less than 5°C lower than the ambient temperature) in different areas of the silage. Often, the temperature at the sides is higher than the lower regions.
2 Silage dry matter (DM)	Squeeze a small portion of silage in your hand. There will be no silage effluent ('juice') if the silage is well-preserved. Silages with DM that are too low will feel wet and effluent will drip out of your hand.
3 Length of feed particles in the silage	These should be > 15 mm and not thread. The grains should be well-crushed without residual fibres.
4 The fibre is distributed equally throughout the ration	Around 1 kg of long fibre/cow/day (e.g. straw) is required for adequate rumen stimulation. Individual cow behaviour (cows not eating fibre) may explain why SARA may occur in spite of the fact that the herd is properly fed
5 Fibre content of the total ration	< 5 mm = 30%; 5–10 mm = 30%; > 10 mm = 40%. Check the proportion of particles that stick to the fingers after a small portion is squeezed.
6 Check faecal quality	Faecal consistency and level of undigested fibres in faeces can be used to assess the presence and/or risk of SARA (< 5% fibres; no grains, no long fibre). For the physical scoring of these parameters: see MONITOR.
7 Cow comfort	Assessment of the following can be used for detection of SARA - frequency of rumination in the herd (at least 75% is expected), shoulder rail height (100 – 110 cm above floor), size of cubicles, % of cows standing idle (< 40% is expected), water supply (water quality; number of troughs; size of troughs), social behaviour (aggression; synchrony; parity 1 position), type of floor in exercise areas, frequency of manure scraping per day (> 4x).
8 Claws	Check for 'waggling' cows; poor back leg conformation and posture; hock lesions; lameness diagnoses; frequency of claw trimming.
9 Clinical examination	Including BCS, Rumen fill, Faeces Consistency, Undigested Fibres in faeces, milk production and composition, feed refusals
10 Barn climate	Smoke powder tests throughout the housing to check ventilation patterns, check for draughts and measure temperature (thermo-neutral zone of cows is from -5°C to +20°C; thermo-comfort zone is from 5 to 15°C).

Herd treatment advisory plan - mastitis

Farm code = **Date =**

Description of the mastitis	Suggested treatment (injectors – Injections)	Withdrawal period (milk)	Withdrawal period (meat)
Mild - clots in the milk, cow is not clinically sick and eating normally. There is a slight decrease in milk yield on occasion.			
Moderate - cow is clinically sick (fever), the affected quarter(s) may be hard, swollen and warm; dry matter intake is reduced. Apparent decrease in daily milk yield.			
Severe - very sick cow, high fever (cold extremities), cow is dull. Cow does not eat. Daily milk yield is significantly lowered. Milk is clotted or watery in nature.	Call your veterinarian!!		

Note: Always complete any course of treatment as prescribed (e.g. three or five days)

Justified use of antibiotics. Eight golden rules

Adapted after www.dier-en-arts.nl

1 Discuss with the farmer how the requirement for antibiotics could be limited

- Propose to the farmer his participation in integrated disease control programmes
- Propose a plan for animal welfare and health

2 Advise him to ensure the isolation of sick (infected) animals.

- Avoid the inappropriate use of antibiotics
- Do not use antibiotics for non-complicated, viral infections
- Limit the use of antibiotics to diseased or high-risk animals only
- Advise the farmer about adequate application of medicines and to finish any therapy duration
- Avoid under-dosage of antibiotics

3 Choose the most appropriate treatments for specific pathogens

- Identify the possible pathogen(s) involved and predict any expected sensitivities
- Design practical (herd or group) treatment protocols for the most prevailing infections based on clinical signs and up-to-date knowledge
- Know the mode of action of antimicrobial products and their pharmaco-dynamic features
- Use narrow-spectrum antimicrobial products

4 Measure the antimicrobial sensitivity of implicated pathogens

- Although the starting point is a clinical diagnosis, the microbial sensitivity should be established so that therapy can be adapted whenever necessary

5 Limit prophylactic applications

- Use antimicrobial products only as a prophylaxis in high-risk animals when it is clear that such a use will reduce morbidity and/or mortality
- Check the impact of such prophylactic use and design appropriate protocols when such application is deemed necessary
- Test the trends in antimicrobial sensitivity (*antibiogram)*

6 Limit the use of antimicrobials around surgery

- Use antibiotics only when deemed necessary and when high standards of antiseptic surgery have been met. Ensure appropriate protocols are in place for such antibiotic use

7 Record and justify all use of antimicrobial products if deviations from designed protocols occur

- Ensure the justification of the choice and dosage of antimicrobial products
- Keep up-to-date records of treatment and therapeutic results in order to evaluate the treatment procedures

8 Report any suspected failure of efficacy of therapy

- Failure of therapy could be the first signal of developing resistance. Report such a phenomenon to the pharmaceutical company and/or the appropriate authorities

Rules of thumb:

- Antibiotics are essential for the treatment and prevention of infectious and zoonotic diseases in both man and animals;
- The risk of resistance development increases with every use of antibiotics;
- Justified use of antibiotics optimizes their effect, while the risk of resistance development is kept as low as possible;
- Justified use of antibiotics means 'using as little as possible to effectively treat the infection'.

Clinical mastitis risk factors checklist

Table 7.14 Checklist for risk factors associated with clinical mastitis

	Risk factors	Target value	Weighing or Scoring result
Cow factors	General disease resistance is low		
	Negative energy balance	< 10% of cows	
	Rumen fill score (1–5)	Average score 3+	
	Hair coat condition		
	Behaviour		
	Teat positioning	Uniform	
	Teat-end callosity scores	85% score < 2	
	Cow hygiene scores	85% score 1	
Environmental factors	Infection pressure	< 15% cows with high cell oount	
	Milking practice		
	Milking machine function		
	Milking hygiene		
	General barn hygiene		
	Hygiene of cubicles		
	Barn climate		
Management factors	Protocol for mastitis detection	"Farm Advisory Plan for Mastitis"	
	Protocol for treatment		
	Drying-off therapy		
	Recording devices/methods		
	Culling strategy		

Claw health

Table 7.15 Checklist for risk factors associated with claw health

	Risk factors	Target value	Score 1 (poor) to 5 (good)
General disease resistance	Negative energy balance		
	Rumen fill score (1 to 5)	3+	
	Hair coat condition & quality		
Floor	Condition and quality	Flat; not slippery	
	Floor hygiene	Dry & clean	
Condition of exercise and rest areas	Exercise and rest areas	Good maintenance; not slippery	
	Number of cubicles or straw yard places	100% availability	
	Size of foot bath	300 x 150 x 15	
	Foot bathing regime	Each month 3 days	
Infection pressure	Hygiene measures		
	Cleaning & disinfection plan		
	Purchase of new cattle		
	Biosecurity plan		
Transition management	Dry Matter intake		
	Concentrates/roughages		
Grassland use	Condition & quality of walkways		
	Walking distance: barn-pasture	< 400 m	

Economics

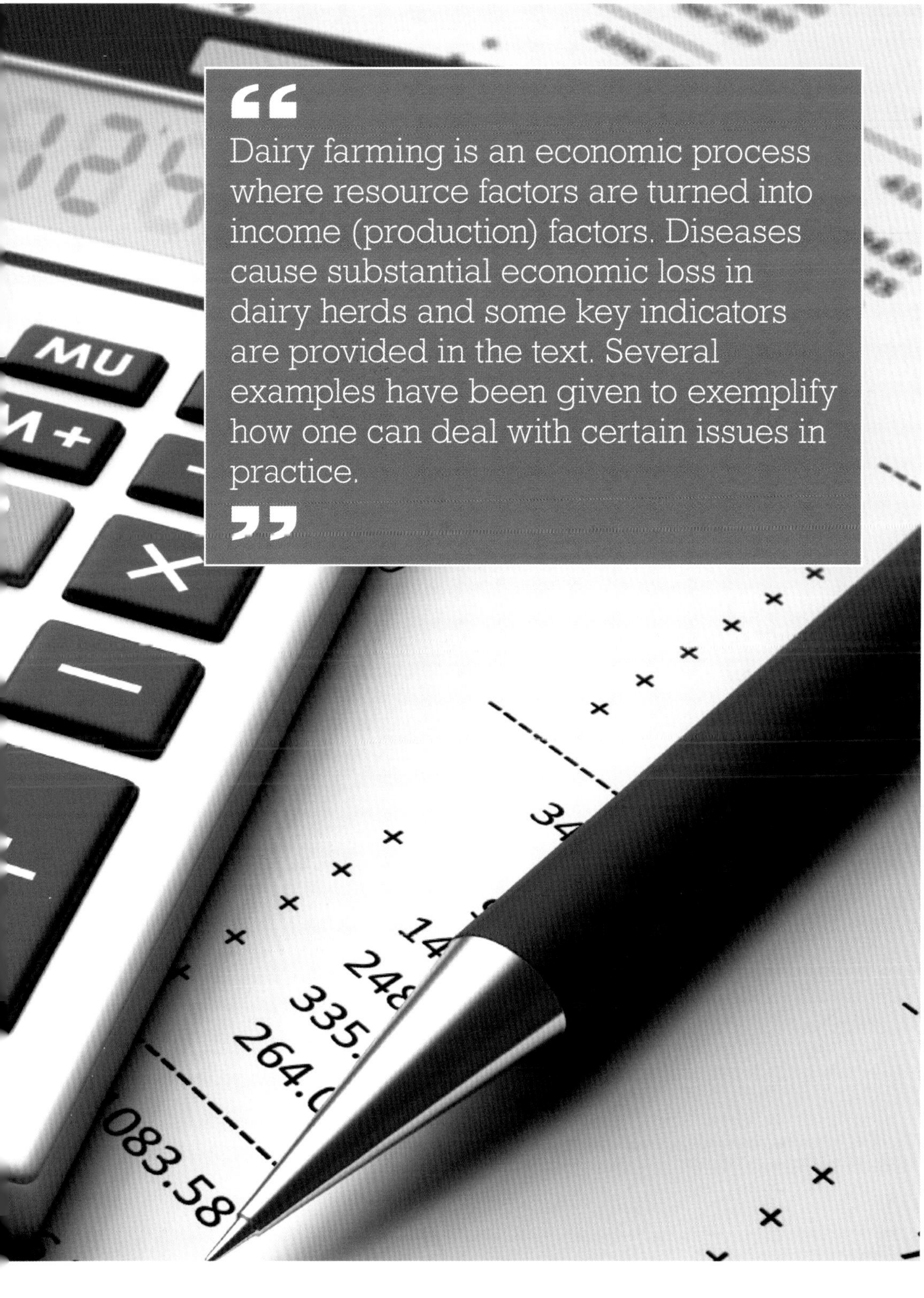

"Dairy farming is an economic process where resource factors are turned into income (production) factors. Diseases cause substantial economic loss in dairy herds and some key indicators are provided in the text. Several examples have been given to exemplify how one can deal with certain issues in practice."

Economic indicators for losses due to disease

Noordhuizen, Cannas da Silva, Boersema, 2007

General

The variation in economic losses between dairy farms generally arises from differences in:

- Efficiency of labour
- Production level
- Costs of bought-in feedstuffs
- Buying and selling cattle
- Geographical location
- Costs of housing and land
- Banking costs, financial margins and resources
- The political situation (national or EU)

When considering the economic losses due to disease, it is important to realise that many of the calculations in literature are based on economic models and mathematical deductions that often use a so-called **normative approach** together with figures originating from a population level rather than that of an individual farm. The situation at farm level is highly variable: the management and husbandry are variable. Therefore, it is difficult to present solutions for all possible situations and the following is more of a general approach dealing with estimates instead of exact calculations.

Udder health

Economic losses due to clinical and sub-clinical mastitis are caused by many factors:

a Loss of milk production (minus the savings in feed costs)

b Costs of drugs -25 to 30 €/case

c Loss of income from non-deliverable milk (mastitis; antibiotics) - 25 €/100 L milk

d Costs of veterinarian - 10-50 €/case

e Additional work (care; treatment)- variable

f Decreased of milk quality - < 60 €/cow/yr

g Premature culling

h Clinical cases - on average 235 €/case

i Sub-clinical cases - on average1 15 €/case

j Other concurrent diseases - variable

The most important factors are: a, b, g h and i.

The average prevalence of clinical mastitis has been estimated at 28% per year, with 1.3 quarters infected, a loss of milk yield of 23% per quarter and a culling rate of 14%. Around 2% of milk losses are calculated for each increase of 100.000 cells/ml of somatic cell count.

Table 8.1 gives the calculation of economic losses due to mastitis.

On average, 71 Euro are lost per cow per year. However, the variation between farms is vast (from 20 to 200 Euro per cow per year)!

With the information presented here, it is possible to (roughly) estimate losses on individual dairy farms.

Health of claws and legs

The yearly incidence of lameness has been reported as 25 to 30%. The majority of these cases relate to claw lesions. The veterinarian treats around 40% of cases, 95% of the cows are trimmed by a professional trimmer or the farmer. The yearly culling rate is approximately 7% of the total number of cows culled (= 2% of the cows present in the herd).

The most important factors used to calculate the economic losses due to claw disorders are shown in Table 8.2 (Berentsen *et al.*, 1999).

This level of loss makes claw disorders the third largest source of economic loss after mastitis and reproductive disorders.

The level of loss in the case of udder health problems varies widely in the literature. Differences between 211 and 342 Euro per cow were noted,

Economic indicators for losses due to disease

Noordhuizen, Cannas da Silva, Boersema, 2007

Table 8.1. Estimation of economic losses due to intra-mammary infections in an average situation (in Euro)

	Losses per cow with clinical mastitis	Losses per cow without udder infection	Total losses per 100 cows (€)	Loss as a percentage of total loss
Streptococci spp.	240	20	2860	40
Coliforms	200	-	1375	19
Staphylococci spp.	275	35	1115	16
Aero pyogenes	285	-	455	7
Bacteriology	235	-	1300	18
IN TOTAL & ON AVERAGE	235	25	7110	100%

(Adapted after Berentsen, Saatkamp, Stelwagen & van Vliet, 1999)

Table 8.2 Losses due to claw disorders

	Per case	Per cow present
Milk losses	55 Euro	15 Euro
Losses in BCS or body weight	5 Euro	1.5 Euro
Costs of treatment & additional labour	20 Euro	5 Euro
Loss of reproductive performance	?	?
Premature culling	225 Euro	5 Euro
Total (per cow present/year)		25 Euro

based on differences in the economic models applied, in regional husbandry methods and in milk price rating, etc. (Østergaard, 2005; Huijps *et al.*, 2008).

Reproductive performance

The ideal calving interval is still around the 365 day period, given the actual milk production level. But under certain conditions, (eg. high production persistency with high level of production) this mean interval could feasibly be extended as losses are largely compensated for by milk income (Berentsen *et al.*, 1999).

Beyond 395 days, every additional day yields economic losses in the region of around 0.5 to 1 Euro per cow per day between 395 and 425 days and 1 to 1.50 Euro per cow per day after the 425 day interval.

If a cow with reproductive problems finally becomes pregnant, the associated economic losses are less than if the same cow was culled. For the latter, losses would be doubled!

In herds with **seasonal calving** only, losses would be around 2 Euro per cow for each day of delay in calving interval (Buckley & Mee, 2006).

Economic indicators for losses due to disease

Noordhuizen, Cannas da Silva, Boersema, 2007

Economic losses due to poor reproductive performance are the second most significant after mastitis.

Metabolic disorders

The losses for each individual case have been estimated at 375 Euro per affected cow. This relates to cases, such as acidosis and ketosis. Incidence and prevalence varies greatly between farms.

Useful reading:

Mastitis in Dairy Production (H. Hogeveen, editor). *Proceedings of the 4th International Dairy Mastitis Conference,* IDF, Maastricht, The Netherlands, June 2005. Wageningen Academic Publishers, Wageningen, The Netherlands

Mortality in calves

Adapted after LAV, 2006

Table 8.3 Mortality in calves

	Loss	Loss (€)	€ per 100 effective calvings	€ per 500 effective calvings
Mortality (peri-natal)	9%	115/day	1035	520
Mortality (pre-weaning)	9%	150	1350	675
Probablitliy of culling cow is calf mortality	11%	950	940	470
Loss of milk porduction	3.5%	0.28/litre	810	405
Additioanl inseminations	0.14	25/ insemination	30	15
Delay in calving interval	13 days	1.10/day	130	65

Source : LAV-LLG-LKV-TSK (Sachsen-Anhalt, Germany, 2006) 'Reducing calf mortality means limiting farm income losses'

Young stock rearing

Adapted after LAV, 2006

Disease losses

The rearing of young stock (replacement heifers) on a dairy farm is a costly business. Not only because it takes, on average, two years to raise a heifer to her first calving but also because of the potential diseases and disorders that may result in economic losses.

There are various different parameters to consider when assessing the economic loss due to these diseases or disorders. The major considerations are give in Table 8.4.

Table 8.4 Overview of 12 major elements concerning the economic losses due to diseases in dairy heifers
Costs of diagnosis (veterinarian; laboratory testing; post-mortem)
Costs of treatment (medicinal products)
Costs of discarded milk due to treatment (with e.g. antibiotics)
Costs of extra labour by farm worker(s)
Costs of retarded growth in calves and maiden heifers
Costs of delayed age at first calving in heifers
Costs of lost body condition and body weight (carcass value)
Costs related to longer dry periods
Costs associated with mortality
Costs due to premature culling
Costs of lost milk production or deviating quality
Costs associated with losses of genetic potential

There is limited information in the literature regarding the estimation of economic losses due to disease in young stock and heifers. This means it is only possible to address certain rearing issues.

The costs involved with rearing young stock

Rearing costs can be considered as the second largest production cost component on a dairy farm (Cady & Smith, 1996). This should be reason enough to pay sufficient attention to young stock rearing and these associated costs.

Different markets have different pricing systems affecting costs and are, therefore, not really comparable. Based on generally accepted target values for rearing (e.g. pre-weaning death rate of calves <10%; first age at calving ±25 months; herd culling rate 25%; average herd calving interval 13 months), it is possible to give an indication of costs, or changes in those costs when the herd performance deviates from the given target value.

A study by Tozer & Heinrichs (2001) showed that rearing costs may be around $32.000 (= €23000) for a 100 cow herd. If culling rate increases from 25% to 30%, there would be too few heifers to replace cows culled. If, however, herd culling rate decreased from 25% to 20%, the rearing costs would decrease by around €5800 (or 24.6%). Additionally, if the age at first calving is reduced from 25 to 24 months, associated rearing costs would also decrease by ~€1000 (4.3%).

The study also calculated that at an average age at first calving of 21 months, any increase in culling rate or in pre-weaning calf mortality rate would cause a shortage of heifers to replace cull cows: up to 10-12 heifers (at a pre-weaning mortality rate of 25% and a culling rate of 35%). Every 1% reduction in culling rate leads to a €720 to €1100 decrease in rearing costs on a 100 cow farm.

Costs associated with dystocia in first-calf heifers

A survey carried out in the UK by McGuirk *et al.* (2007) demonstrated that economic losses of mild dystocia could be estimated at £110 (€125), while for severe dystocia this figure increased to around £350 to 400 (€400 - 450).

The most important determinants of these costs were labour, followed by increased number of days open, increased mortality and premature culling.

Young stock rearing

Adapted after LAV, 2006

Comparing the feeding of pasteurized, non-saleable milk with commercial milk replacer.
Cows with mastitis and/or cows undergoing treatment with antibiotics for whatever reason are always present on dairy farms and there is always the question regarding what should be done with the milk from such animals. A study by Godden *et al.* (2005) reported the comparison of feeding to calves such milk (following pasteurization) with milk replacer.
The calves fed milk replacer exhibited:

- Less growth per calf per day (-120g)
- A lower body weight at weaning (5.6kg)
- A higher probability of being treated for any disorder (Odds Ratio 4)
- A higher probability for premature death (Odds Ratio 30)

Therefore, pasteurizing discarded or antibiotic milk can be an attractive option for feeding calves. The authors calculated a saving of €0.50 per calf per day. However, this system would only be feasible for (large) herds, which can feed at least 23 calves per day. Moreover, investment into the pasteurization equipment has to be made.

Estimation of economic losses due to respiratory disease in young stock
A study by VanderFels-Klerx *et al.* (2001) addressed both pneumonia in calves younger than three months and a seasonal outbreak of respiratory disease in maiden heifers up to 15 months of age. The components of economic losses were:

- Treatment costs
- Mortality
- Premature culling
- Reduced growth
- Lowered reproductive performance
- Milk yield loss in first lactation.

The calf pneumonia model showed an average loss of €31.2 per maiden heifer (range = €18 to €57), while the seasonal outbreak caused in average a loss of €27.1 per maiden heifer (range = €17 to €43).

Intervention costs

Intervention is defined as: "any strategy or action that should lead to recovery from disease, to premature culling, or to prevention of a certain disease in young stock". Such strategy or action could comprise treatment and surgery, diagnostic testing, risk identification and risk management, or wait-and-see.

Ideally, the costs associated with different intervention options should be estimated before deciding which option to choose. This estimation can be carried out via *partial budgeting* and through *cost–benefit assessment* (or cost–effect assessment), while a *decision-tree analysis approach* can be helpful in making the most appropriate choices. It is outside the scope of this text to elaborate in detail about the inputs and outputs, or methods of such estimations. Given the parameters described in Table 8.5 and those related to the various intervention options, it should be possible to assess both costs and benefits for a certain intervention.

In the case of neonatal disease, commonly infectious in nature, the economic loss elements are diagnostic costs, treatment costs, retarded growth, increased probability of mortality and increased probability of secondary disease.

An example intervention could be to wait-and-see, while another intervention could comprise of oral and/or injection-based antibiotic treatment, as well as physiological salt infusions intravenously for three or five days.

The first intervention costs nothing but has an increased risk of calf mortality. The second intervention requires daily farm visits from the veterinarian for the duration of the antibiotic treatment, as well as the costs of treatment and additional labour. However, the second intervention has a higher probability recovery of the calf, although the recovery of daily weight gain afterwards may be retarded. By assigning costs to potential benefits and losses it is possible to calculate the expected total economic loss and benefits for a given intervention.

Milk losses associated with poor herd reproductive performance

Adapted after Nordlund 1991;Grusenmeyer 1983; Rogers 1988; Cassell 1990

Table 8.5 Summary of losses (€) from tables 1 to 5.

	EXAMPLE 1 (Herd of 500 cows)	EXAMPLE 2 (Herd of 600 cows)	Your farm
Losses 1 (€)	148.920	14.892	
Losses 2 (€)	12.075	1.380	
Losses 3 (€)	46.000	4.800	
Losses 4 (€)	15.000	1.000	
Losses 5 (€)	27.600	2.070	
TOTAL(€)	249.595	24.142	

ACTION PLAN:

Economic losses associated with mastitis in first-lactation heifers

(After: Huijps, de Vliegher & Hogeveen (2007). Economic losses associated with mastitis in first lactation heifers. In: Proceedings of the Heifer Mastitis Conference. Ghent, Belgium, June 24-26, 2007; de Vliegher, editor)

A clinical case of mastitis costs, on average, €209. This means an average of €14 , with a range from €0 - €180.

Losses comprise of:

- Loss of milk production €68
- Non deliverable milk €17
- Premature culling €118

Costs associated with a case of sub-clinical mastitis in heifers are around €25, with a range from €0 to €250.

Losses comprise of:

- Premature culling €20
- Loss of milk production €3

The total losses per average heifer are €55, with a range from €1 to €321.

Costs of mastitis

Jos Noordhuizen January 2008;
Adapted after Nordlund 1991 Grusenmeyer 1983;Rogers 1988; Cassell 1990

Table 8.6 Variation between farms

Costs	Mean	Minimum	Maximum
Expected (€/cow/year)	52	0	161
Production loss (€.kg)	0.07	0	0.12
Visits(€)	24	0	100
Drugs (€/treatment)	33	5	110
Labour (€/hour)	19	0	200
Culling (€/cow)	383	0	750
	Results (€ per cow per year):		
Milk loss sub-clinical infections	36	0	105
Milk loss clinical infections	10	0	32
Veterinarian	0.6	0	3,9
Drugs	10.6	1	37.7
Non-deliverable milk	7.9	2.1	21.9
Culling costs	17.9	0	86.4
Reductions	0.30	0	3.8
Extra work	3.8	0	19.0
Total	€ 78	€ 17	€ 198

Costs of mastitis

Table 8.7 Clinical and sub-clinical mastitis (€/cow)

Cost	Clinical mastitis	Sub-clinical mastitis	TOTAL
Production loss	22.80	76.60	99.40
Veterinarian	0.30	0	0.30
Drugs	6.00	0	6.00
Non-deliverable milk	9.00	0	9.00
Culling	21.60	0	22.00
Additional labour	3.30	0	3.30
TOTAL	63.00	76.60	140.00

(Kirsten Huijps, Theo Lam, Henk Hogeveen (2007). Costs of mastitis underestimated. Veeteelt, The Netherlands December issue 2007)

Table 8.8 Economic impact of most important disorders in dairy cattle

Recorded disorders	% of total	Economic impact	Losses (€)	Control costs (€)
Mastitis (SCC)	33	78	52	26
Reproduction	18	40	31	9
Metabolic/digestive	14	27	19	13
Around calving	12	27	19	5
Calves	11	24	15	9
Claw/leg health	5	11	9	2
Heifers	4	9	xx	9
Other	3	8	xx	8
Parasitism	1.5	3	xx	3

(Fourichon Ch, Seegers H, Bareille N, Beaudeau F (2001) Estimation des pertes et de l´impacte économique consécutifs aux principaux troubles de santé en élevage bovin laitier. 8ème Rencontres Recherches Ruminants INRA, Institut de l´élevage, Nantes, France)

Management measures

Adapted after Nordlund 1991; Grusenmeyer 1983;Rogers 1988; Cassell 1990

Cow Comfort comprises different domains, such as housing & barn climate, feed & feeding management, water management, health Care.

In the two figures below, the effect of given management measures in the domain of cattle housing are qualitatively highlighted by comparing the on-site situation (milk yield/cow and health status) before and after the implementation of the management measures.

Apparently, the management measures for improving cow comfort do pay off economically (van Eerdenburg *et al.*, 2005).

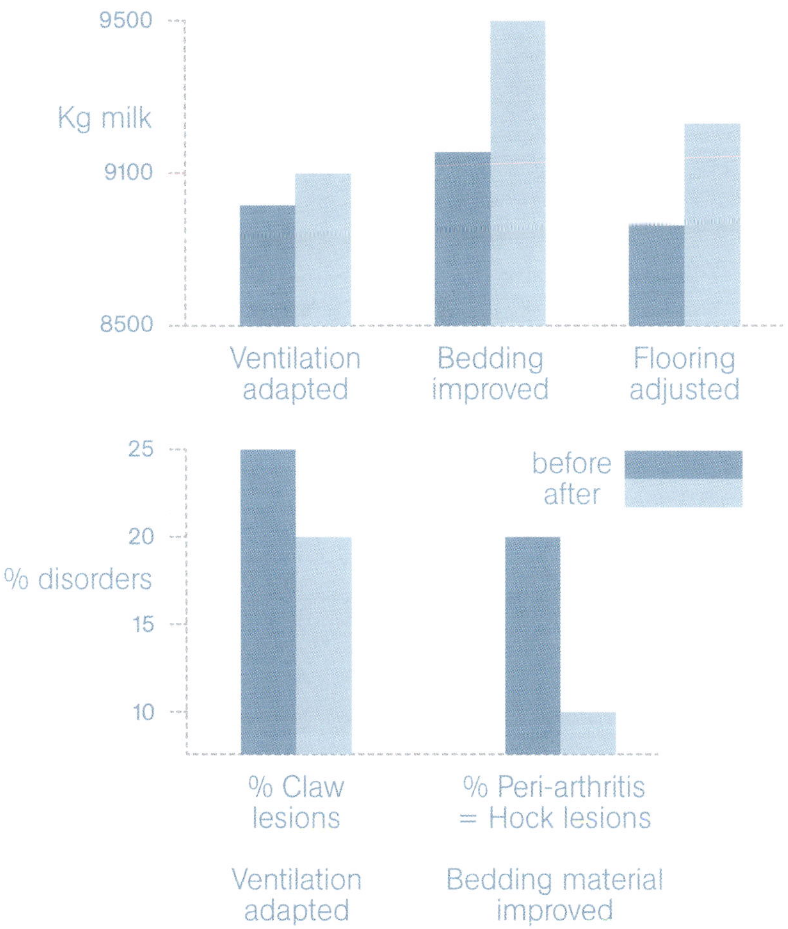

Figure 8.1 The economic benefits of comfort-improved husbandry (ventilation, bedding and flooring) expressed as relative values of milk yield and health before and after adjustment

Husbandry-related costs

After Johan Achten, Veeteelt April 2011

Husbandry costs comprise the following basic elements:

- Veterinary costs
- Medicinal products
- Milk recording organization
- AI
- Bedding material
- Other costs, such as teat dip.

The variation between farms is substantial from around €105 to €260 per cow. Table 8.9 gives the husbandry (CC) costs (€) for four different classes of dairy farms (LIBA, 2011).

Table 8.9 Husbandry costs (CC) per cow and per L of milk (€), excluding calf-rearing costs

	Low CC costs	Average CC costs	Average CC costs	High CC costs
Veterinary	46	67	80	102
Milk recording	8	12	15	19
AI; semen	21	30	43	51
Bedding material	12	21	20	30
Other	22	30	37	51
CC costs per cow	109	159	195	258
Net revenue per cow	1047	1107	1121	915
CC costs per L milk	14.07	13.29	13.08	10.34
L milk per cow	7423	8279	8497	8743
Replacement rate	27%	29%	27%	28%

Economics - robotic vs. conventional

Source: ABAB, 2008

Table 8.10 The economics of robotic versus conventional farms

Costs (€)	Conventional	Robot farms
Net revenue per 100 kg milk	28,82	27,90
Fertiliser	0,46	0,35
Energy	1,22	1,31
Personnel	0,33	0,24
Depreciation	7,90	9,19
Lease costs / quota	0,51	0,59
Other	7,10	7,75
Financing	4,52	5,11
Totals (including machine costs)	6,78	3,36
Depreciation - machines/installations	1,92	3,45
Maintenance, small materials	1,19	1,52
Other machinery	0,23	0,32
Fuel	0,59	0,58
Calculated interest (machinery)	0,59	1,20
Costs of machines/installations	4,52	7,07

Recently a benchmark study has been performed by Lely on 72 UK dairy farms in order to investigate the economic benefits of installing a robot (A.Stewart, 2012). Average milk production level per cow per day at the start-up of installation was compared to milk production at 6 to 8 weeks, and 8 to 12 months after installation. The results can be summarized as follows: At start, the average milk production was 24.7 kgs, at 6-8 weeks after start 27.1 kgs and at 8 to 12 months after start 28.6 kgs per cow per day. In economic terms this means that the average annual farm income increased with 12.370 pounds (or 9.7% after 6-8- weeks) and with 20.102 pounds (or 15.6% after 8-12 months) respectively for farms of 65 cows and 305-days lactation in average (26 pence per kg milk). A farm management support team may have contributed largely to the success of these farmers.

Management measures

Costs and efficacy of various management measures to improve udder health on Dutch dairy farms

The objective of this study was to estimate the costs and efficacy of different management measures to improve udder health and related intra-mammary infections due to classical and environmental microorganisms. Simulation modeling was carried out using data from literature and interviews with experts. Table 8.11 gives the effects of such measures on mastitis incidence and somatic cell count of the bulk tank milk (effects have been ranked in order of importance and the five most influential measures have been highlighted).

Table 8.11 Effects (ranked in order of importance – 1 being most important) of varying management measures on mastitis incidence and bulk tank SCC on Dutch dairy farms *(after Huijps K., Hogeveen H., Lam T.J.G.M., Oude Lansink A.G.J.M* (2009) Journal of Dairy Science 93:115-124).

Measure	Effect on mastitis incidence		Effects on bulk tank milk SCC	
	Environmental	Contagious	Environmental	Contagious
Milking mastitic cows last	15	7	5	5
Milking sub-clinically infected cows last	17	5	2	2
Individual udder towels	10	6	18	15
Cleaning of dirty quarters	9	15	16	16
Eliminating first milk jets	16	17	6	7
Wearing gloves	18	18	17	14
Post-milking teat dip	1	1	1	1
Cluster & teat liner flushing after milking a mastitic cow	12	9	12	8
Cluster & teat liner flushing after milking a cow with sub-clinical mastitis	12	9	12	8
Replacing teat liners regularly	11	11	14	12
Existence of a treatment protocol	12	12	11	10
Drying-off treatment	5	4	4	3
Blocking access to cubicles after milking	7	14	15	17
Optimal mineral supply at drying-off	3	3	3	4
Avoiding high stocking densities	4	8	8	11
Cleaning the resting area	6	13	9	13
Ration optimization	2	2	7	6

First calving at different ages

Johan Achten, in Veeteelt, October issue, 2011

Table 8.12 Economics effects of first calving at different levels

Age at first calving	24.1 mo.	25.1 mo.	27.0 mo.
Costs of heifer calf (€)	1231	1318	1336
Milk per cow (kg)	9028	8877	8336
Net revenue per cow (€)	1811	1774	1682
Net revenue/100 kg milk	20.06	19.99	20.17
Age at culling (mo.)	65	65	65
Milk per hectare (kg)	20121	17978	16170
Milk per farm (kg)	1.032.693	932.781	889.587
Milk per cow (kg)	9028	8877	8336
Culling in 1st lactation (%)	10%	5%	15%

(Study on 150 dairy farms)

- Heifers calving at a younger age will produce during 3 months more than elder herd mates which calved for the first time at older age.
- Farms that want to intensify more, should restrict young stock rearing costs, and hence reduce the age at first calving.
- Heifers calving at 27 mo. age produce less milk and are culled at a higher rate than younger calving heifers.

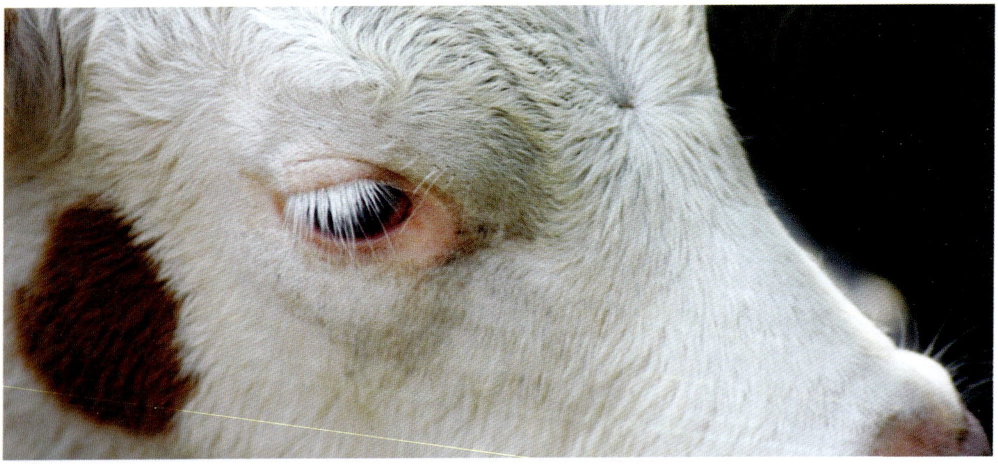

Biological vs. conventional

Michel de Haan, WUR Wageningen, Veeteelt 2 November 2009

This comparison is based on the following:

- Milk prices of 0.32€ and 0.40€ for conventionally- and organically-produced milk, respectively.
- Absence of cost for removal of manure in organic farm situation (assumes that all farm-yard manure is used within the system).

Situation 1 shows the best options in economic terms. Situation 2 is severely punished because it concerns soil that is sensitive for dryness. Sandy soil of good quality is highly appropriate for biological farming. Changing from conventional to biological farming is rather easy for extensive farms, especially when the milk production level is not very high, when forage is sufficiently available, and number of cows limited. Biological farming requires more labour than conventional (about 30 hrs/cow more per year). If one needs to employ a farm worker in the Bio system, the advantages rapidly diminish (see situation 3 in the Table 8.13).

Table 8.13 A comparison of the benefits of 3 different biological (= organic) farming situations with the economics of conventional farming as a reference.

Parameter	Situation 1	Situation 2	Situation 3
Quota (kg of milk)	500.000	500.000	750.000
# of cows	67	56	100
Kg of milk per cow	7500	9000	7500
Expected milk loss	600	1400	600
% culling rate	30	30	30
Intensity of farm (kg milk per ha)	10.000	15.000	11.000
Soil	Peat	Sandy (dry)	Clay
Surface (hectares)	50	33	68
Proportion of mais %	0	15	20
Surface of mais (ha)	0	5	14
Surface of grass (ha)	50	28	54
Grazing system	L+6--- R+3	L+8---L+6	L+6---L+6
Costs of manure delivery (in euro per m^3) Conventional versus Bio	10 / 0	10 / 0	10 / 0
Advantage Bio	YES	NO	YES
Additional income for Bio (euro)	18.500	-14.600	6700

L= limited grazing (ration-grazing); R= rotational grazing over different smaller plots. L+6 = limited grazing during 6 days. www.biologischondernemen.nl

Feed & Feeding Management

"Cattle nutrition is one of the pillars of cattle production. Evaluation sheets and checklists are presented designed to help fine-tune feed and feeding management on the farm. Lists highlighting areas for risk assessment have also been included. Maize silage has received particular attention, not only because it is a good ration component, but also because it has sustainability features. Finally, this chapter contains a simple monitoring tool that includes scoring rumen fill, faecal consistency and undigested faecal fibre, which can be used as part of routine monitoring of rumen function, rumen health and productivity.."

Key points on nutrition

E.Gruys, Dier & Arts, 2009 #4

Points to consider before evaluating the nutrition of dairy cows

1 The most important feed components:

- Water
- Energy (glucose, carbohydrates, cellulose and other fibres)
- Proteins
- Minerals and vitamins

2 The most frequent disorders:

- Lack of energy in the ration (NEB / ketosis / acidosis)
- Excess energy in the ration (fatty liver in dry cows leading to steatosis and infertility after calving)
- Lack of protein (milk yield and protein drop)
- Deficiency in minerals and/or vitamins (e.g. hypomagnesaemia, hypocalcaemia)

3 Conditions that require an increased nutrient intake:

- Parasitic infections
- Cold stress (< -10°C; < 5°C with strong wind)
- Loss of appetite (chronic diseases)
- Pregnancy
- High level of milk production

4 The interaction between disease, nutrition and stress factors:

- A basic level of feed is necessary for immune competence
- Stress factors cause a drop in dry matter intake (DMI) and metabolism, which, in turn, may cause immune suppression
- Exposure to unknown antigens may lead to disease
- Decreased DMI during disease challenge
- Several diseases have a direct effect on absorption in the intestines
- Disease states may be a source of stress for the animal

Field evaluation

After Paul Dobbelaar, January 2008

Table 9.1 Points for a field evaluation when feeding the dairy cow

General aspects
(1) The inspection of cows

	Early lactation	End of lactation	Dry cows
Rumen fill score			
Hair coat quality			
Manure (consistency, residual fibre/grains)			
Appetite and intake			
Rumination (frequency)			
Locomotion score			
Oedema			
Abnormal behaviour; stereotypies			
Other behaviour (dominance, antagonistic)			
Milk yield of older cows			

→ see also at www.vacqa-international.com

(2) The inspection of feedstuffs

	Grass silage	Maize silage	Other
Amount fed (kg DM/d)			
Dry matter (g/kg)			
Homogeniety			
Colour			
Nutritional values:			
Crude protein (g/kgDM)			
Neutral detergent fibre, NDF (g/kgDM)			
Energy value (MJ or Mcal/ kgDM)			
Starch (g/kg DM)			
Other			

Field evaluation

Length of particles:			
Feed bunk			
Manure			
Visible moulds			
Soil contamination			
Odour			

(3) Feeding management

	Lactation	Dry period	Close-up period
Feed refusals (should be < 5% per day)			
Water (quality, availability)			
Forages (available; refusals)			
Forages and concentrates:			
Offered separately			
Offered together			
High and low quality N feeds:			
Offered separately			
Offered together			
Feed changes (sudden/gradual)			
Concentrates distributed over the day (yes/no)			
Concentrates distributed via:			
Milking parlour			
Automatic dispenser in the barn			
TMR			
Increases in concentrates (level (kg/d) and duration of increase (d))			
Concentrates used			
Cows eat after milking			

→ See also www.vacqa-international.com

Field evaluation

(4) Environmental conditions and management			
	Early lactation	End of lactation	Dry period
Housing:			
Barn climate			
Exercise area			
Feed bunk			
Hygiene of feed areas:			
Cubicles (size, quantity, quality)			
Bedding (type, hygiene, comfort)			
Cob webs (ceiling or walls)			
Hours of daylight			
Hours of darkness/reduced light			
Management			
Manure scraping (method, frequency)			
Storage of feedstuffs			
Storage of bedding material			
General hygiene			
Disease detection			
Treatment of disease			
Behaviour towards animals			
Lactating animals split into production groups			
Milking practice			
Milking machine maintenance			
Claw and hock lesions			

Field evaluation

Analysis of strong points and areas for improvement in the milk recording sheets

The average at herd level

-1.1- Averages per day for subsequent recording dates (L milk, milk fat and protein). Put them in a table → easier to see trends.

Reference values at herd level:

Milk fat: > 4.20% when housed; > 4.00% at pasture.

Milk protein: > 3.50% when housed; > 3.30% at pasture.

Stage of lactation as a reference for the herd = 160 days (year around calving)

Compare the results of the milk recording with those of the milk processing plant - sometimes there are differences!

-1.2- Production over 305 days (L milk, milk fat and milk protein). Compare these figures with those of the population or region (be sure to take breed differences into account!).

Remarks:

- Between 60 and 120 days, milk fat is at its lowest
- During the first two months of lactation, milk protein is at its lowest
- Peak milk production occurs before 60 days
- If milk fat is high before 60 days: suspect ketosis, look at BCS...
- Note the number of cows per group.
- Parity 1 cows should be managed as a separate group!!

Use of pasture- issues for consideration:

- Estmate at what stocking density a 24 hour, rotational grazing would become no longer feasible and estimate which restrictions to grazing should be implemented.
- Change cattle between plots at 1700 kg DM per hectare.
- Assess when a field is considered 'grazed' and should a change of field be considered.
- One could consider pre-grazing when incidence of abomasal displacement is high.
- Be aware of the fact that grass does not grow more rapidly in spring in all the fields, because there are e.g. biological differences between fields.
- Trying to ensure good quality grass silage for winter by taking a cut from all plots at the same time could be an option, but calculate the consequence for grazing cows.
- There is often a plan for grazing rotation. However, there are also variations in grass composition over both a 24 hour period and during the seasons.

Table 9.2 The averages at the level of groups (per lactation stage and or parity)

Reference Values :				
Stage of lactation :	L milk/day	milk fat (%)	milk protein (%)	This farm:
< 61days	38.5	4.27	3.22	
61 – 120 days	36.0	4.14	3.33	
121 – 180 days	31.1	4.33	3.51	
> 180 days	23.9	4.68	3.77	

Field evaluation

Grazing systems:

1 Rotation
2 Rationing (restricted pasture)
3 Use of electric fencing to "strip graze" (= restricted pasture).

Points for consideration when grazing cattle:

- Drinking troughs (source of water; number of troughs; cleanness)
- Dry cows (summer and autumn: avoid lush pasture)
- Evaluation of pasture use ~~ analysis of forages
- Urea in the bulk tank
- Grazing-related diseases

Pasture improvement:

- Annual soil analyses (laboratory)
- Application of manure (quantity per hectare; quality; consider a potential association with the goal of that respective field: is the grass for cutting/silage or for grazing? metabolic disorders: vitamins; minerals)
- Maize and hay
  - * Assess how much maize is needed or beneficial during grazing.
  - * Estimate the effect of giving hay while grazing.
  - * Define the maximum level of concentrates beneficial for cows in early lactation.
- Dry period
  - * Grazing dry cows in lush pastures (spring—summer—autumn) can be a catalyst for milk fever and other disorders (ketosis; acidosis). Housing of dry cows is indicated.
- Weather
  - * Prevent or at least reduce the effects of heat stress.

The calculation of rations

The basic ration serves

1°= to accommodate energy demands
2°= to accommodate protein demands
3°= to synchronise the two above

Calculate the maximum and minimum amount (kgs) of energy and protein required; then work between these two values to synchronise the ration. The quality and composition of protein sources should be taken into consideration and the crude protein level used as a guide only.

Balance the ration taking into consideration other parameters:

- Sugars (g/kg DM) (70; 60 – 140)
- Starch (g/kg DM) (175)
- NDF (g/kgDM) (420 – 500)
- ADF (g/kgDM) (240 – 290)

Grazing: yes or no?

In Holland, there are farmers who prefer grazing, while others prefer zero-grazing. Grazing is still carried out on 80 percent of the farms in Holland.

The usual reasons for zero-grazing are:

- The lack of grazing close enough to the farm (29 percent of farmers)
- That robots do the milking (23 percent)
- That it requires too much labour (17 percent)
- Legislation problems (5 percent)
- Herd size related (5 percent)
- Milk production related (3 percent)
- Related to level of forage production (3 percent)

Herd size is regarded as not being relevant for deciding on graz or zerograzing.

Scoring the dry period to early lactation

Adapted after Nutreco--Hendrix, 2006

Table 9.3 Scoring paramters for evaluating the transition of dairy cows from the dry period to early lactation

Parameter	SCORE 1	SCORE 2	SCORE 3
Pregnant heifers show udder oedema (%)	>30	10-30	< 10
L of colostrum at 1st milking after calving	> 10	6-10	< 6
Milk fever in cows of parity > 2 (%)	> 20	5-20	< 5
Retained afterbirth cases (%)	>15	5-15	< 5
Abomasal displacement cases (%)	> 10	5-10	< 5
Clinical mastitis cases (%)	> 15	5-15	< 5
Milk protein level at 0–60 days in milk (%)	< 3.1%	3.10-3.25	> 3.25
Ratio between milk fat and milk protein at 0–60 days	> 1.4	1.40-1.25	< 1.25
Loss of BCS units in the 1st 30 days after calving	> 1.0	0.5 – 1.0	< 0.5
Forced culling (%) in the first 3o days after calving	> 10	5-10	< 5

NOTE: the last column at the right shows the reference value for that particular class.

Number of scores 1 =	x 0 points =	points
Number of scores 2 =	x 5 points =	points
Number of scores 3 =	x 10 points =	points
Total number of points for transition management =		points

If the total score is < 50 points, then the economic losses are estimated at being between €5000 and €12000.

If the total score is between 50 and 80 points, then the economic losses are estimated to vary between €2000 and €5000.

If the total score is > 80 points, then economic losses will be few (estimated at a maximum of €2000) and often due to complications or other factors.

In the first two cases (points < 80) there is apparently a need to reconsider the management of cows in the transition period: feed quality, feed quantity, feed composition, cow comfort issues, body condition scores, scores of rumen fill. Possibly, there is also a need to reconsider the management at late lactation and in the early dry period.

Maize silage particle length

GD Ruminants, September 2004, Deventer NL

Table 9.4 The optimal length of maize silage particles relative to the proportion of maize silage in the ration

Proportion of maize silage in the ration	Fibrosity of other forages	Length (cm)
< 4 kg DM	High Low	6 8
4–8 kg DM	High Low	8–10 10-12
> 8 kg DM	Not applicable	13–19

Table 9.5 The proportion of three different particle lengths in the ration at three different lengths of chopping

Length at chopping (cm)	Smallest particles (%)	Middle-sized particles (%)	Long particles (%)
6	67	32	<1
8	47	51	2
17	29	67	4
% distribution as advised for rations with 50% maize silage	35–45	45–65	3–8

These figures have been assigned following the Penn State particle separator test.

For good rumen function, there should be at least 20% of particles with a length of 2.5 to 5cm in the ration. The frequency of rumination per minute increases in this case. This results in increased saliva production and subsequent better rumen pH buffering.

Reduced feed intake risk factors checklist

Adapted from Interact Agrimanagement BV, 2004

Table 9.6 Checklist for risk factors associated with reduced feed intake

	Risk factor -Note Score* 1= poor; 5= good	Target value	Score*
Dry period management	Preparation		
	Conditions at close-up		
Feed	Palatability of grass silage		
	Palatability of maize silage		
	Fibres in grass silage		
	Fibres in maize silage		
	Other feed (by)products		
Feeding management	Feeding according to targets		
	Fresh, quality rations fed		
	Conserved forage analysis		
	Mixing & mixing time	< 10 min	
	Supply	24 hr/day	
	Increase in concentrates after calving	300 g/d	
	Competition among cows for cubicles (100% availability needed)		
	Competition of cows for feed places (100% availability needed)		
	Competition of cows for escape routes (3 routes for every 60 m)		
	Feed refusals and quality (< 5%; homogenous)		
Claw/leg health			
Cow comfort	Ventilation		
	Light (at level of cows shoulders)	150 lux	
	Space available behind feed rack	350 cm	
	Size of cubicles (depends on type)		
	Softness of bedding (knee-test)		
Water provision	Availability (additional troughs in heat stress periods)	6 cm width per cow (summer: 9 cm)	
	Trough positioning in barn		
	Cleanness (fresh; no sediment; no odour)		
	Water distribution system		
	Water distribution for calves separate from dry and lactating cows		
Production persistency			

Palatability of forage risk factors checklist

Table 9.7 Checklist for risk factors associated with palatability of forage

	Risk factors	Score for primary forage source 1 (poor) to 5 (good)	Score for secondary forage source 1 (poor) to 5 (good)
Feed harvesting & stock	Particle length		
	% DM in silage		
	Additives		
	Silage density		
	Height of silage hump		
	Feeding speed of the silage hump per week		
	Silage made in layers		
	Silage made in vertical portions		
	Heating found in silage hump		
	Covering of silage correct		
	Temperature in the silage hump is > or < ambient temperature		
Silage-face cutting method	Remainders		
	Time interval between cutting and feeding		
	Quality losses		
	Feeding speed of the silage hump per week		
	Time taken to mix TMR (norm < 10 min)		

TMR mixer wagon checklist (Total Mixed Ration)

After Tom Oelberg, Diamond V, USA www.dairyherd.com 2010

From the cow's point of view, a TMR from a mixer wagon is the best way to provide an optimal ration. However, there are several considerations associated with a mixer wagon.

The appropriate feeding system for a farmer is determined by the:

- Husbandry conditions/practices
- Work load
- Costs involved
- Working or management style of the farmer
- Number of feedstuffs in a ration
- Goals of the farmer: health, production, fertility etc

The ration should be based on good quality forages and any deficiencies must be corrected by additional feedstuffs.

So, **it is essential to weigh the feed ingredients accurately**.

The ration that comes out of a mixer wagon must be **homogenous**.

Too often, the mixer is run for too long (< 10 min) or the machine is poorly calibrated and there is a lack of long particles (leading to a lack of fibres in the ration and the possible risk of acidosis).

The cows should not be able to sort their ration. If necessary, water or molasses should be added (especially during periods of heat stress and/or too dry maize silage).

A mixer wagon can be highly beneficial for creating rations for dry cows and (maiden) heifers.

A fresh ration should be mixed every day due to the development of moulds resulting in a loss of feed intake and/or development of butyric acid spots. If only one meal is provided to the cows, it should preferably be delivered to them at morning milking time.

A **dosing machine** is not a true mixer wagon. Cows can too easily select feed components and leave refusals for their herd mates.

In a forage-based ration, if sufficient eating space is available at the feed rack and at least two different types of concentrate feeders are present, then it is possible to offer a well-balanced ration without the use of a mixer wagon.

GOAL= to create a homogenous TMR every day to reduce variability

1 Are the cutting blades sharp?

2 Is the pushback panel well-maintained?

3 Does the TMR wagon empty itself rapidly and completely?

4 Have the scales for the ingredients been calibrated recently and regularly and are they in good working order?

5 Is the TMR mixed well and for a sufficient duration following the addition of the last ingredient?

6 Is the TMR wagon stood on level ground when it is filled with ingredients?

7 Are there any threats of iron parts, plastics or moulds around the augers or anywhere else in the wagon?

8 Are there remains from the previous TMR on the augers?

9 Have you installed super-magnets at the outlets of the wagon?

10 Do you conduct daily or weekly checks of the wagon and, if needed, do you take the appropriate measures?

Good rumen function

After AABP 2004; Dickrell and Hutjens

When the rumen functions well, around 80% of the energy demands and 60% of the amino-acid demands are supplied by the rumen microbes.

Problems at the rumen level can result in acidosis, ketosis, claw problems, reduced reproductive performance, loss of milk production and mastitis.

Table 9.8 Parameters and associated criteria for assessing good rumen function

Parameter	Specifications
Milk fat and protein%	Look for inversions (protein more than 0.2 higher than fat %). Check for cows that are > 1 unit lower than herd average (query acidosis)
Forage particles (TMR)	Excessive particle length should be avoided. If necessary, re-chop the ingredient.
Ration composition	Fibre length can be monitored using a scoring system (see elsewhere in this series) or using the PennState particle separation method. If the DM content in the TMR is > 40% then add 3-4 L water; another liquid feedstuff could also be added (0.5 to 1kg). Feed refusals must be removed
Monitor claw lesions, wiggling, poor posture, hock lesions, locomotion	Every 3-4 months, trends may be observed in these parameters. The rate of clinical lameness should not exceed 15%. Locomotion scoring has its own reference values (see elsewhere in this series)
Ruminal acidosis & SARA	Monitor rumination frequency in the herd (> 75% is desired) and in the individual cow (~60 chews per cud)
Monitor faecal quality	Faecal consistency and undigested fibres can be scored from 1 to 5 (see elsewhere). Long particles in the faeces are undesirable
Grains (maize)	About 5% grains in faeces are acceptable. Grains should be crushed.
Feed efficiency	Calculate the ratio between litres of fat-corrected milk and kg DMI per cow. Adult cows should be at > 1.6 (heifers at > 1.5). If a TMR is fed, then this ratio should be around 1.4. If it is < 1.3 this indicates potential problems with feed efficiency)
Fibre-starch balance	NDF should be ~28–33% of the ration; effective NDF should be20–22%; ADF should be 19–21%
Milk urea nitrogen level	Values of < 250 mg/L and > 350 mg/L indicate potential problems. Variation within a herd will be around 1 to 3 units. Always check milk urea nitrogen levels in groups of cows after any feed changes
Straw	If ration fibre level is insufficient, straw may be added to the ration. For example, if effective NDF is marginal, NDF-digestibility is > 60% or faecal consistency scores are very low (1-2). For lactating cows, 0.5 - 2kg straw per cow per day can be added (length 4–8 cm)

Maize silage heating

Milchpraxis, G (2008)

The degree of heating in maize silage clamps largely depends on how the clamp was filled and whether there was adequate consolidation. The better compacted a silage is, the lower the risk of heating due to the removal of oxygen. High DM silages are at the same time at risk of heating. Silage temperature can be measured by a long, 100 °C thermometer, to be put in the silage clamp horizontally. Ensure the right (heavy) machinery is employed to fill and compact the clamp. Table 9.9 gives the relative air penetration at varying densities.

Table 9.9 Relative air penetration (cm) of maize silages at varying densities (kg DM/m3)

Density of maize silage (kg DM/m3)	Depth of air penetration (cm)
120	50 – 100
150	45 – 80
180	30 – 60
210	25 – 40
240	20 – 30
270	15 – 20

DM= dry matter

The greater the opportunity for air to penetrate the silage clamp, the higher the risk of **mould** development (grey–white–bleu–green spots) and/or of **butyric areas** to be found. The latter affects the taste for cows negatively.

A **critical moment** is when the clamp is opened for the first time. Another critical factor is the rate of feed-out. If the rate of feed-out is too slow, there is the risk of spoilage.

A known phenomenon is also the situation when a farmer puts several blocks of silage in front of the cows beforehand for e.g. the whole weekend, to be at ease himself. These blocks may continue to heat and develop butyric areas, and even moulds, on the spot, which may impact on feed ingestion later the day. The same may occur when large quantities of feed refusals are left in the feed bunk overnight.

Two groups of dry cows

R. Knook, De Heus Voeders, NL, Veeteelt Oct 2008

Objective = a maximal dry matter intake (DMI) before and after calving

Duration = two months (50 to 60 days)

The cation—anion balance should be decreased by 200 mEq/kg DM in the last two weeks of the dry period. Concentrates can be used to decrease the balance.

Differences in animal health status as found when comparing 1 group and 2 groups of dry cows:

In the **situation of 2 groups** of dry cows, there were 42% less dull calves; 17% less endometritis cases; 20% less mastitis cases; 32% less rumen acidosis; and 31% more abomasal displacements (caused by excessive concentrates offered after calving!)

Maize silage stimulators

Amelie BELAN, PLM, France March 2005

Table 9.10 Maize silage stimulators

Product and concentration	Doses TMV (L)	Crop	Advantages	Disadvantages
Formic Acid 85% (FA)	2.3 – 3.5	Grass; legumes 25% DM	Can be used under poor harvest conditions	Corrosive. Increases effluent production.
Propionic Acid 99% (PA)	5 - 10	All silages	Can be used in non-sealed silos exposed to air.Inhibits moulds.	Less effective than FA
Ammonia 82% N or 24% N	13.3	Maize silage	Improves crude protein content in silage	Volatile and caustic (caution!). Not useful for grass silage.
Urea 45% N, or in solution 22.5% N		Maize silage 35% DM	Improves crude protein content. Better than ammonia and less dangerous.	Less effective than ammonia for fermentation.
Molasses (beet), 45% sucrose	14 - 28	Grasses; legumes	Increases levels of sugars available for fermentation, decreases N.	Favours the butyric acid bacteria.
Enzymes		Young plants and DM < 40%	Increases sugar content and digestibility of silage	Effectivity is variable
Lactic acid bacteria		Grasses with DM > 25% Legumes with DM > 30%	Rapid action: the final silage pH will be lower than without.	Need sugars for optimal bacterial growth

(N= nitrogen)

Quality of maize silage

Bruno Bertheloz, 31840 AUSONNE, France Michel Vagneur,39300 CHAMPAGNOLE, France

It is possible to evaluate on-site the quality of maize silage thanks to some simple and practical criteria. This enables any adjustments to be made to dairy cow rations to ensure their suitability.

The criteria are:

- visual observations
- pH
- length of particles
- density
- temperature

Laboratory analyses can be used to complement the above. Climatic changes, as well as the impact of humans and technology mean that practical evaluation of silage clamps is variable between farms and years. The latter source of variation means it is important to take measurements on a yearly basis.

Introduction

Maize silage plays a key role in many rations for dairy cows and represents about 70% of dry matter intake in some markets. It also represents 10% of the land dedicated to forage cultivation (1.4 million hectares) and 20% of the total dry matter intended for ruminants. Maize is an interesting feedstuff, rich in starch, half-forage and half-concentrate. The yield per hectare can be above 20 tons (DM). Achieving such yields requires careful management. With increasing herd sizes, the increase in mechanisation and speed of labour, the **quality of maize silage becomes a critical** component on certain farms.

The quality of maize silage is indirectly associated with animal health, production and fertility. Good quality forages are also demanded from chain-based quality assurance systems.

Laboratory analyses can complement the criteria noted above and allow for more, in-depth ration composition calculation. For example, the composition of acids (lactic, aceto-acids, butyric acid), the percentage of soluble nitrogen, ammonia and other nutritional values.

Poor conservation may cause production losses (kg milk; milk fat and protein %), metabolic disorders (thin cows; ketosis; acidosis) and/or poor reproductive performance (poor oestrus expression). Additionally, the risk of a butyric fermentation increases, which can result in penalties for cheese-making and, hence, impact the price paid for the milk. The risk of listeriosis also increases, as well as the potential risk associated with mycotoxins (this is harder to evaluate but could be important in some cases).

A thorough understanding of forage conservation and how to evaluate it using simple criteria allow the veterinarian to accurately advise the farmer.

Remember

Ensiling is a method of conserving forages by promoting or restricting fermentation of residual sugars in the plant material. The pre-requisites for good fermentation are:

- A stable and acidic pH (below 4 for maize silage)
- The absence of air, which corresponds to good consolidation, can be partly detected by assessing the density of the silage.

The cow needs long fibre in the ration. As maize silage is often the only source of fibre in the ration, the farmer must focus on a chop-length that is sufficiently long and appropriate to the use of the maize in the ration. The minimum chop length should be 10 mm; a chop length of 15 mm could be considered an optimum between rumination-inducing capacity and ingestion capacity. Depending on individual situations, a shorter (5mm) or longer (25-30mm) chop length may be advised.

The energetic value of the maize silage is mainly derived from rumen fermentation processes. Grains

A practical approach

contribute to the energetic value and must be crushed to increase the surface area available to rumen microbes. Good fermentation of long-chain carbohydrates (fibre) is equally important if the farmer wants to maximise the energetic value of the forage.

The criteria for evaluating maize silage are:

- a pH below 4
- a good density (compaction)
- a reasonably long chop length
- crushed grains
- a stable temperature - close to ambient temperature

Direct observations

Moulds

Conditions for growth

In contact with air, endogenous yeasts will develop and allow the fermentation of residual sugars. This phenomenon is amplified if the temperature is high. If this situation continues, moulds will appear. The presence of moulds is a criterion for poor conservation, loss of nutritive value and health risks for cattle. Although the presence of moulds does not imply contamination with mycotoxins, they do increase the risk. Conversely, the silage can be mould-free but heavily contaminated with mycotoxins. Endophytic fungi/moulds are present on all plants but do not develop in an anaerobic environment, therefore, well-conserved silages generally have low presence of moulds. Certain conditions, such as a high DM > 40% or an insufficient density may enhance their development.

Problems due to moulds

The main issues with regards to moulds are loss of DM, nutritive value and palatability. Moulds generally lead to losses of refused forage, a loss of nutritive value and loss of appetite. The majority of moulds cause little harm, however, their respective mycotoxins can cause significant losses and care should be taken to avoid feeding contaminated/ mouldy forage. A particular mould, Aspergillus spp., can cause pulmonary problems and allergies in farmers/workers via inhalation.

Identifying moulds

The following may e identified on opening a silage clamp:

- *Monascus purpureus* white in colour then turning red (red on maize). Is mainly localised on the outer sides of the clamp but can also be found in the middle and develops in any air pockets in silages with a high DM content (poor density). It increases appetite and poses no risk; however, it can produce a toxin, Monokalin K, which has a antibiotic effect on cellulolytic bacteria in the rumen, reducing the value of the silage, as well as the efficiency of the rumen.
- *Fusarium roeum or graminarium*- this mould is white followed by rose coloured and has the appearance of cotton. It is localised in the main body of the silage. It is a mould of field origin but can remain and be visible in the resultant silage. It decreases appetite and presents risks via its mycotoxins, **zearalenone** (oestrogenic activity) and the trichothecenes Dioxynivalenol (DON) and T2 toxin (diarrhoea, intestinal haemorrhages).

Moulds of storage/conservation origin are:

- *Aspergillus fumigates* - a dark green mould, this can be found localised on the front of the clamp. It develops on the face of the clamp when exposed to air, developing slowly in autumn (at temperatures around 20°C). It does not modify appetite but presents an abortion risk.
- *Pénicillium roqueforti* - a blue-green mould found on the face of the clamp. It is generally well visible one month after opening silage that has been poorly compacted. It reduces appetite, does not represent a major risk of toxicity but can synthesise Roquefortine C, Mycophenolic acid, PR toxin and Patulin, all of which have antibiotic effect and can have health consequences when the animal is faced with a multiple challenge.

Quality of maize silage

Bruno Bertheloz, 31840 AUSONNE, France Michel Vagneur,39300 CHAMPAGNOLE, France

- *Mucorales* - mucor is white turning grey with black points. It is localised in the areas exposed to oxygen and appears at feed-out, especially in silages with high DM and poor density. It causes reduced appetite and is not toxic.
- *Byssochlamys* - persistantly white and compact. Localised in the body of the silage, it appears one month after opening. It does not affect appetite but represents a risk due to the potential production of Patulin (affects rumination; meteorism).

In all cases, the contaminated areas must be removed and disposed off.

Ustilago maydis

Spores are always present in the soil and develop on stressed plants (heat periods; drought). In the case of heavy infection, conservation problems and appetite loss can be encountered but this parasite is non-toxic in normal conditions.

Odour

Lactic acid has little odour, so a good silage does not smell. Butyric odours are related to the presence butyric acid and indicate poorly fermented silage. A vinegar odour indicates the presence of acetic acid and, again, poorly conserved forage.

Compaction

A simple test is to try to push your fingers straight into the silo. Ideally, the silage should be solid enough to stop your fingers from penetrating. Poor compaction is evidenced by the ability of your fingers pushing into the silage.

Aspect

The end of the leaves must have been clean-cut by the harvesting machine. If they are frayed, there might have been a problem of countercutters in the machine. Grains must have been crushed - a whole grain is likely to be found in the manure representing a waste of feed energy.

Density and compaction

The last 15 years the machinery for harvesting has evolved to 8 rang harvest and silage machines. The speed of work has increased too. Hence it is more difficult for the farmer to obtain a long cutting, crushed grains, and optimum compaction of the silage hump. Moderate compaction yields the risk of repeated fermentation.

When the DM content is high (> 40%), the residual air in the hump may be 3 to 5 times more important.

Between 30 and 35% DM the ideal density is just above the 240 kg DM / m3; many silage humps do not achieve this threshold. A study on 97 silage humps in 2004 (Pioneer), showed a density below 240 kg DM / m^3 in 50% of the cases.

As a paradox, the long cut silages are not necessarily the less compacted. Even in case of a high DM content, much better care must be taken of compaction. In spite of all, it is not necessary to go down to a cutting below 10 mm.

The less a silage clamp is compacted, the more and the deeper there are entries for air, the ph is higher, and the percentage of acetoacid is higher.

Yeast develops: alcohol odour and heating are primary signs. Thereafter, the moulds take over. Nutritive value diminishes: for silage at 30 % DM, an increase of 10 °C means losses up to 2 or 3 % per day in the affected zones.

Economic losses are very high: direct losses (DM; nutritive value), development of butyric acid spores, decreased feed intake. Conservation products (biological stabilizers or acids) allow improving the conservation and hence having a positive economic contribution.

To assess the density, one should know the DM content of the silage. It is sufficient to take a sample with a given volume auger, weigh the obtained sample. The DM content in kg DM per m^3 = weight of sample x DM % / volume. The goal is to stay below 220 kg DM per m^3.

A practical approach

To evaluate the DM content

- below the 28 % DM, the silage leaks fluid when pressed in the hand;
- between 28 and 32 % DM, the silage leaks a few fluid drops when hand-pressed;
- over 32 % DM, the silage is dry on touching.

Succesful silaging

It is possible –even with high DM levels- to achieve the correct density with complying to good silage practices and silo (clamp) handling.

Some tips to maximize compaction of the silage:

- Compaction by using tyres with an air pressure > 2 bars or filled with water. Remember that new, low pressure tyres have been developed for not compacting the soil, so the compaction of silage by tractors with these new tyres must be avoided!!
- Put sand bags for compacting the top and the front of the silage clamps; they put 5 times more pressure on silage than tyres.
- Increase the speed of feeding out the silage by taking less deeply into the clamp and by coming back to the same spot more rapidly. It is recommended, in general, to progress by 20 cm per day in summer and 15 cm in winter time.
- Do not open the silage clamp in advance; a maximum of 2 to 3 days is acceptable.

Avoid altering the front structure. A clean-cut front is needed at all times: a milling machine or rotor yields such results. Other machines leave openings through which air can enter the clamp and allow heating.

Do not leave silage remainders at the foot of the silage clamp; they easily re-ferment.

Size of particles.

In the 80's

In the eighties, the ITCF (currently: *Arvalis Institut du Végétal*) has developed a shaker-sieve; sieving permits a simple, rapid and practical evaluation of the particles in silage. Recommendations to limit the proportion of particles to a certain volume are presented in the Table 9.11 below:

Table 9.11 Silage Particle Size

Size of particles	Recommendations in volume
> 2 cm	Less than 2%
1 to 2 cm	5-10%
< 1 cm	90-95%

ITCF - devenu Arvalis Institut du Végétal – Recommendations in the 80's

The goal was:

- Reduce the large particles in the ration which are hardly consumed and which hamper compaction of the silage:
- Limit the particles of intermediate size (1-2 cm) which facilitate mastication:
- Obtain a large proportion of small particles to facilitate compaction, and increase consumption, while having the maximum of grains crushed.

Current recommendations

Advice has evolved over time toward a much longer particle size because maize silage is often the greatest source of fibres in the ration and the feed intake has increased considerably. The required particle size must evolve likewise.

The goal is now to obtain many particles of > 5 mm, crushed grains and few very long particles. Good maize silage cannot be obtained without the precise setting of the crusher.

Quality of maize silage

Bruno Bertheloz, 31840 AUSONNE, France Michel Vagneur,39300 CHAMPAGNOLE, France

Table 9.12 Particle size of maize silage as recommended by Pioneer Sémences

	DM assessed	Length in mm	Setting the crusher	Length without crusher
Milky – Paste	< 28%	15-17	Open > 4 mm	10
Pasty – Hard paste	28-35%	12-14	Tight 2-3 mm	5
Hard paste - Glassy	> 35%	10	Very tight 1-2 mm	5

To obtain lengths of 15 mm, which is recommended for standard maize silage, the crusher must function well. If there are ravels found, the crusher must be reset, but should not cut too fine particles.

In a study where silage cuts at 0.95–1.45–1.90 cm were compared, the authors concluded that a cutting at 1.90 cm is recommended with crushed grains (cows showed 25,3 kg DM ingested and 38,2 kg standardised milk yield).

Table 9.13 Maize silage cutting length as recommended on the basis of findings with the ITFC shaker sieve

Size of particles	Recommendations
> 2 cm	Less than 2%
From 1 to 2 cm	20-30%
From 0,5 to 1 cm	40-50%
< 0,5 cm	20-30%

Table 9.14 Particle size recommendations based on the Penn State shaker sieve (note that holes are round, not square)

Size of particles	Maize silage	TMR
> 1,8 cm	3 - 8%	2 - 8%
0,8 - 1,8 cm	45 - 65%	30 - 50%
0,18 - 0,8 cm	30 - 40%	30 - 50%
< 0,18	< 5%	< 20%

Overall, one should talk about particle size in mm, as that is the internationally accepted measurement. For the Pennstate shaker-sieve a conversion table has been proposed to convert the results of sieving into average particle size.

A study comparing crushed grains with non-crushed grains showed that 1.2 kg milk and 1.5 kg DM could be gained by crushing (study cows ingested 25.3 kg DM and produced 38.2 kg standardized milk).

A reduction of particle size from 13 mm down to 6 mm reduced the mastication activity with 30%.

Temperature

Measuring silage temperature is done at the face and at depth, while comparing these temperatures with ambient temperature. A difference of 5°C is still acceptable; beyond this point the silage becomes too hot.

A practical approach

When silage is created during hot weather conditions, the plant mass and the air pockets may retain warmth for months, especially when the silage clamp/hump is high and large. One should not confuse "hot silage" with silage in the process of fermentation.

Table 9.15 Maize silage temperature

DM of silage	Elevation of temperature				
	5°C	10°C	15°C	20°C	25°C
	Daily loss of DM% in the heated silage mass				
20%	1.6	3.2	-	-	-
30%	1.2	2.3	3.5	-	-
50%	0.7	1.5	2.2	2.9	3.7

H. Honig & M.K. Woolford 1980

The pH

The pH is measured at several points in the clamp; the reference value is a pH of 4. The closer you get to a mould zone, the higher the pH gets. This shows the farmer the changes around affected areas.

Prevention of undesirable fermentation in maize silages

- Poorly conserved parts of the silage and dispose of them away from the clamp to avoid the risk of recontamination;
- Air entry facilitates repeated fermentations; entry must be limited as much as possible;
- Avoid shaking the silage when taking sections for feeding; It is best to use vertical, clean cuts (e.g. a shear grab) to minimise air exposure at the face; Ensure that the face is 'refreshed' at a minimum of every three days;
- Spray the front of the silage with propionic acid solution if needed and if indicated as urgent;

Use of conservation products.

- Accelerate the lowering of pH;
- Preserve the maximum of nutritive value to the forage;
- Limit the multiplication of butyric acid-producing organisms;
- Improve the value of the forage per animal.

* Or:

- Limit undesirable fermentation when opening the silage clamp;
- Avoid nutritive losses;
- Prolong the duration of good conservation;
- Limit the synthesis of unpalatable and toxic components;
- Improve the ingestion of forages;
- Improve the sanitary quality by reducing the butyric acid-producing organisms, listeria and toxic substances.

Conclusion

"When a problem has been well-defined, you are halfway to the solution ". S Covey.

During a farm visit, silage quality must be evaluated. The following criteria are the most important in the field: (a) Visual observation; (b) The pH; (c) size of particles; (d) level of compaction; (e) temperature.

If there is an issue with one or more of these criteria immediate advice must be given to the farmer, as well as discussion issues raised for the next harvest season.

Heating in silages

Inge van Drie, Suzan Nicolasen,Tymen van Zessen, Franck Lensink, Veeteelt 2011, April

Heating in silages seem to be more common every season. The higher the dry matter content of the silage, the higher the probability of heating is. A **threshold of 45%** dry matter is considered the critical limit; 40% is the optimum with a density of 220 kg DM/m^3 grass silage.

A high dry matter content reduces the potential for good compaction in the clamp resulting in the incomplete removal of oxygen from the clamp. Moreover, the water-activity is lower in such silages, which reduces the desired drop in pH. It is postulated that reduced nitrogen (N) fertilizer use contributes to this phenomenon (less N results in less protein and more sugars for conversion into lactic acid).

Heating in silage is defined as the temperature between the upper layers and the lower layers of the silage clamp differs by 8°C or more.

Heating in silages causes an increase in pH, development of moulds and spoilage.

A (very) high protein content in silage also causes heating via buffering of acid produced and preventing the required rapid drop in pH.

Laboratories that analyse forage quality parameters,, such as energy, protein and dry matter, can now provide an indication of the stability of the silage with regards to heating. The heat susceptibility index (HIS) is composed of various parameters that are selected by laboratories following the results of the standard analyses, usually dry matter content, acetic acid content and silage pH.

Table 9.18 Silages can be qualified according the HIS

0 to 20 points	16% of silages	Little susceptibility
21 to 50 points	-	Heat susceptible
> 50 points	37% of silages	Very susceptible

Whether a heat susceptible silage (for example in class 21 – 50 points) will actually heat and spoil is dependent on the **speed of feeding** (**20 cm per day in summer** time is standard) and the thorough **covering** and sealing of the clamp (soil; sand; tires).

Heating seems to be a localized farm problem: there are repeat offenders who find themselves in this category year after year. This suggests heating is related to grassland, harvest and ensiling management.

Heating in silage causes **losses** in overall nutritional value (20%), sugars (30%) and protein (10%), as well as an increase in ammonia N and reduced palatability. According to field surveys, economic losses vary from €3000 (15% heating) to €12.000 (55% heat) (*Hendrix UTD Agro, NL, 2010). Lactobacillus buchneri* in silage supplementation can sometimes reduce the risk of heating, however, there are usually DM losses associated with using this organism.

Special feedstuffs

Table 9.19 Special feedstuffs and associated risks

Product	Maximum daily intake (kg product/day)	Risks/remarks
Potatoes (crude)	10 – 15	Acidosis
Potatoes (steamed)	10 – 20	Acidosis
Apples	15	Acidosis
Gherkins	30 - 35	
Bulbs	none	
Bread	2 – 3	Acidosis
Spinach; salad	30	NO_2
		Add fibre!
		Little feed value
Green/white cabbage	15	NO_2
		End of lactation
		Bitter
Green beans	25	
Tomatoes; cucumbers	50	Tomatoes (K)
		Cucumbers (NO2)
		Not for dry cows
		Little feed value
Onions	10	Milk odour!
		1 kg per day and then increase
Chicory	10 – 15	Wash first
Carrots	20 – 25	Milk palatability
Brussels' sprouts	10 – 15	

Nitrate toxicity in dairy cattle

Managing high nitrate feed

Table 9.20 Cattle feeding guidelines for forages containing varying levels of nitrate[1]

Method of Reporting Nitrate Level			
Nitrate (NO_3)	Nitrate Nitrogen (NO_3-N)	Potassium Nitrate (KNO_3)	Recommendations for feeding
Percent of Forage Dry Matter			
0.0 - 0.44	0.0 - .10	0.0 - 0.73	Safe to feed in all situations.
0.44 - 0.66	0.10 - 0.15	0.73 - 1.10	Safe for non-pregnant animals. Limit to 50% of diet dry matter for pregnant animals.
0.66 - 0.88	0.15 - 0.20	1.10 - 1.47	Limit to 50% of diet dry matter.
0.88 - 1.54	0.20 - 0.35	1.47 - 2.57	Limit to 35-40% of diet dry matter. Avoid feeding to pregnant animals.
1.54 - 1.76	0.35 - 0.40	2.57 - 2.93	Limit to 25% of diet dry matter. Avoid feeding to pregnant animals.
over 1.76	over 0.40	over 2.93	**DO NOT FEED**

To convert from parts per million (ppm) to percent, move the decimal point four places to the left (i.e. 8800 ppm = 0.88%).

[1] Source: Sniffen, C.J. and L.E. Chase. 1981. Nitrates in Dairy Rations, Department of Animal Science, Cornell University.

much nitrate to kill a ruminant when the nitrate comes from forages that are eaten over a long period and have been gradually introduced as compared to a feed that is consumed quickly and to which the animals have not been previously exposed.

Stored Feeds: Test the stored feed for the presence and amount of nitrate. Frequent intake of small amounts of a high nitrate feed increases the total amount of nitrate that can be consumed daily by livestock without adverse effects, and helps livestock adjust to high nitrate feeds. For example, feeding limited amounts of long-stemmed, high nitrate feed several times a day, rather than feeding large amounts once or twice daily, is a good strategy. Animals should not be allowed to consume feeds containing more than 0.5 per cent nitrate if they have not been exposed to such feeds prior to this.

To help animals safely make the transition to high-nitrate forages, these feeds should be physically mixed with low-nitrate forages so that the overall nitrate level is less than 0.5 per cent. This can be accomplished easily in feedlot rations where grain is fed and forages are chopped and mixed as a complete ration. Putting out one bale of high-nitrate feed along with one or more bales of low-nitrate feed is not acceptable as it would be possible for some animals to eat only from the bale containing high levels of nitrate. Therefore, the feed must be

Nitrate toxicity in dairy cattle

physically mixed together to blend the nitrates down to a safe level. Be sure to make the transition to the questionable feed over a period of one to two weeks and to avoid moving cattle on and off high-nitrate feeds.

If mixing is not possible, put out low-nitrate feed at the beginning of each feeding and follow up with the higher-nitrate feed to ensure each animal has consumed the low-nitrate feed first.

A balanced ration tends to reduce problems from nitrates in the ration. Feeding adequate levels of energy, vitamins (A and E) and trace minerals reduces the risk of toxicity. Feeding grain in combination with high-nitrate feeds helps reduce the effect of the nitrate content. Energy from the grain apparently helps complete the conversion of nitrate to bacterial protein in the rumen. This can be another tool in managing high-nitrate forage which cannot be physically mixed with low-nitrate forage.

If the animals are in good body condition, are receiving a balanced ration and have been introduced to high-nitrate feed over a period of time, it is possible for them to maintain normal levels of growth while consuming feeds with nitrate levels of one per cent or higher. Livestock should have access to clean water at all times.

Do not feed damp hay, straw or fodder suspected of being high in nitrate. Damp feed seems to be more toxic because some of the nitrate has already been converted to the more toxic nitrite.

Don't panic when you get a high nitrate level in your forages. Proper sampling and a nitrate test will allow you to make use of the forage through proper feeding management.

Pasture: Allowing livestock to graze pastures suspected of having high nitrate levels is not without risk. Implementing one or more of the following management practices will reduce the risk of livestock losses to nitrate toxicity.

- Don't overstock suspect pastures to limit consumption of lower stem material which is highest in nitrate content.
- Don't strip-graze suspect pastures to limit consumption of lower stem material which is highest in nitrate content.
- Provide other feeds that contain little or no nitrate during grazing.
- Graze suspect pasture during the day and remove the animals at night during the first week to reduce the amount of pasture consumed and to acclimate cattle.

Observe livestock frequently when they begin grazing a new pasture that is suspected of having high nitrate levels to detect any signs of nitrate toxicity. Recent research indicates that beef heifers can safely graze sudangrass that contains high levels of nitrate (0.93 per cent nitrate NO_3-), but cattle grazing these pastures tend to selectively graze the leaf portion of the plant that is not high in nitrate, and slowly adapt themselves to these feeds. If stocking rates are heavy for a pasture suspected of high nitrate concentrations, it may force the cattle to consume high-nitrate plant parts, increasing the possibility of nitrate poisoning.

Causes of High Nitrate Levels in Forage

Nitrate taken up from the soil through plant roots is normally incorporated into plant tissue as amino acids and proteins. The primary site for this conversion is the actively growing green leaves. Nitrate accumulates in the stalk or stem of plants when factors interfere with normal plant processes.

All plants contain some nitrate, but excessively high amounts are likely to occur in forages that have been grown under stress conditions such as shade or low light intensity, detrimental weather (including drought, frost, hail, low temperatures), herbicide applications, insect predation and diseases. The amount of nitrate in plant tissues also will depend on the species, stage of maturity, part of the plant and nitrogen fertilization.

Nitrate toxicity in dairy cattle

Shade. Conversion of nitrates into amino acids and proteins is linked closely with photosynthesis. Light is the energy source for these activities, so shaded plants or lower leaves may be higher in nitrates than plants grown in full light.

Weather. Not all drought conditions cause high nitrate concentrations in plants. Some moisture must be present in the soil along with the nitrate to permit absorption and accumulation. If the major supply of nitrate for the plant is in the dry surface soil, very little nitrate will be absorbed by plant roots. In plants that survive through drought, nitrate levels are often high for several days following the first rain.

The nitrate level in plants can rise and fall. If the source of stress is drought, and timely rain allows the plant to recover, the nitrate level falls as the plant uses the nitrate in the production of new plant tissue. If drought persists, so does the nitrate.

Frost, hail and low temperatures all interfere with normal plant growth and can cause nitrate to accumulate in the plant. Frost and hail may damage or completely destroy the leaf area of the plant. A decrease of leaf area limits photosynthetic activity, so nitrate absorbed by the roots accumulates in the stem or stalk.

Most plants require temperatures above 13°C for active growth and photosynthesis. Nitrate can be absorbed quickly by plants when temperatures are low, but conversion to amino acids and protein occurs very slowly during times of low temperature.

Herbicides. While herbicides, such as 2, 4-D tend to disrupt normal plant processes and can result in temporary high nitrate content in plants; however, spraying pastures and silage crops to control weeds may actually reduce the nitrate hazard of these feeds, especially when weeds high in nitrates are killed.

Disease. Plant diseases interfere with normal growth and development. This can cause nitrate to accumulate by interfering with nitrate reduction, protein synthesis or manufacture and translocation of carbohydrates.

Insect Predation. Leaf damage caused by insect predation can result in reduced photosynthesis and may result in accumulation of nitrate.

Plant Species. Plants vary in the amount of nitrate that accumulates in various tissues. Certain weeds, such as pigweed, kochia and lambsquarter, often are high in nitrate. Oats and millet cut for hay at an immature stage can also have high nitrate concentrations. Sorghum and sudangrasses often store high amounts of nitrates. Brome and orchard grass store very little nitrate under normal growing conditions. Legumes generally do not contain high nitrate concentrations. Alfalfas, vetches, trefoils, peas and clovers generally do not accumulate nitrates. A list of common crop and weed species that can accumulate nitrates is shown in Table 9.21.

Stage of Growth. Stage of growth markedly changes the nitrate content of forages. Nitrate concentrations usually are higher in young plants and decrease as the plant matures. However, plants grown in soils with excessive nitrates or grown under stress may be high in nitrate content at maturity.

Plant Parts. Plant parts closest to the ground contain the most nitrate. Leaves contain less nitrate than stalks, and the seed (grain) and flower usually contain little or no nitrate. Most of the plant nitrate is in the bottom third of the stalk.

Nitrogen Fertilization. Nitrates in the soil are the source of nitrate in plants. While a positive relationship exists between soil nitrate and nitrate in the plant, the effect of nitrogen fertilization appears to be less important than the conditions listed previously in causing high nitrate content in forages.

Harvest Method and Nitrate Levels

Harvest as Hay. If hay in the windrow remains damp for several days or is rained on, nitrate levels

Nitrate toxicity in dairy cattle

Table 9.21 Common crops and weeds that have potential to accumulate nitrate.

Agricultural Crops		Weeds	
Oats	Sorghum	Pigweed	Wild Sunflower
Canola	Pearl and German Millet	Bull Thistle	Canada Thistle
Barley	Corn	Fire Weed	Kochia
Wheat	Sudangrass	Lambs Quarters	Mustards
Rye	Rape	Nightshade	Millet
Sugar Beets	Soy bean	Russian Thistle	Smartweed
Flax	Sudan grass	White Ragweed	Canada Thistle
Alfalfa		Johnson grass	Jimson weed

(T.L. Provin and J.L. Pitt; Texas A&M University, Cooperative Extension Service, USA, 2011)

will decrease slightly. Hay dried quickly will lose very little nitrate. Nitrate concentration in hay bales does not change appreciably over time.

Cut hay when the weather is most favourable. Bad weather has destroyed more hay than nitrate. After the hay is baled, test for the presence and the level of nitrate. Forages with high levels of nitrate can be diluted with other feeds.

Harvest as Silage. Ensiling tends to reduce the nitrate content of forages. Forages high in nitrate can lose from 40 to 60 per cent of their nitrate content during fermentation.

Harvest forages suitable for silage at the stage of optimal quality and quantity. Excessive nitrate in forages will not always be reduced to safe levels during ensiling. If silage is suspected to be high in nitrate, analyze the feed for nitrates before feeding. The analysis will help in designing rations to prevent livestock losses from nitrate poisoning.

Harvest Near Maturity. Crops normally have lower nitrate levels at maturity, so for crops suspected of containing nitrates, silage should be harvested as near to maturity as possible. If the field being harvested as silage has been identified as or is suspected of having high nitrate concentrations, raise the cutter head to avoid stalk bases that have the highest nitrate concentration.

In crops such as oats for hay, it may be wise to harvest the crop at a more mature stage than desirable for maximum yield of digestible nutrients.

Harvest as Green Chop. Some forages are harvested in an immature stage and fed to cattle in the form of green chop. High nitrate feeds that are piled in mounds and allowed to heat before feeding are especially dangerous to livestock. Heating causes nitrate to convert to nitrite, making the feed about 10 times as deadly. An analysis of nitrate will help in designing feeding programs to prevent livestock losses from nitrate poisoning.

Nitrate in Water. Mature livestock can tolerate higher concentrations of nitrate in their water supply than young livestock. In the case of calves, however, much of their fluid intake is derived from nursing and very little nitrate is secreted in milk.

Nitrate toxicity in dairy cattle

Livestock drinking well water are not likely to experience nitrate toxicity. Nitrate toxicity from water is more likely to occur when livestock drink water from ponds, road ditches or other surface depressions that collect drainage from feedlots, heavily fertilized fields, silos, septic tanks or manure disposal lagoons.

As with feed, frequent intake of water containing nitrates appears to increase the overall nitrate tolerance; however, when evaluating possible toxic situations, the nitrate in both the feed and water must be considered because they are cumulative.

Nitrate toxicity is not likely to occur from water containing less than 443 ppm nitrate NO_3, provided that animals are fed a balanced ration that is not high in nitrate and that sound feeding, watering and management practices are followed.

Treatment

Nitrate poisoning can quickly become fatal. When high nitrate is suspected, remove the contaminated feed and provide a high-energy feed, such as barley. A veterinarian should be called immediately to confirm the tentative diagnosis. Because death is a result of oxygen shortage, handle cattle as little and as quietly as possible to minimize their oxygen needs.

Administer a 4% methylene blue solution intravenously at a dosage rate of 4.55 mg. per 0.45 kg. of body weight to treat cattle with nitrate poisoning. Treatment may have to be repeated every six to eight hours because the rumen may be full of forage or feed that contains nitrate and nitrites will continue to move from the rumen into the blood stream. Mineral oil may be given orally to protect the irritated mucous membranes.

(Canada – Government – Agriculture: www.agriculture.gov.sk.ca & aginfo@gov.sk.ca)

Examples of nitrate field tests

A rapid test is the **diphenylamine spot test**. The diphenylamine test is comprised of diphenylamine salt (0.1 grams) dissolved in sulphuric acid (30 ml-36N). A single drop of this acid reagent is placed on a freshly split plant stem. If a dark blue color develops immediately (in the first 5 seconds) nitrate is present. If there is no immediate change in colour, there is no nitrate. The diphenylamine field test indicates only the presence or absence of nitrate. It does not determine the actual nitrate concentration and, as with any field test method, it should be used **only as a screening tool**. Any positive result from the spot test should be followed up with a laboratory analysis for quantification.

Another option may be found at www.nitrate.com/store/ NECi, Lake Linden, MI, USA

1. Dry forage nitrate test (11 US dollars) (e.g. hay):
2. Testing range is from 0 to 9000 ppm nitrate, or 0 to 2000 ppm nitrate-N, or 0.9% N Feedstuff has first to be grindled.
3. Livestock wet feed nitrate test (11 US dollars) (e.g. silage, green chop or standing hay):
4. Testing range as above (1). Use a garlic crusher or equivalent to squeeze out the excess liquid.
5. Soil test (11 US dollars)
6. Water test (30 US dollars)that animals are fed a balanced ration that is not high in nitrate, and that sound feeding, watering and management practices are followed.

Feeding replacement heifers

After FISHER (2011) International Dairy Topics 10(2)

It is beneficial to graze heifers prior to their first calving (Benson, 2011).

Exercise while at pasture increases appetite and prepares them lactation stress. Moreover, grazing represents a lower exposure level to pathogens than when housed. The welfare of grazed heifers is also improved (Galindo & Broom, 2010) with regards to social interaction - significantly fewer aggressive interactions occur. Animals also show more behavioral synchrony and, hence, less stress. In general, grazing heifers is economically beneficial (grazing; manure spread; soil quality, erosion decline; better health status; lower veterinary costs; better growth).

The physical benefits are as follows:

- Increased bone density
- Improved blood circulation
- Improved calving ease
- Increase heart and lung size
- Increased oxygen levels in tissues
- Reduced incidence of pneumonia
- Reduced time to placental expulsion
- Fewer open days after calving
- Enhanced appetite (carried over into period when housed)
- Correct structural development of legs and claws

Setting up a grazing system:

Small heifers are selective eaters and will not consume the same amount of pasture as adult cattle. In general, small heifers need twice as much pasture per kg of bodyweight as adult cattle.

1 **Continuous grazing**: Heifers have access to the whole field throughout the year. This causes a decline of preferred pasture species, uneven nutrient deposition, compaction and erosion.

2 **Set stocking**: Fields grazed almost continuously, with some reserved for re-growth or forage harvesting. Disadvantages similar to those noted in system 1.

3 **Rotational grazing**: Heifers are moved from field to field based on the amount of forage present. This option is better for heifer growth and for pasture quality than systems (1) and (2).

4 **Intensive rotational grazing**: Using numerous smaller fields at a high stocking density for a short period of time. Good for pasture quality and for heifers, but more labour-intensive. This option yields the best heifer growth rates and is beneficial in terms of pasture quality and species survival. Heifers are moved daily or every two days. Very young heifers can be followed by older heifers to bring pasture down to the normal residual level. If *Johne's disease* is present or suspected, then the surface grazed by young cattle should not be used for adult cattle. Fields should be set aside for young stock up to 12 months of age. Moreover, manure should not be spread on these fields.

➲ Carry out a rough estimate of the **feed budget** and consider:

A The number of calves the farmer needs to graze
B The average body weight over the time they will be at pasture and the desired growth rate
C The percentage of the year they will be on that pasture
D The months of the year there will be active pasture (re)growth
E The likely amount of kg of grass per ha of pasture
F The level of feed supplementation with which the farmer is planning to provide the heifers.

This will give a rough estimate of the number of hectares needed for grazing the heifers. Choose an

Feeding replacement heifers

appropriate site with easy access for people and trucks, with shade for summer and sufficient drinking space supplied by good quality water distribution systems. Water troughs should not be shared between two fields, nor should they be fenced by an electric fence system. Choose the most appropriate pasture species, considering soil type, climate and the use of the fields. Ensure that the fencing around each field is adequate (electric; barbed wire).

Management issues:

1 Make sure that the transition of youngest heifers to pasture is done gradually. The rumen takes two to three weeks to adapt to a new diet. Hence, for the first week at pasture the heifers must still be offered the ration they were fed prior to going out to grass.

2 Until six months of age, young heifers do not have the rumen capacity to survive on pasture alone. The grazing must be supplemented with additional feed.

3 When pasture quality is poor or during cold weather, even older heifers may need to be supplemented with additional feed (concentrates).

4 Set average daily gain targets for the young stock and check this at least once a month on one animal per age group.

5 Cooperate with the nutritionist to develop accurate feed budgets and rations for different heifer groups; develop feed supplementation strategies.

6 Take feed samples for quality testing before ration changes; if needed, supplement them with concentrates and or silage.

7 Learn pasture assessment skills: pasture yields can be assessed using a pasture ruler, a rising plate meter, capacitance probes, echo-sounder or laser quad-bike-mounted pasture meters.

8 Keep records of pasture exploitation: pasture growth, feed analyses, feed supplements, heifer-growth and pasture maintenance measures. They are useful for evaluation and planning.

9 Consider the first grazing seasons as a learning experience when you start. Do not become disheartened or give up when it fails to turn out the way the farmer (or vet..) expected.

Potential problems:

A Internal parasites - carry out regular faecal egg counts followed by treatment if necessary.
B Diarrhoea with or without meteorism– this is a risk in pastures with > 50% clover or luzerne, but also in immature pasture with high protein levels.
C Clostridial disease – often in heifers grazing lush, immature pastures. Consider vaccination programmes.
D Escaping heifers - repair fences; check for presence of a neighbouring bull.

Excess vitamin E & the risk of mastitis

Ruth Bouwstra, PhD thesis, UGCN, Veeteelt, Nov 2, 2010, Netherlands

In many commercial concentrates and mineral mixtures, feed mills add vitamin E to improve the udder health situation on dairy farms. Scientific publications generally show that this supplementation is beneficial to the cow's disease resistance.

However, a recent field study was conducted in which cows with differing basal vitamin E blood levels were exposed to different levels of supplemented vitamin E.

On five dairy farms, half the dry cows were supplemented with 3000 IU per cow daily, while the other half only received 135 IU per cow per day. The rations for both groups were the same. Cows were sampled at drying off, four week and two weeks before calving, within the first 24 hrs after calving and at the first milk recording. Body condition scores (BCS) and rumen fill (RF) scores were also collected, as well as information about udder infections up to 100 days lactation. The study period was nearly one year and the selected farms all had a mastitis problem.

Results

Firstly, the mastitis situation improved on all farms (possibly due to the frequent visits by the veterinarian, more attention given to the cows and/or better dry-off procedures). Data analysis showed that the groups with the highest vitamin E supplementation had the highest incidence of mastitis (the risk was 1.5 times higher than the other levels of vitamin E).

> *The dairy cattle at highest risk are the* ***dry cows*** *and the* ***(pregnant) maiden heifers****.*

Cows at pasture will consume grass, which is high in vitamin E. This poses little risk.

Much earlier research showing an improvement with high levels of vitamin E supplementation used cows with an already low blood level of vitamin E due to ration composition. In these animals, additional supplementation would prove beneficial. In situations where the basal blood vitamin E level is adequate, there should be some assessment of current ration and status to determine if supplementation is really necessary.

In 2005, another study demonstrated that the mean blood level of vitamin E was 6.4 IU in lactating cows without supplementation and 8.9 IU in dry cows with supplementation. In 2007, cows at drying off had a blood vitamin E level of 17.9 IU (probably due to the fact that feed mills had started to include additional vitamin E in their concentrates). From a level of 14.5 IU further supplementation is not necessary. In 2008, around 81 dairy farms were revisited to assess vitamin E status again. It was clear that mean vitamin E levels had, yet again, increased.

> *The threshold above which supplementation is indicated is 7.4 IU*

Supplementing dairy cattle with a blood vitamin E level of above 7.4 IU may increase the risk of mastitis in the first 100 days of lactation. High risk groups must be monitored more closely.

Excess vitamin E decision diagram

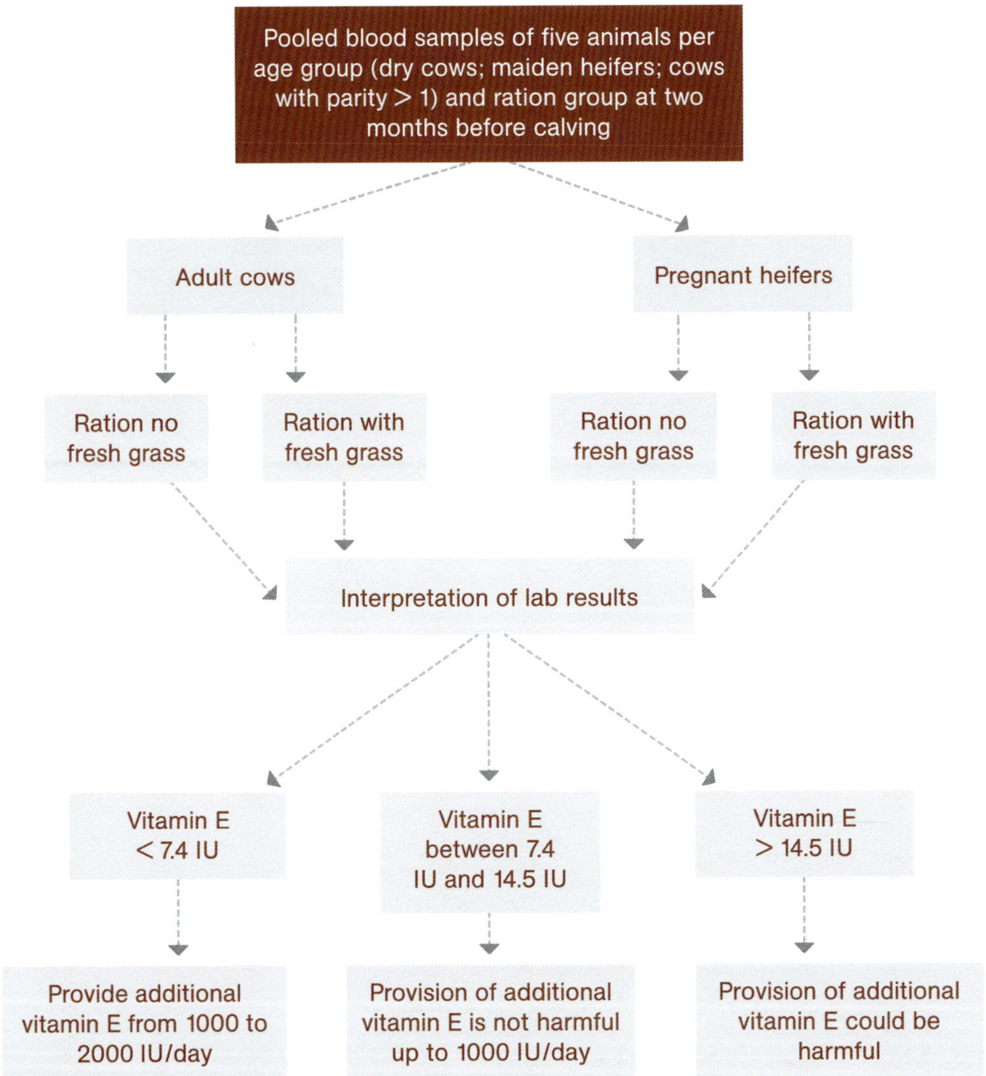

Figure 9.1 Decision diagram for vitamin E inclusion in the ration

Risk factors affecting mycotoxin occurrence

1. Field crops: Field mycotoxin contamination

- Biological factors
- Susceptible crops
- Compatible toxigenic fungi
- Environmental factors
- Temperature
- Moisture
- Physical damage
- Mould contamination
- Insect/pest control
- Crop rotation
- Tillage

Goal: Adapt sustainable practices to reduce mould growth and mycotoxin production.

2. Harvesting: Field mycotoxin contamination

- Crop maturity
- Season
- Temperature
- Moisture
- Physical damage
- Cross-contamination
- Chop length
- Fast fill and seal of silo with no compromise on packing density

Goal: Harvest the crop at the correct stage and ensure optimum dry-matter for the forage or grain being harvested.

3. Silo storage/By-products: Storage mycotoxin contamination

- Temperature
- Moisture
- Relative humidity
- Mould control
- Storage duration

Goal: Maintain uniformly clean and dry grain.

4. Feed processing: Storage mycotoxin contamination

- Temperature
- Moisture
- Season
- Mould control
- Hygiene
- Pest control

Goal: Watch moisture levels of finished feed to avoid contamination post-production.

5. Local farm: Storage mycotoxin contamination

- Temperature
- Moisture
- pH
- By-product use
- Storage duration (covered or open)
- First in - first out practice
- Lower leg swelling is a mycotoxin symptom.

Goal: Maintain clean feedstuffs and pitch any spoiled material.

6. Forage management

- Target packing density > 17 lbs/ft3 of DM
- Quick clean fermentation (silage inoculant)
- Face management: Keep it clean and tight
- When removing, always apply downward pressure.
- Pitch the waste on the shoulders and the floor
- Maintain face at <85 degrees Fahrenheit to reduce yeast/mould activity

Goal: Minimize the growth of yeast and mould in stored forage.

7. Farm environment

- Season and moisture
- Speed of feed consumption (number of feeding times per day)

Risk factors
continued

· Avoid feed residue (clean troughs daily or weekly as needed)

Goal: Minimize growth of yeast and mould in TMR during feedout.

8. Mycotoxin symptoms on the farm

· DON: reduced milk production and increased SCC

· T-2 Toxin: Hemorrhagic Bowel Syndrome (HBS), diarrhea

· Zearalenone: enlarged udders, poor conception rates

· Fumonisin: poor milk production and immunity

· Aflatoxin: aflatoxin M1residues in milk, reduced immunity

· Silage mycotoxins: altered rumen function, retained placentas, abortions

Goal: Observe symptoms carefully to introducepreventive and therapeutic measures. Feedadditives known as mycotoxin adsorbents are an effective tool for producers to use for prevention and treatment of challengeson farm.

Vitamin E

Vitamin E is essential for all cattle but especially important for dry cows, sick cows and young stock, as well as when diets high in unsaturated fat are fed.

Table 9.22 Reference values for serum vitamin E (mmol alpha-tocopherol/L) levels in dairy cows

Vitamin E (individual results)	Vitamin E (group results)	Interpretation
< 4.0	< 4.9	True deficiency
4.0 – 7.4	4.9 – 7.9	Sub-optimal
> 7.4	> 7.9	Adequate

(source: GD, Deventer, The Netherlands, 2009)

It is advisable to check the **selenium** levels at the same time, because vitamin E and selenium are closely associated and complementary.

Aflatoxin analysis & control

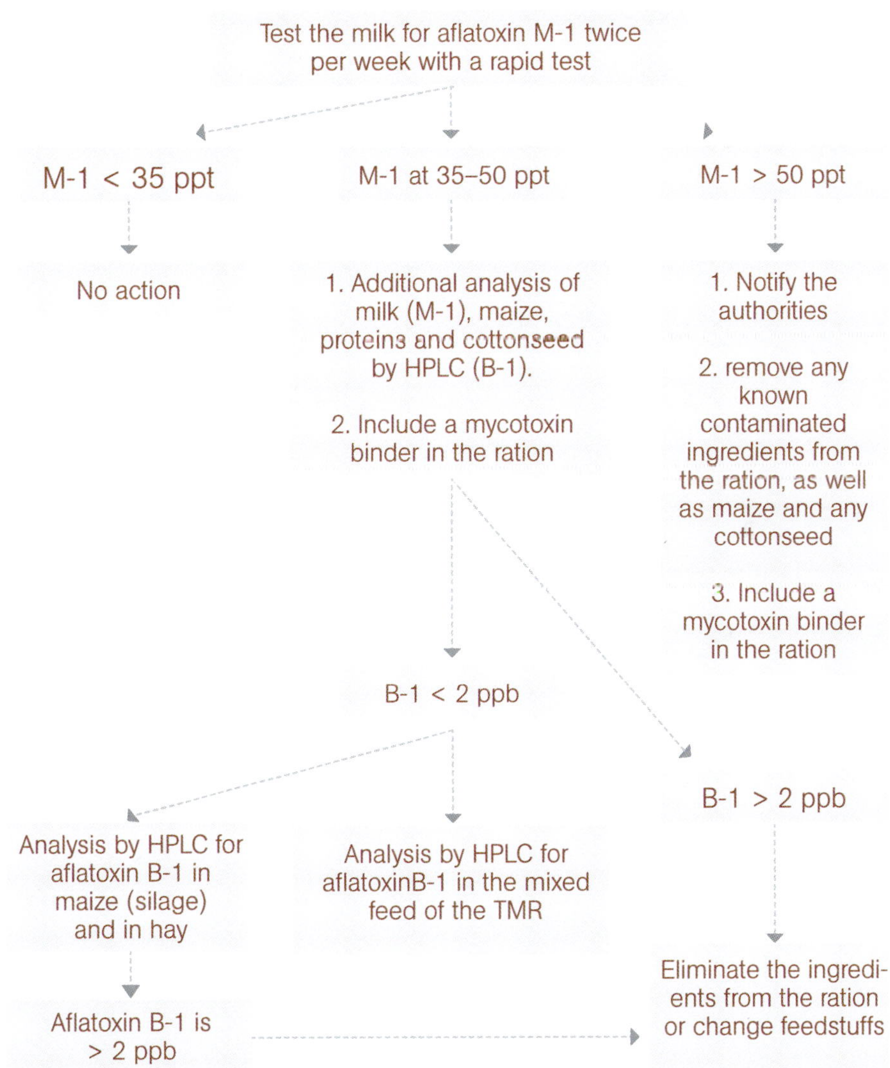

Figure 9.2 Protocol for aflatoxin analysis and control on the farm

Control of aflatoxin

After M. Zambianchi et M. di Benedetto, IDT vol 8 no 6, Novembre 2009

The degradation of aflatoxin in the rumen is lowest in bovine animals, however, they are still less sensitive than swine.

Aflatoxin is most frequently found in imported feedstuffs (countries where such feedstuffs are cultivated under humid, hot conditions, cotton seed, for example).

Aflatoxins in cattle can cause immune suppression, lowered feed intake, liver lesions, loss of milk yield and poor reproductive performance.

In Europe, a rise in aflatoxins is often seen towards the end of autumn-beginning of winter (November–December) but can also be seen in the middle of winter (January–February) or spring (April). The incidence of aflatoxin contamination of rations depends on the ration composition and changes made relating to the choice of feedstuff.

To eliminate aflatoxin as quickly as possible, a natural mycotoxin binder can be used, such as Mycosorb (Alltech Inc., KY) or a non-digestible clay. Suitable mycotoxin binders should be effective at low inclusion rates, have a high affinity for a wide range of mycotoxins, a high capacity to bind mycotoxins at high concentrations and be stable across a range of pH.

Metabolism:

Aflatoxin B-1 is rapidly absorbed in the rumen transported to the liver where it is metabolized to Aflatoxin M-1.

Aflatoxin M-1 is carcinogenic and hepatotoxic and has a relatively high rate of transfer into milk.

The peak of Aflatoxin M-1 occurs in milk at ~24 H after ingestion.

There is large variation between the concentration of M-1 found in the milk compared with that in the ration and the relationship is not linear. A transfer of 1 to 2% has been reported but often it is ~5 to 6%. The latter means that milk can often go above the legal limit of 50 ppt (0.02 mg/kg = ppm in Europe).

- This 50 ppt can occur when the ration, e.g. a TMR, contains more than 5 ppb of aflatoxin/kg (calculated at 12% humidity) in spite of the fact the individual ingredients contain less than 5 ppb each !!

- Action is necessary when the concentration of aflatoxin goes above the limit of 35 ppt in milk and or 2 ppb in, for example, maize.

See also the protocol scheme below for the analysis and control of aflatoxin on the farm.

Mycotoxins in bovine nutrition & related disorders

After Schuh. Isr. Journal of Vet. Med.1997; 52 (2-3) 93-97

Zearalenone causes infertility due to an oestrogen-mimic effect. At ambient temperatures of 27°and a humidity of 23%, *Fusarium* spp. grows well. For toxin production, the margin of temperature is from 12 to 24°C and sudden temperature changes are necessary. At dosages of 25 ppm in cereals, vaginitis, delayed oestrus, reduced feed intake and milk production loss can be observed. Additionally, infertility following a dosage of 14 ppm from hay or grass silage has also been noted.

The rumen represents a substantial detoxification, which can neutralise aflatoxin, ochratoxin A and toxins of *Fusarium*, mainly by the action of protozoa (rather than bacteria). However, in ruminal acidosis, the activity of the protozoa can be significantly reduced and, hence, the detoxification function is largely lost. Stress factors in general (social stress disease oestrus high milk production transition period) contribute to the occurrence of disorders.

Table 9.23 Subclinical effects of mycotoxicosis (Mike Watkins, Biomin USA, 2010)

Around calving	More ketosis, abomasal displacement, mastitis, retained placenta.
Examination postpartum	More metritis, pyometra, ovarian cysts
Treatments	Lower response, repeated treatments
AI-period	More irregular oestrus, anoestrus, false oestrus
Pregnancy diagnosis	More non-pregnancies, abortions
General	Longer lactations, more culling

The **diagnosis of mycotoxicoses** in cattle is very difficult. Moreover, it is very expensive.

Firstly, all clinical signs must be noted, as well as data from rations and feedstuff. Even laboratory reports that state 'no mycotoxins detected' are not completely reliable with regard to the actual situation on farm – the tests may not be sensitive enough to detect low levels and/or the contamination may be coming from another source e.g. bedding material. Taking samples in hot spots (suspect areas) in silage is often needed.

Secondly, the climate should be considered - season, sudden temperature changes between day and night, storage conditions of hay and silage (grass; maize) and of the concentrates, as well as the weather at the time of harvest and/or ensiling. Mould spots should be looked for in silage clamps.

Antibiotic **therapy** is not effective.

Bacterial and viral diseases have to be excluded from the differential diagnosis list. Post-mortem investigations can help with diagnosis.

It is possible to add **mycotoxin binders to the rations** to negate the negative effects of mycotoxins either by conjugation or deactivation by enzymes.

An example is Mycosorb by Alltech (Alltech Inc., KY). Many other products exist but they differ largely in efficacy. Mycotoxin binders should be effective at low inclusion levels but high mycotoxin levels, stable across a wide range of pH, able to bid a wide range of mycotoxins and be supported by scientific literature.

A true **therapy** does not exist. Change the ration and the feedstuff when suspected/proven to be contaminated.

In order to alleviate the symptoms, treatment may include:

- Blood transfusion
- Vitamin C and Ca-gluconicum
- Sedatives, vitamin B complex
- Transfaunation of the rumen, propionic acid

Mycotoxins continued

- Antihistaminics, glucocorticoids, glucose infusion
- Destroy the contaminated feedstuff

Prevention.......

Immediately after harvest, turn the land over (if plant remains stay at the surface, they may be contaminated and increase the risk of mycotoxin contamination by four times).

If the same plant/crop is grown in subsequent years then the risk of mycotoxin production is also increased. Crop rotation in the fields should be employed e.g. follow grass with maize.

Grass or maize should not be harvested too late in the season, and not in wet or very dry conditions.

Forage and humid cereals should be dried following harvest to a standard moisture content of 14%.

As always, good silage and storage management is essential to reduce the risk of mycotoxin contamination.

Applying organic acids (propionic acid 0.5 percent) to cereals can be an option.

Continuous inspection and monitoring is required.

Public health issues

Milk consumption represents a far smaller risk than the consumption of cereals and fruits.

However, there is a risk related to Aflatoxin-B-1 in cattle feed. Cattle metabolise Aflatoxin-B-1 into Aflatoxin-M-1, which can be subsequently found in milk.

VOC in silages

Y.A.Aragon, Internatl. Dairy Topics 10 (6) 27-29, 2011

Reduction of VOC in silages using inoculants

Volatile organic compounds (VOC) mostly originate from plants, far less from animals or bacteria. VOC are very well-known in silages and comprise many different types. Among these are: isoprene C_5H_8; monoterpenes $C_{10}H_{16}$; oxygenated VOC; sesquiterpenes $C_{15}H_{24}$; alcohols (ethanol, methanol, propanol, butanol); esters of acetate and hexanoate; aldehydes; ketones; carboxyl compounds.

The most important for cattle silages are the alcohols (especially ethanol), aldehydes and the esters of acetate and hexanoate.

Alcohols such as ethanol are produced by yeasts, and may result in a loss of feed energy in the silage of up to 40%.

Yeasts may develop when *good dairy farming codes of practice* (GDF) are not well followed. Elements in this GDF are for example: harvesting at the right time; good compaction of the silage hump; a proper silage sealing; putting sufficient weight on the covers; the proper silo design; the feeding speed per week; a clean cut surface when using the silage. These elements together can keep the level of yeasts in the silage low and hence the ethanol level low.

Ethanol emission from a silage hump may also occur, depending on air speed, temperature, exposure time, porosity of the silage. The latter is determined by e.g. its density and its structure. A lower density and poorer structure will lead to greater porosity of the silage.

When there is sufficient level of acetic acid in the silage, the ethanol level will also be low.

These VOC may represent human health hazards too.

Inoculants with lactic acid bacteria will diminish the hazards for higher ethanol contents of silages.

Mycotoxins on a dairy farm

Adapted from www.knowmycotoxins.com Alltech 2010

In cases where **signs** of a mycotoxin problem have been detected, the following checklist may be used, together with a visit from the vet, to discuss measures to be taken. Typical signs may include cows not milking as usual; lower than usual milk fat content; increase in somatic cell count; reduced reproductive performance; reduced feed intake; poor rumen function; lethargy; lower leg or teat swelling; impaired immune function. BEDDING material, such as straw has also been found to contaminate the milk or affect the cows if ingested.

Moulds can be detected on maize crops, grass and other (fermentable) plants, as well as in silages. Moulds can easily develop on plant remainders in the field, therefore, after harvest, the land must be ploughed. If crop is harvested late, the product may be too dry and is then much harder to consolidate well resulting in air pockets and mould development.

The definitive **diagnosis of** "mycotoxins" or "mycotoxicosis" is very hard; counting spores is invalid because moulds occur in hot spots and the presence of moulds does not necessarily mean the presence of mycotoxins. Moreover, the signs are very non-specific and vague. If the farmer suspects the feed as source of disorders, he must immediately remove this feed from the ration.

The economic losses are related to milk production, reproduction and the health issues named above.

Strategies for prevention & control:
Planting insect- and disease-resistant varieties and practising crop rotation. Minimise plant stress. Apply timely harvesting to avoid a high dry matter crop that is difficult to pack. Clean silage bunkers thoroughly prior to harvest; close silage clamps immediately after harvest. Check regularly for holes in silage clamps ad bales; close them as quickly as possible. Eliminate affected parts of the silage from the clamp and from the feed bunk.

1. Ensure aerobic stability in silage clamps (even weight on top e.g. sand) by good compaction at ensiling. Inoculants can be applied for a rapid fermentation. Dry feeds must be kept free from moisture and rodents. If heating does occur in the silage, try to feed smaller portions in the feed bunk more frequently, or add propionic acid to the TMR while mixing.
2. Eliminate any suspected feedstuff from the ration. Mycotoxin binders could be added to the ration but their efficacy is highly variable from highly effective to little or no effect. Organic acids have no beneficial effect and can even have a negative impact!

Table 9.24 Assessing the risk of mycotoxicosis on the farm

Questions	Applicable on your farm?
Forage & feed:	
Are you feeding high dry matter forages?	
Is there any soil contamination of the forage?	
Is there evidence of spoilage or heating in any feeds?	
Were there unusual pre-harvest and/or post-harvest weather conditions?	
Has poor herd performance coincided with ration changes?	
Herd health & performance:	
Has there been any inconsistency in milk production?	
Has there been an increase in disease incidence?	
Are there cows showing signs of ruminal acidosis?	
Have you noticed any inconsistency in faecal consistency?	
Are there any other issues of concern regarding the herd?	

Levels of UFA & omega-3 fatty acids

Agrifirm Holland, 2008

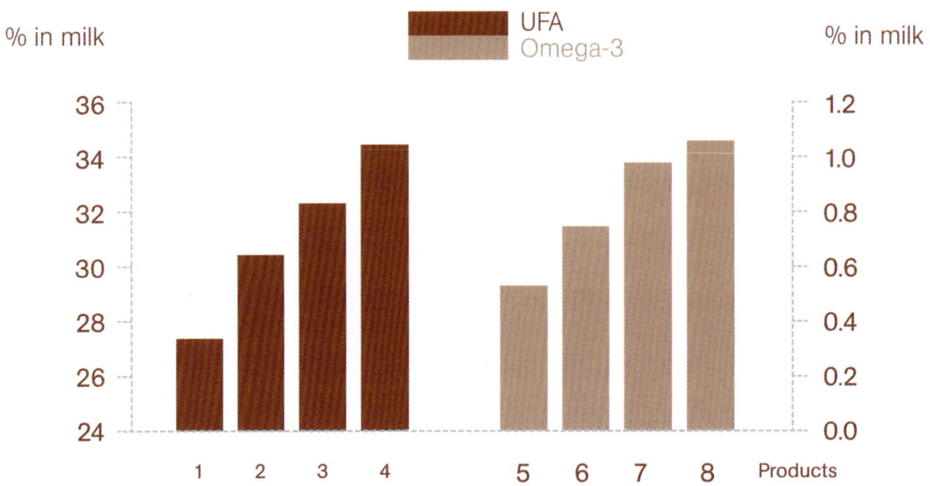

Figure 9.3 The percentages of UFA and omega-3 in milk, following a supplementation of one out of eight linseed products in the ration of cows respectively following dietary supplementation of one of eight linseed products

Levels of unsaturated fatty acids (UFA) and total omega-3 fatty acids in milk following use of different products of linseed

The so-called "healthy milk" represents an increasing trend in dairy marketing and consumer demands. The most important issue is related to unsaturated fatty acids and omega-3 contents of milk.

Therefore, a field trial was set up to study the effect of UFA's and omega-3 fatty acids supplementation to feed on their transfer to milk. The ration was the same for all groups of cows, except for the supplements. The results are shown in the Figure 9.3 above.

Product 1	no omega feed supplied
Product 2	UFA - untreated linseed
Product 3	UFA - extruded linseed
Product 4	UFA - grounded linseed***
Product 5	no omega feed supplied
Product 6	Omega FA - untreated linseed
Product 7	Omega FA - extruded linseed
Product 8	Omega FA - grounded linseed***

***= ground in purest form

Linseed content in the ration was 1.4 kg.

There is a large variation in the fatty acid content and profiles in different grass silages. This range corresponds with a 0.5 kg of linseed.

Conclusion: The products 4 and 8 resulted in the highest transfer to milk, with products 7 and 3 as second and third best.

Clostridium spp. contamination

Preventing contamination of silage by *Clostridium spp.* - including silages of good quality.

Clostridium spp. (or butyric acid bacteria) in silage cause economic losses, either by penalties for quality failure, or by loss of DM (up to 15%). In the latter case, the digestibility, feed intake (about 1.5 kg per cow per day) and the metabolisable energy also decrease. When the temperature of the silage is 5°C higher than the ambient temperature, the silage is at risk of developing Clostridium spp. These high temperatures can be found at depths of 50 to 100 cm from the silage face.

It was recently shown that, despite fresh silage analysis showing very low *Clostridial* bacteria count, the silage can become heavily contaminated once the clamp is opened. This phenomenon is caused by an aerobic instability which begins with yeasts and continues with moulds. This may happen **even in silage of good quality** and has been demonstrated in maize silage and sorghum silage.

After harvest, at ensiling, there can be around one hundred spores, however, after opening and feeding (exposure to air), the number of spores increases up to millions per gram silage after five days.

Treatment of silage with *L. buchneri* (NCIMB 40788; anti-mould product) could stabilise the silage for at least 300 hours (or 5 days) maintaining the pH at a level below 5. The spores of *Clostridium* did not develop under air exposure. However, there are significant DM losses often associated with application of this micro-organism.

In normal situations, ensiling maize reduces the pH rapidly (rapid acidification) and spores cannot develop. But at the moment of feeding (opening, cutting and feeding) there is ample exposure of the silage to air and a reactivation of aerobic fermentation occurs, starting with yeasts and followed by moulds. The pH increases and the temperature increases up to 45°C. The increase in pH causes the *Clostridium* spores to develop and so contamination begins.

A= with product *L. buchneri* ; B= pH in maize silage without this product; C= number of *Clostridium* spores without treatment

Another way to protect the maize silage against the losses due to *Clostridium spp* is to use **poly-ethylene_tissue** to cover the silage clamp.

However, recently it was determined that **black-on-white oxygen-barrier-films** (125 microns) were far more effective than poly-ethylene tissue.

The silages most susceptible to losses in DM, digestibility and aeration are:

- maize silage (contains already yeasts in the field)
- silage with high sugar contents and high DM content
- bales of silage with high DM content
- maize silage in a TMR wagon.

Maize silage should not contain more than 3 x log10 spores per gram of silage. If it is possible to restrict the concentration of spores to below 5 x log10 per gram, then the concentration in the bulk milk tank will be lower than 3 x log10 spores per litre of milk.

Clostridium contamination

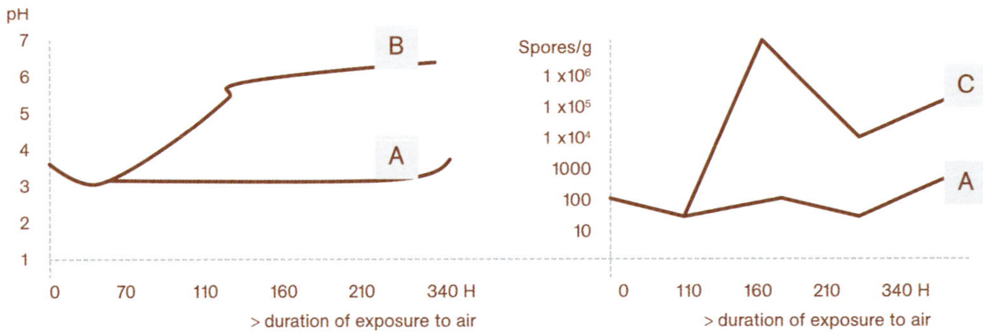

Figure 9.4 The formation of *Clostridium* spores in maize silage following exposure to air. For explanation of A, B and C see text.

Useful reading

Anonym. 2009. Preventing Clostridia contamination at feed out – even in good quality silage. International Dairy Topics 8 (5) p.11

Borreani G, Tabacco E, Cavallarin L. 2007. A new oxygen barrier film reduces aerobic detoriation in farm-scale corn silage. J. Dairy Sci. 90 (10) 4701-4706

Borreani G, Tabacco E. 2008. Low permeability to oxygen of a new barrier film prevents butyric acid bacteria spore formation in farm corn silage. J. Dairy Sci. 91 (11) 4272-4281

Holzer M *et al.* 2003. The role of Lactobacillus buchneri in forage preservation. Trends in Biotechnology 21 (6) 282-287

Vissers *et al.* 2007. J. Dairy Sci. 90 (7) 3278-3285.

Production process

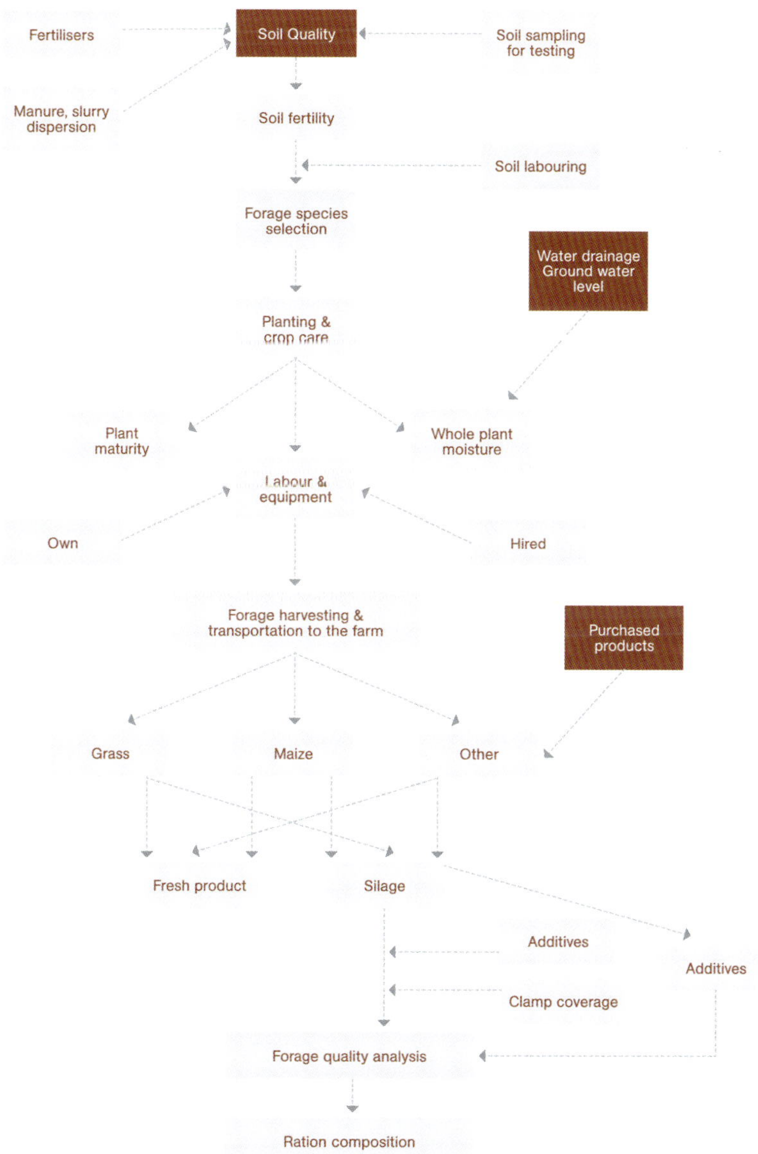

Figure 9.5 Production process diagram - pasture and crop exploitation and harvesting

Production process

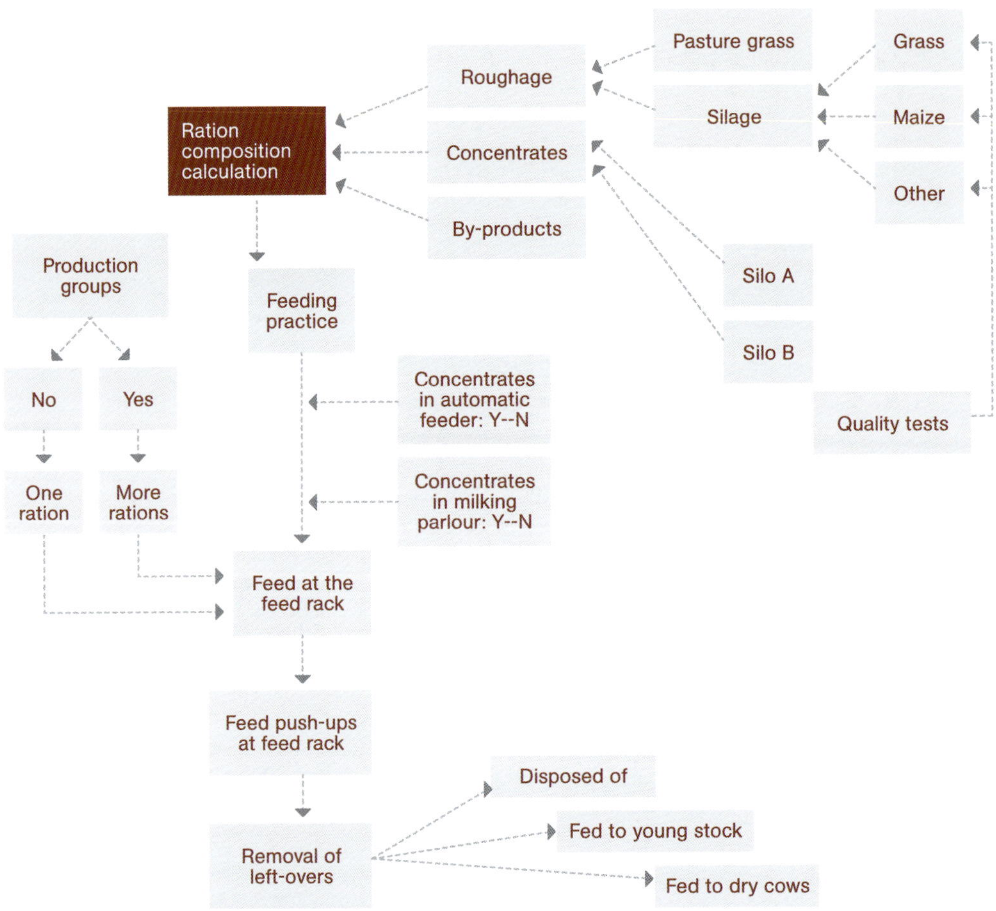

Figure 9.6 Production process diagram - feed,

Production process

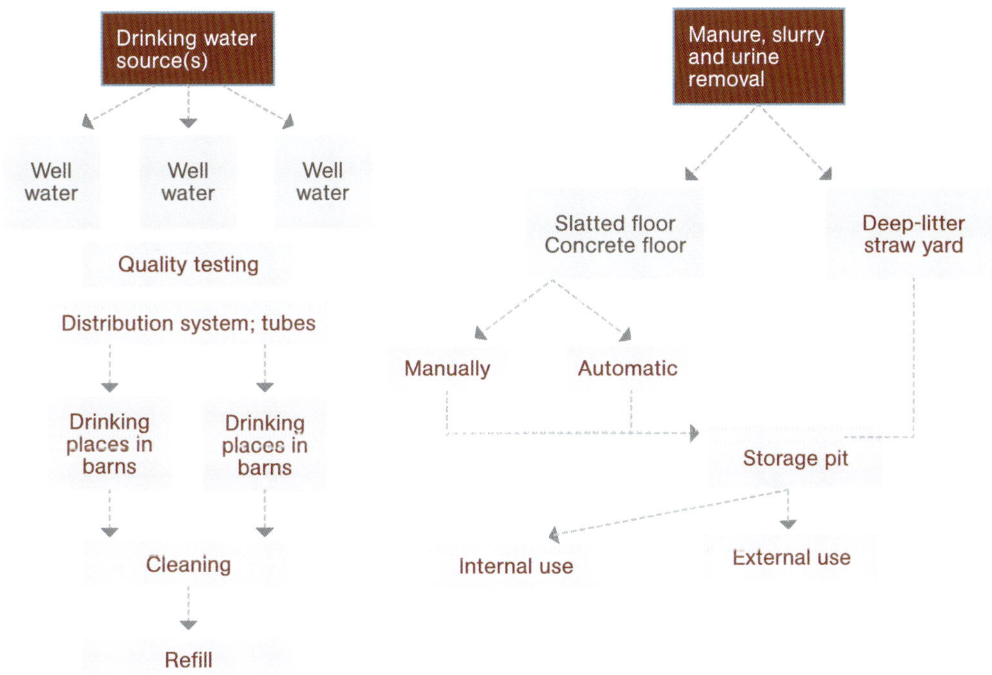

Figure 9.7 Production process diagram - water and manure

Estimating stage of maturity of maize for silaging

There are different parameters for estimating the stage of maturity of maize in the field. These parameters jointly determine the final dry matter content of the total crop and, therefore, whether the maize is fit for ensiling.

Table 9.25 The percentage of maize-ears in the crop

The percentage of maize-ears	Interpretation
40	Poor (possibly due to severe drought, soil structure problems, poor soil fertilization, early flat-out of crop; or due to a too high crop density)
50	Average (at normal culturing conditions and crop density)
60	High (very favourable culture conditions; or low crop density; or poorly developed hybrids)

Table 9.26 Estimation of the dry matter content (%) in leaves and stem

DM% estimated in leaves and stem	Interpretation
21%	If you cut a stem, fluid is flowing; leaves are still green
24%	If you cut a stem, the cutting edge is humid but no fluid flows; the leaves are yellowish
27%	There is little or no fluid on the cut edge or in the stem; leaves are brown

Table 9.27 Estimation of the dry matter content (%) in the maize-ears

DM% estimated in maize-ears	Interpretation
35%	Milk line in the grains is at ¼ ; grains crush under cutting or under fingernail scratching
40%	Milk line is at 1/3 ; milk flows from grains when crushed or scratched
45%	Milk line is at ½ ; only the lower part of grains contains milk
50%	Milk line at 2/3 ; no milk left; grains easy to scratch with fingernail
55%	No milk line; hard grains; grains hard to scratch with finger nail
60%	Black spot visible; hard grain; no scratching

Estimating stage of maturity

Table 9.28 To determine the right stage for harvesting

% DM in whole plants	Other characteristics	Harvesting decision
< 28%	DM% increases by 1 point every 3 days (under average conditions)	Do not harvest
Between 28% and 35%	DM% of grains is > 50%	Harvest
	DM% of grains is < 50%; DM% of leaves and stem is > 24%	Harvest
	DM% of grains is < 50%; DM% of leaves and stem is < 24%	Wait until DM% of leaves and stem is at 24%
> 35%		Harvest immediately

Practical recommendations:

1 (1) If the DM% of leaves and stem is > 24% (leaves and stem are desiccated) → chop more finely in order to safeguard and improve conservation.

2 (2) If the DM% of grains is > 60% (black spot visible) à use a harvester that can crush the grains.

3 (3) If the DM% of the whole plant is < 28% and, if the maize must be harvested, then adjust the cutting from 20 to 30 cm.

4 (4) Under conditions of stress, such as severe drought (leaves and stem are desiccated; DM > 38%), the actual values of the different parameters are often underestimated.

Note this approach can also be found in a "slide rule" format for practical handling.

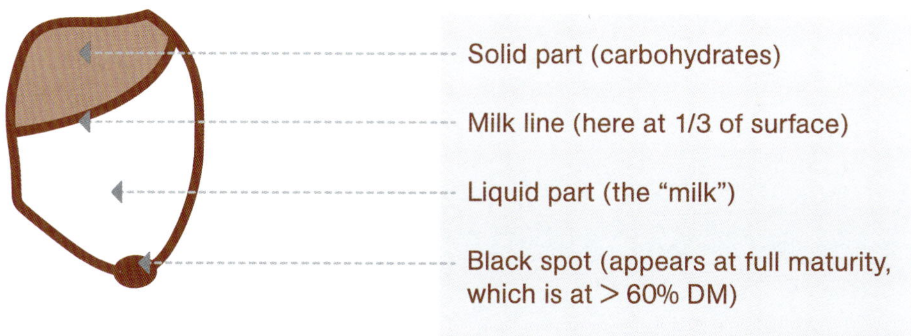

Figure 9.5 Structure of the maize grain

Minimizing the risk of heating in silage clamps

1 Harvest grass at an estimated dry matter content of 40%.

2 Harvest maize at a dry matter content of 28 to 32% (the grain is hard-pasty) and set the cutter at 6 to 8 mm.

3 Distribute the harvested grass or maize in layers of 30 cm height maximum and compact it thoroughly.

4 Define the height of the silage clamp according to the expected feeding speed: feed a minimum of 100-150 cm clamp depth per week in winter and a minimum of 150-200 m clamp depth per week in summer season.

5 Seal the clamp immediately after ensiling with optimal cover material and ensure it is airtight.

6 Cover the entire clamp with at least a 10 cm thick sand or soil layer.

7 When using a clamp silo, line the sidewalls with plastic sheeting prior to ensiling. The plastic sheeting should provide at least 2 m cover over the top of the forage once the clamp is full.

8 Do no open maize silage that has been ensiled for less than six weeks to allow adequate conservation and cooling.

9 When feeding silage from a clamp on soil (not on concrete), it is necessary to put sand bags on the plastic in front of the clamp in order to keep out the air as much as possible.

10 Make sure that the silage face is neat and straight; do not leave loose feed in front of the clamp, and close the front as good as possible with the cover.

Monitor-scoring rumen fill

Table 9.29 Legends to the card of scoring rumen fill

Note that Rumen Fill scores may change over a couple of days, while e.g. Body Condition scores will only change over weeks or months! It makes the Rumen Fill score a day-to-day management tool.

Score 1	can be observed in cows with calving problems, like dystocia and milk fever, but also in sick cows (e.g. acidosis). This score should be avoided at all times
Score 2	can often be observed in freshly calved (high yielding) cows which show some problems to consume enough feed; it can also e observed in cows with a more or less severe negative energy balance and ketosis. It would be acceptable for a short while till the cow reaches a higher consumption level within a few weeks after calving (score 3)
Score 3	at least this score should be aimed for before and around peak milk yield and halfway lactation, although at the end of lactation it would be a good score too
Score 4	cows around peak milk yield, halfway and at the end of lactation and in the dry period should show this score level, under optimal conditions
Score 5	cows showing this score have a high feed consumption level, but is most often observed in dry cows too

1

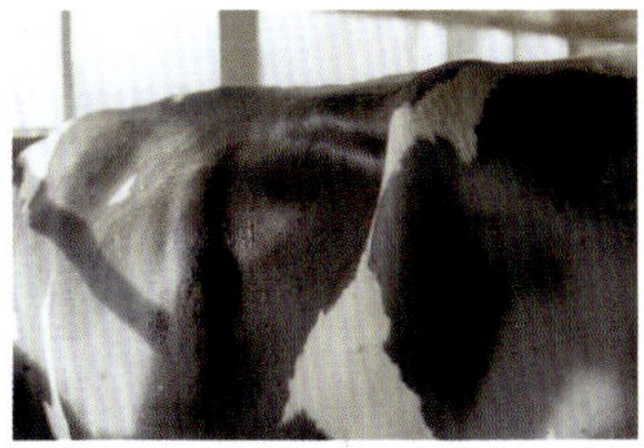

2

3

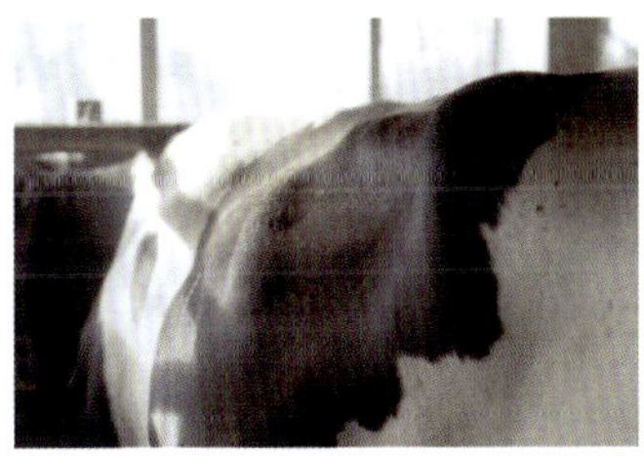

4

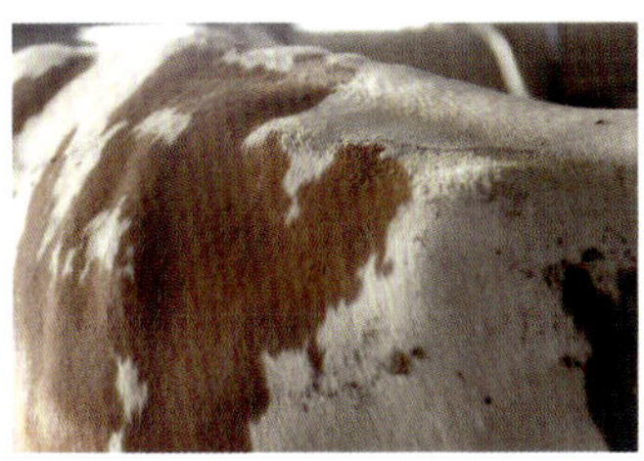

5

Monitor-scoring undigested fractions in the faeces

Table 9.30 Legends to the card of scoring undigested fractions in the faeces	
Score 1	this score indicates a satisfactory digestion and is strived for in general in lactating cows
Score 2	when some fibre particles are showing, there is a tendency towards more poor digestion. But it still can be acceptable for early lactating cows
Score 3	too much particles are observed, parts of these may be grains. It indicates the (start of) digestion and or fermentation disorders, or a deficient ration
Score 4	here there are truly too many fibres and or grains. Possibly a deficient ration or a feed mixer not properly working or roughages chopped too long
Score 5	too many fibres and or grains visible. See under score 4

1

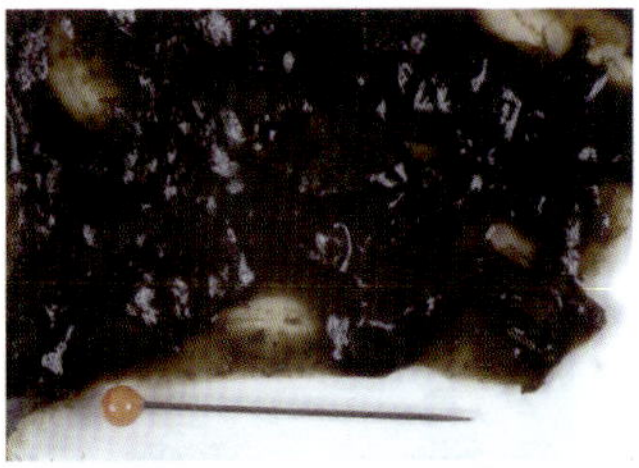

2

3

4

5

Monitor-scoring faecal consistency

Table 9.31 Legends to the card of scoring faecal consistency

Score 1	is always to be avoided because it shows disturbed rumen fermentation. It can also be observed in diseased cows (acidosis; infectious diseases with fever; certain severe coliform mastitis cases; etc)
Score 2	can be observed when cows are in lush pastures (protein rich; few fibres) or consume much energy-rich concentrates (high yielding cows) or low fibres level
Score 3	this score is in general what we are looking for in lactating cows. It all depends –of course- of the composition of the ration supplied
Score 4	thick faeces like these can be observed in late lactation and dry cows
Score 5	these faeces are too thick, even for dry cows. It indicates a disturbed fermentation and or digestion, or water intake is deviant

1

2

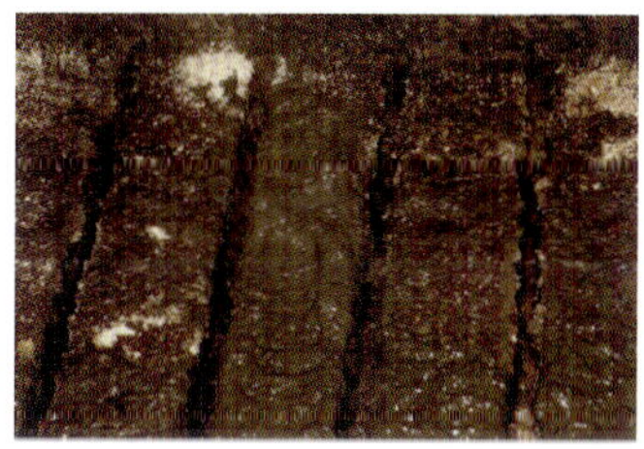

3

4

5

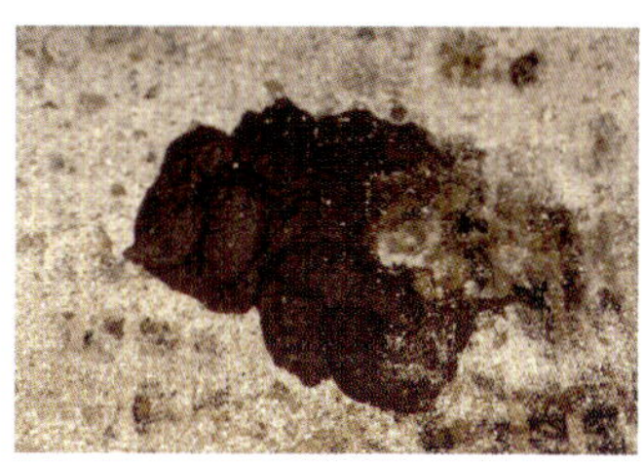

Housing

> Cattle barns are a basic component of the cattle environment. The current principle is that the barn design should follow the cows' needs rather than human needs or human welfare. Many errors exist in today's cattle barns. Practical elements are provided in this chapter together with checklists for evaluation.

Good housing practice

Adapted after Vin H & Vin-Dekoker J, JN GTV, Nantes, France, 2007

Clinical diagnosis:

1 hygiene parameter

2 climate parameters

7 major parameters in total

Hygiene

1. **Hygiene score of cows (see scoring chart elsewhere)**
   On a scale of 1-4, where 1-clean and 4-dirty, scores should ideally be < 2.

2. **Straw yard bedding temperature must be < 41°C at 15 cm depth of straw, or 40°C at 10 cm depth.**
   Above 41°C, the straw should be cleaned out and replaced with fresh.

3. **The surface area per cow → includes the passageways, exercise areas, areas close to drinking troughs and at the feed rack.**
   To calculate the quantity of clean straw needed, one should eliminate the non-occupied areas and dirty (manure) areas. Standard surface needed is 6 m2/cow and 1 kg of straw per m2 surface. If the straw on exercise areas is used for manure in the field, then 600 gr straw would be sufficient (2 kg straw per cow for 3 m2).

   Carry out the observations at the start of the farm visit, with as little disturbance to the cows as possible!

Barn climate

4. **The markers of a poor barn climate (rotten wood; rusty poles; cobwebs; damp coat of cows; rain falling directly into the barn).**

5. **Verify the following elements:**
   The quality and quantity of air through inlets and outlets:
   - building double-roofed, closed or mono-roof with open space-boarding or chimneys.
   - closed building with ventilation through vertical outlets.
   - half-open, double-roofed building.
   - half-open, mono-roof building with large outlets.

   → tests with smoking powder = in less than three to four minutes the smoke must have dissipated (in winter < four minutes; in summer < three minutes)

   Check the size of outlets (space boarding) (Table 10.1)

   Check the volume of fresh air available per cow.

6. **Additional markers of barn climate: hygrometer; air speed (equipment!)**

7. **Other relevant parameters:**
   Water availability and access (effective width of drinking places: 4 ½ m for 100 cows increased to9 m width in total for 100 cows in hot conditions 1 drinking place for every 10 cows.

   The light in the barn: 10% of the roof surface as windows, or 8% of the floor surface as windows (or 2 neon-lights per 10 calves). A good, practical test is, whether there is sufficient light to read a newspaper in the barn.

Other aspects

8. **Wild animals (foxes, badgers) should be prohibited from entering the barns; birds and pigeons (contamination of feed!) should also be discouraged.**
   Dead animals (and their fluids) should be removed from the barn to a specific site away from live animals, feed and bedding material.

Good housing practice

Sick animals must be kept in a **sick pen**, separated from healthy herd mates and from the **calving pen**. The sick pen should have a surface area of 4 x 4 m with a drinking trough.

9. **There must be a place set aside where professionals can easily handle and work with the cows, for example a contention box (inseminator; veterinarian; claw trimmer).**

10. **The microbiological and chemical quality of water should be tested twice yearly in the laboratory (for zoonoses; contamination)**
    On-farm equipment must be available for the cleaning and disinfection of barns, boots, clothes.

11. **Pollution of the environment by manure and urine must be avoided**
    (nitrogen; nitrates; phosphate; heavy metals, other chemicals) → Directive Nitrate 1991 CEE.

12. **Emission of polluting gases containing micro-organisms**
    (salmonella; brucella; coliforms etc) by aerosols during cleaning must be avoided.

Also, emission of **NH_3, H_2S_2, methane** should be addressed. To reduce gas output, turn and stir more frequently the slurry and compost.

Table 10.1 Recommended surface area (m^2) of air inlets and outlets for adequate ventilation based on liveweight (kg) of cattle and type of building

	Weight of cattle (kg)	Surface of air inlet (m^2/cow)	Surface of air outlet (m^2 /cow)
Closed barn, double roof, or mono-roof with open space-boarding or chimneys	50	0.02 – 0.1	0.025 - 0.05
	100–200	0.04 – 0.12	0.06 – 0.08
	250–350	0.15	0.075
	400–500	0.18	0.09
	600–700	0.20	0.10
Closed barn with ventilation through vertical openings	50	0.075	0.075
	100–200	0.135	0.135
	250–350	0.18	0.18
	400–500	0.21	0.21
	600–700	0.275	0.275
Half open barn, double roof	50	0.025	0.013
	100–200	0.06	0.03
	250–350	0.075	0.035
	400–500	0.09	0.45
	600–700	0.10	0.05
Half-open building. Totally open building.	50	0.15	0.15
	100–200	0.30	0.30
	250–350	0.40	0.40
	400–500	0.46	0.46
	600–700	0.55	0.55

Straw yards & potential contaminants

Jean-Luc MENARD March 2000 ; Institut d'Elevage, Pays-de-la-Loire, France

Factors influencing the microbiological contamination of the bedding material:

1 Type of bedding material
* straw favours the development of *Strep. uberis*
* wood shavings favour growth of *E. coli.*

2 Humidity in the barn. For example RH > 60%

3 Temperature of the surface of the bedding (must remain < 25°C, max 30°C)

4 Aerobiosis (for example, *E.coli* develops well under aerobic conditions); when the volume of straw increases, so does the temperature.

Some experimental results and field studies:

The temperature in the bedding increases by **5°C** (up to 40°C in some cases) if ~1.6 kg straw per cow per day is added.

The **temperature at a depth of 10 cm** in the bedding gives the best predictive results. That temperature must be **below 35°C.** If the temperature exceeds 35°C, the bedding should be removed and replaced with fresh straw.

On 41 farms:

The longer the straw remains in place, the lower the temperature of the bedding (with the exception of the aforementioned situations).

The level of milk production is only of minor importance regarding contamination of bedding.

Adding ~ **1.2 kg straw per m²** (or adding less than 2 kg straw/m² after cleaning out), the temperature in the bedding remains lower for the next three weeks compared with the addition of, for example, 1.6 kg.

The addition of large amounts of straw after cleaning can increase the temperature of the bedding by at least 5 °C over three weeks!

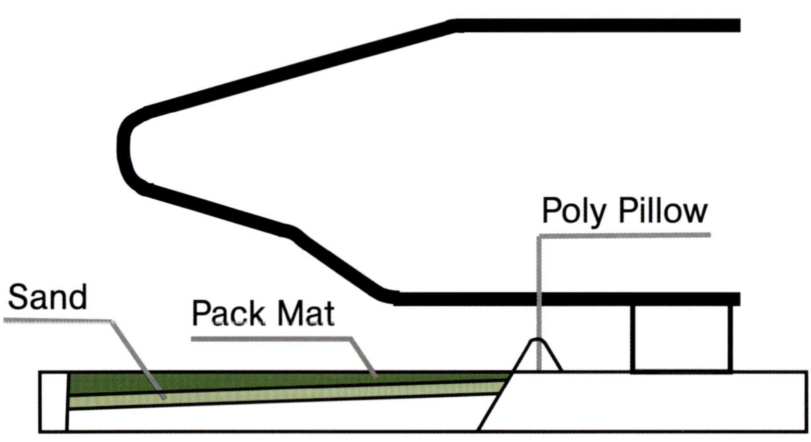

Figure 10.1 Pack mat with sand divider (Wisconsin)

Checklist for housing quality

Loose housing with straw yard	Total surface of the straw yard; resting area; exercise area; slats and slots in the floor; drinking troughs; hygiene; stale air
Loose housing with cubicles	Number of cubicles; cubicle sizes; shoulder rail height and position; brisket rail; width of area between rows; hygiene; bedding material;
Barn climate	Air inlets (place; size; surface); air outlets (idem); air streams; ventilation; humidity; temperature; dust; cobwebs; ammonia odour; light; over/under-ventilation; draughts
Surface quality	Slippery floors; poorly maintained flooring; abrasive; dirty; damp floors; poor bedding or topping;
Areas at risk for trauma	Irregular surfaces; holes; fixtures and fittings sticking out of material; step height > 20 cm; too many 90° angles in passages; too much inclination/ ramps;

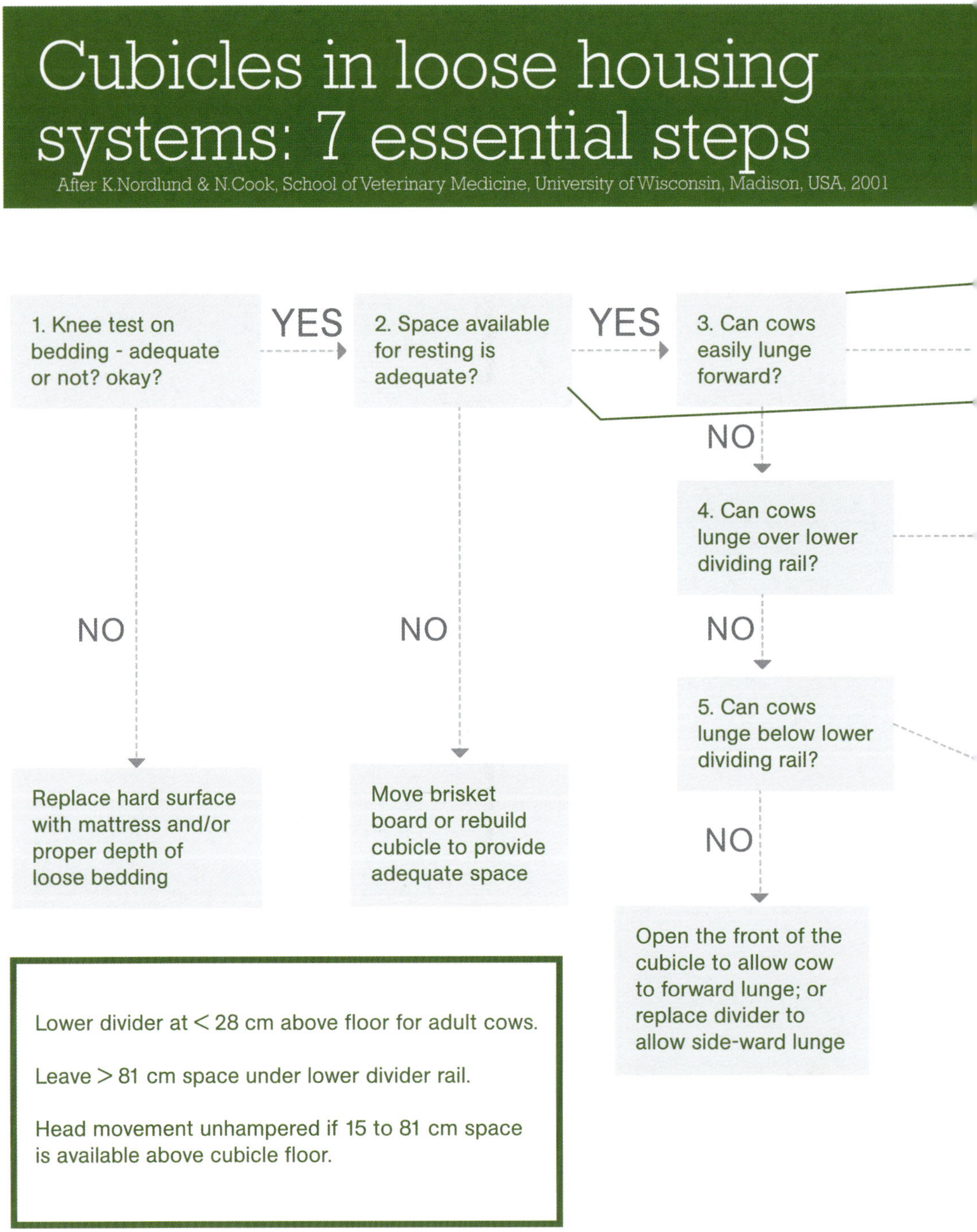

Figure 10.2 Flowchart for assessing cubicles in loose housing systems: 7 essential steps

Cubicles in loose housing systems: 7 essential steps

After K.Nordlund & N.Cook, School of Veterinary Medicine, University of Wisconsin, Madison, USA, 2001

YES

YES

Recommended dimensions of cubicles

Kg Body Weight	Cubicle width	Resting length	Total length	Neck rail height
180	75 cm	110 cm	145 cm	81 cm
275	84	122	165	89
360	91	132	185	94
540	112	155	229	109
630	119	168	249	114
720	130	178	269	122

YES

6. Can cows lower their muzzle at end of the lunge?

YES

NO

7. Can cows rise without hitting the neck rail?

Remove obstructing object (remove bars; hang dividers from vertical posts, etc)

NO

Reposition the neck rail to recommended height and distance from curb

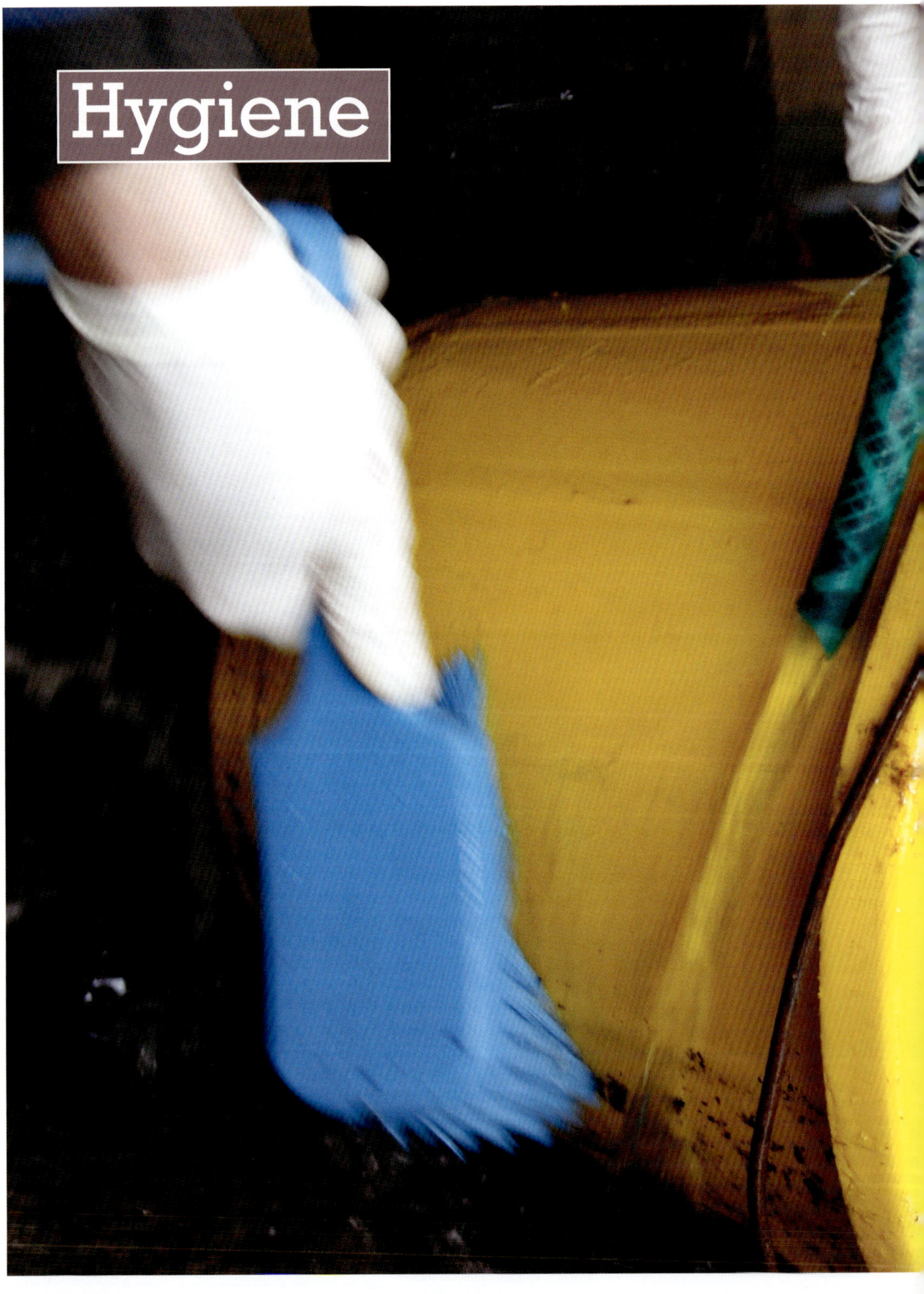
Hygiene

> "When attempting to improve animal health status and product quality, the application of strict hygiene practices is a prerequisite. This chapter provides a number of golden rules for hygiene and disinfection."

Housing hygiene

Maintenance measures in the straw yard	Frequency of straw renewal; quantity of straw added per day; straw distribution; straw quality; cleaning out frequency; manure scraping frequency; straw bedding temperature at 10 cm depth;
Maintenance measures in cubicle housing	Elimination of manure; manure scraping frequency; quantity of bedding renewed each day;
High risk areas (hygiene)	Areas with stagnant water or manure/urine; Manure heaps/slurry pits; areas unprotected against bad weather;
Passageways, roads to pastures	Poor maintenance; gravel paths; dirty; soiled with manure and water;

Conclusions:

Actions:

General hygiene

After Bioprotex CIDLines, Belgium; Internatl.Dairy Topics, vol.9 nr.3, 2010, pp 11-12

DATE:

Farmer:

Address:

Phone:

Veterinary consultant:

Address:

Phone number:

Area	Description	Already executed Y/N	Planned for future Y/N
1	Cleaning of the cattle houses		
2	Disinfection of the cattle houses		
3	Cleaning & disinfection of the calving area		
4	Cleaning and disinfection of the calf hutches and housing		
5	Drinking water treatment		
6	Pre-milking udder hygiene		
7	Post-milking udder hygiene		
8	Cleaning & disinfection of the milking parlour		
9	Cleaning & disinfection of the milking machine		
10	Claw hygiene		
11	Cleaning & disinfection of the dead cattle house/area		
12	Cleaning & disinfection of the trucks and lorries/machinery		
13	Boot tubs		
14	Personal hygiene		

Example protocol for cleaning a cattle house:

1. Remove all manure
2. Soak with water for six hrs under low pressure
3. Apply **cleaning** product (gel-forming alkaline product 3-5% solution; 3L/10 m2 surface) and allow a contact time of 10 to 60 min; apply **disinfectant** (quat and aldehyde based; 0.25 – 0.5% solution; use 3 L/m2 surface) and allow contact time of at least 15 min. Disinfect at least three times a year.
4. Rinse with water under high pressure (50-150 bar; 12-30 L/min)
5. Allow to dry completely

(Note: full use of safety equipments by humans is imperative; observe any safety indications for cattle and other animals)

Cleaning & disinfection

B. van den Brink & M.Stam, VEIP & Dier-en-Arts nr. 10, 2009, pp 310-313; Biotechniek 1, February 2001

Before undertaking any cleaning and disinfection of a barn, the animals must be removed from the animal house.

There are three basic groups of cleaning products available:

1 **Acids** → to eliminate residual urine, calcium and rust. Contact time should be no longer than 20 minutes, otherwise deposit remains or corrosion occurs.

2 **Neutral** → most often used for daily cleaning of floors, walls and equipment. These are mild products for man and material and, therefore, contact time can be longer than 20 minutes.

3 **Alkali** (mild and strong types of products) → to eliminate organic material, such as faeces, food remains, blood, milk and faeces. Products with a high NaOH content (> 5%) are potential more effective but also more aggressive for wood and metal materials; contact time must not exceed 20 minutes. These products are also available as foam, which is practical for walls. In this case, application can be with high or low pressure equipment. Masks and gloves should be worn for protection. It is sensible to alternate between this type of cleaning and an acid cleaning once every two weeks.

Disinfection is meant to lower the bacterial or viral load in houses, surfaces and air. Almost all products are required to be officially licensed for use in practice, the only exemption being products based on **alcohol**. The latter, however, are only useful for **small surface areas**. Table 11.1 details several groups of disinfectants based on their active substance.

It is advised to alternate quaternary ammonia compounds with chloro-compounds or aldehydes in order to prevent resistance development in microorganisms.

Table 11.1 Groups of disinfectants based on their active substances and their efficacy against various pathogens and organisms

Active substance	Bacteria Gram+	Bacteria Gram-	Moulds	Virus lipophilic	Virus hydrophilic	Spores	REMARKS
Alcohol	++	++	+	++	+	--	Small surfaces
Aldehydes	++	++	++	++	++	+/±	Formaldehyde; glutaraldehyde
Chloro compounds	++	++	++	++	++	+	Can be corrosive
Iodine compounds	++	++	+	++	++	±	For skin disinfection
Peroxides and acids	++	++	++	++	++	±	Often used in closed systems
Quartenary ammonia compounds	++	±	+	+	--	--	May cause resistance

++ = good, + = sufficient, ± = moderate to poor; – poor

The cleaning & disinfection cycle

1. Remove the animals from the house
2. Remove the macroscopic dirt
3. Rinse with water
4. Prepare cleaning solution (follow the label for proper dosage)
5. Clean surfaces with a brush and/or low pressure equipment and cleaning solution
6. Rinse with water
7. Prepare disinfection solution (follow the label for proper dosage)
8. Apply disinfecting solution and allow contact time
9. Rinse with water
10. Let the surface(s) air dry completely.

Creating an effective barrier room design

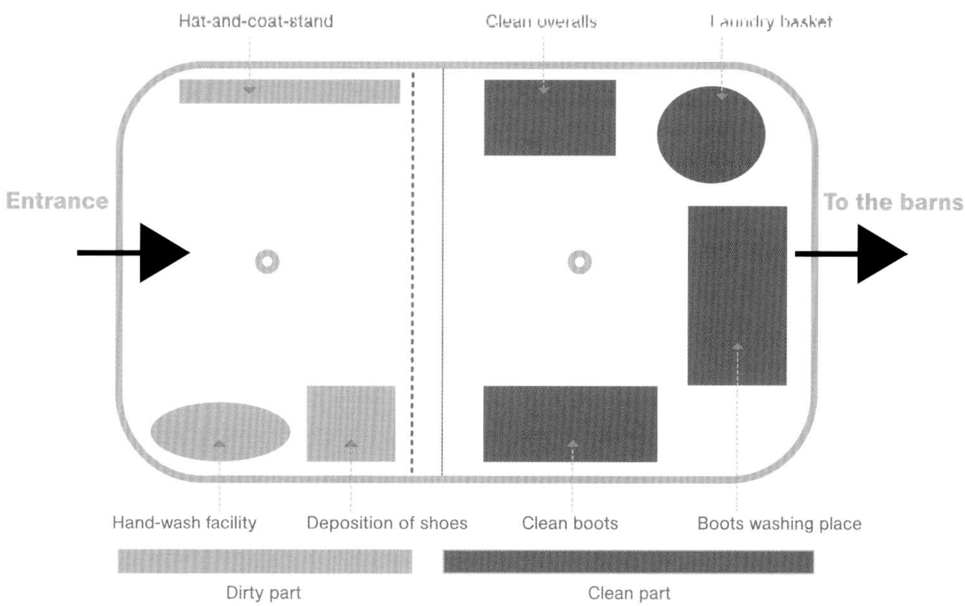

Figure 11.1 Creating and effective barrier - room design

Hygiene & management in the milking parlour

After S. Klimpel & B. Maassen-Francken, GEA Group, Internatl Dairy Topics vol.10, nr.3

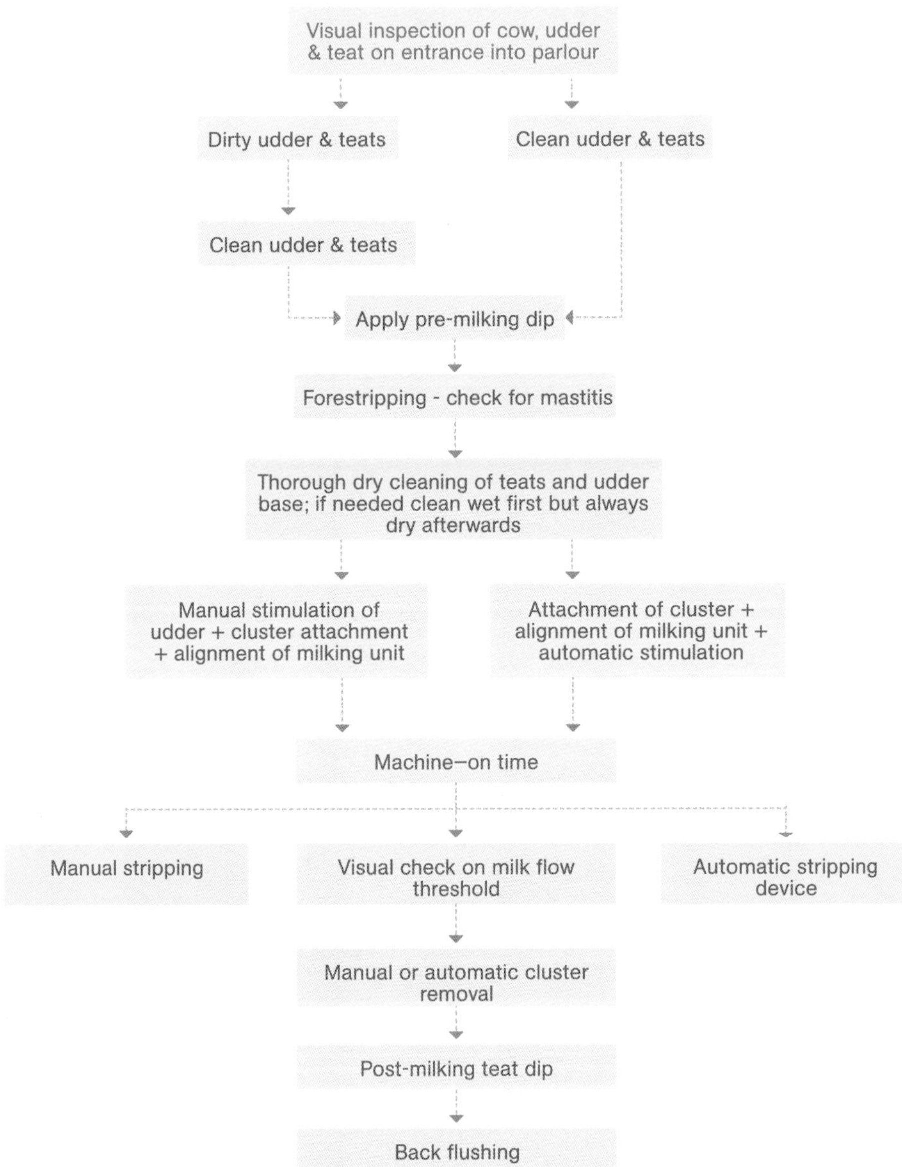

Figure 11.2 Hygicnc and management in the milking parlour

www.dairymaster.com

Management

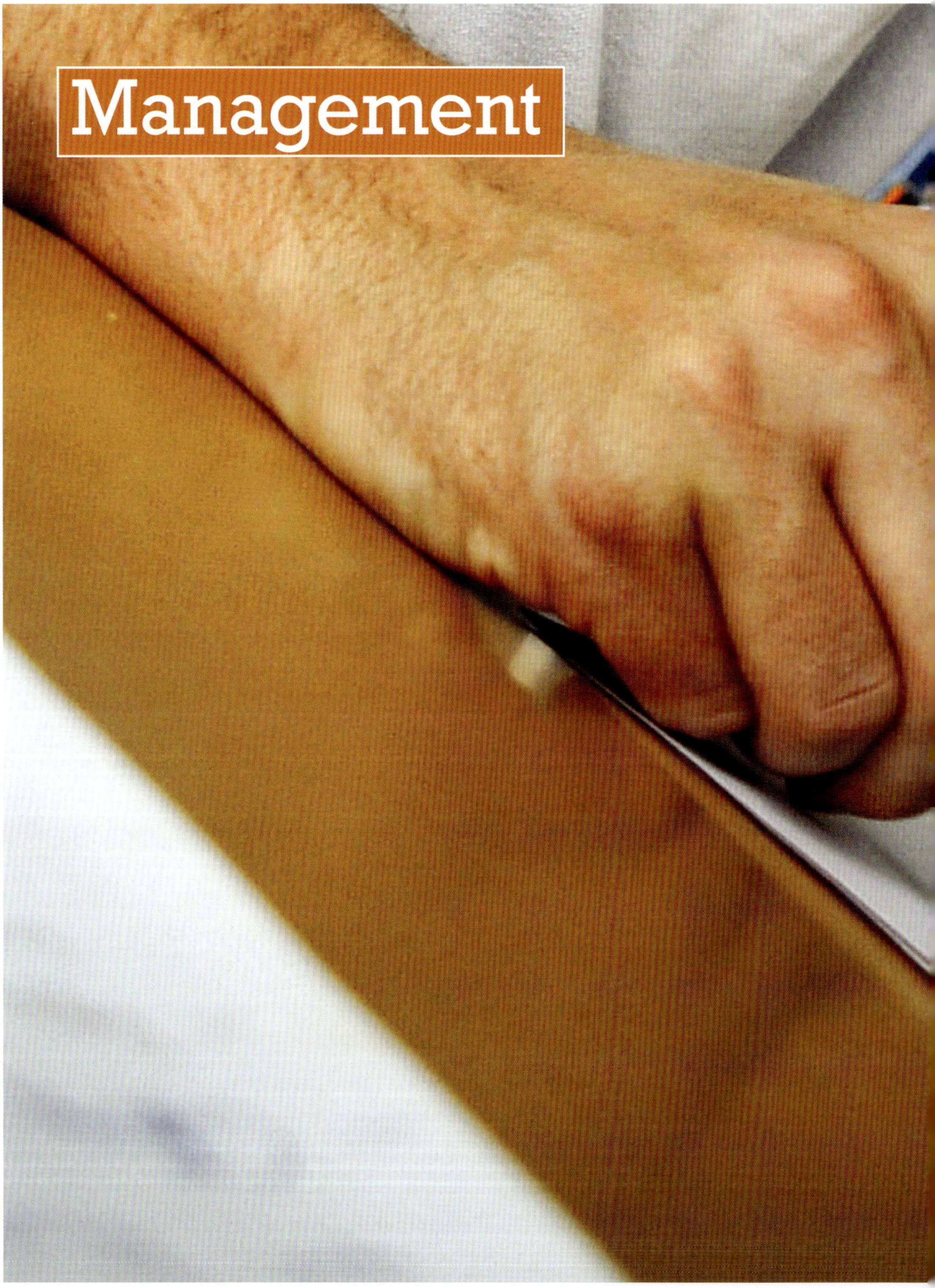

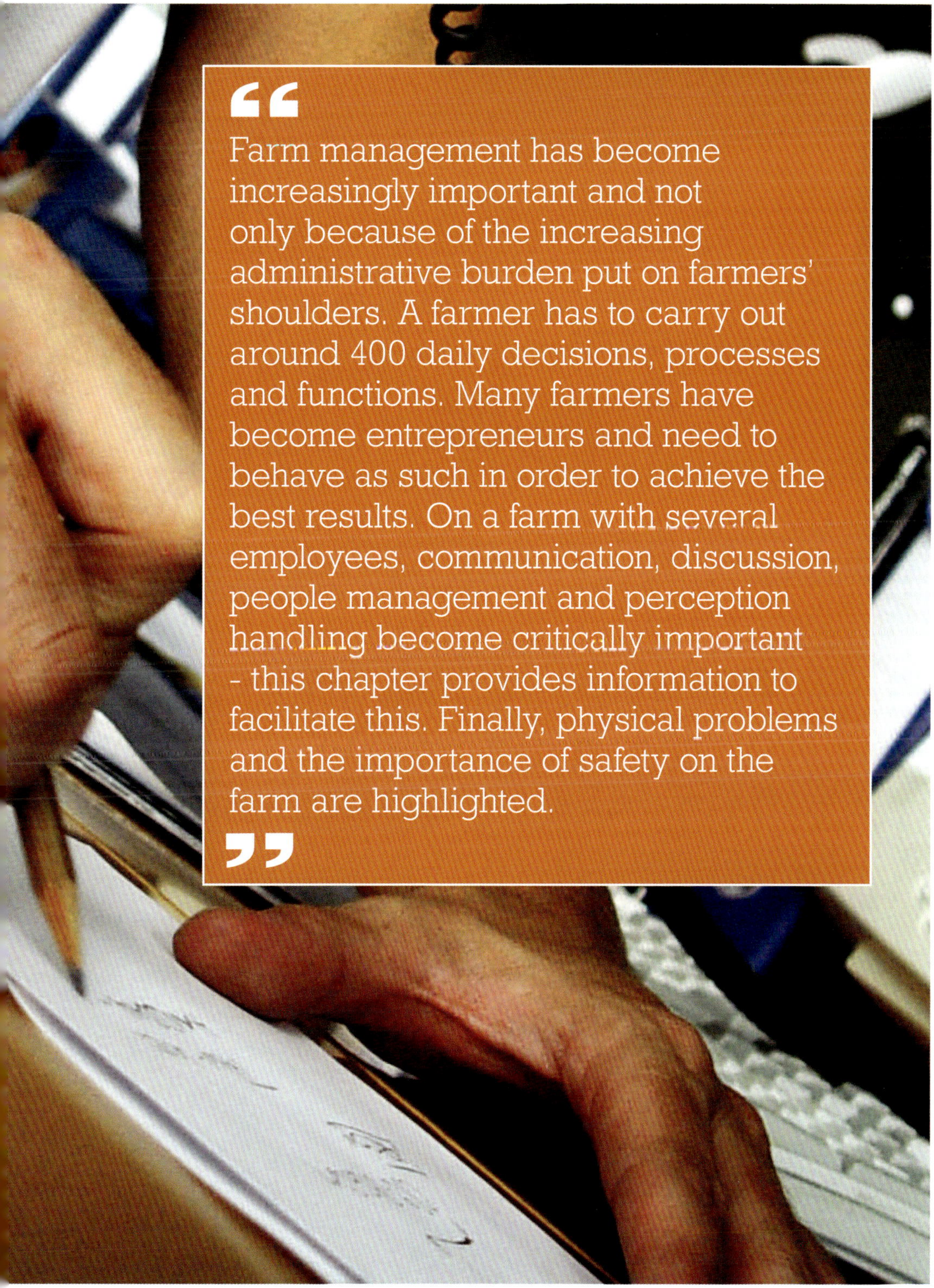

“Farm management has become increasingly important and not only because of the increasing administrative burden put on farmers' shoulders. A farmer has to carry out around 400 daily decisions, processes and functions. Many farmers have become entrepreneurs and need to behave as such in order to achieve the best results. On a farm with several employees, communication, discussion, people management and perception handling become critically important - this chapter provides information to facilitate this. Finally, physical problems and the importance of safety on the farm are highlighted.”

The entrepreneurial wheel

A true entrepreneur has three basic characteristics, which are well-balanced.
DREAM - DO - THINK

The factors contributing to this balance are numerous and they describe the features of the entrepreneur (see below). Commonly, an entrepreneurial farmer will "show" a mixture of these factors.

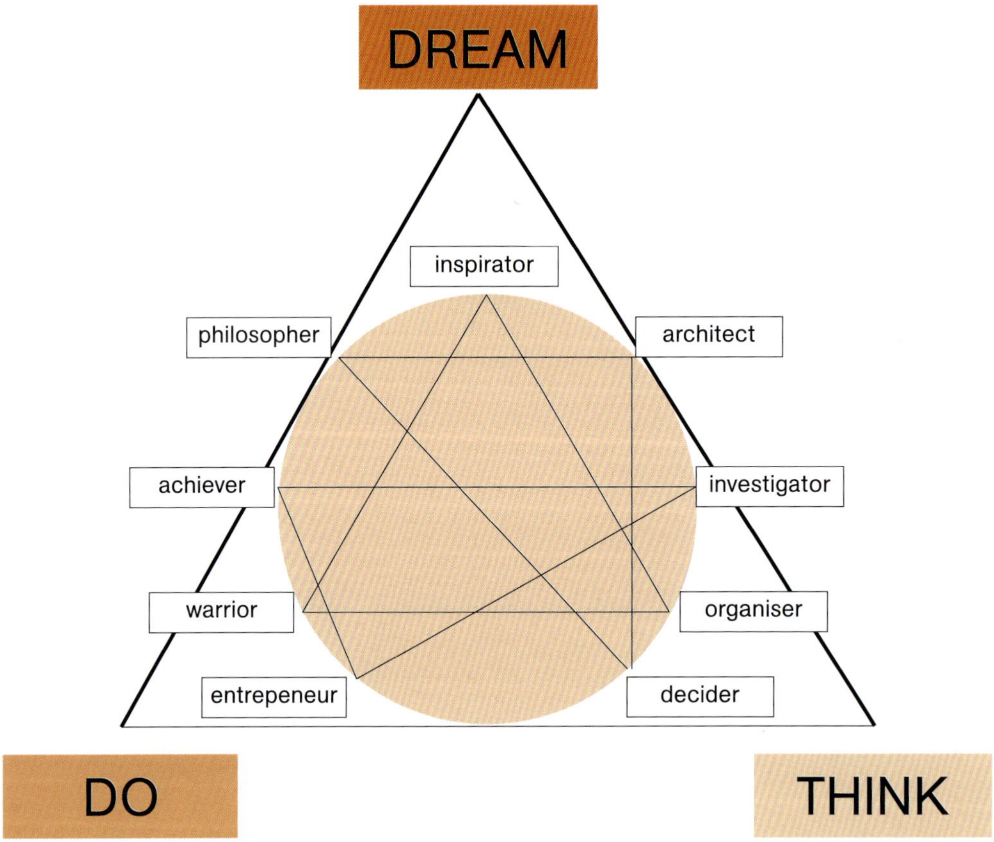

Figure 12.1 The entrepreneurial wheel

Veterinary vs. farmer perceptions

After Lievaart *et al.*, 2003, Tijdschrift voor Diergeneeskunde

Table 12.1 Perceptions of farmers regarding the veterinary practitioners

Strong Points	Weak Points
Competent in veterinary medicine	Too dominant
Often competent in (some) zootechnologies	Doesn't know when/how to listen
Relationship of trust with farmer	Arrogant
Strong veterinary network	Not very proactive
Has many contacts	Not really structured
Reliable	Does not work on planning
Always available, day or night	Knows little about cow nutrition
good at handling calvings & fertility	Knows little about genetics
good at handling emergencies	Knows little about managerial issues
Remains unbiased	Has no (not always) time for discussion
Works on the basis of trust and respect	Does not do economic assessments
Can be a professional partner, close to the farmer	Does not like to write reports
	Does not give working instructions
	Shows too little empathy sometimes
	Communication is often not very professional

Table 12.2 Perceptions of vets regarding the farmers

Strong Points	Weak Points
Knows his farm	Poor at recording data
Knows most often his animals	Has too much information in his head, little written down
Knows how to do his job	Is more of a manager than an entrepreneur
Is often capable to handle most of the 400 functions and processes that are part of the farm	Is not very exact in his answers
Represents a certain socio-economic category	Does not consider the HHPM vet as an investment
Likes to discuss with his veterinarian	Considers veterinary costs as losses
	Sometimes hard to communicate with

Marketing concepts

Molhoek AWI, Endenburg N; 2009; TvD 34 (1) 4-10

Table 12.3 What a client perceives as important in a veterinary clinic.

Degree of importance for the client	Subject	The six P's of the marketing concept	Ranking (order of importance)
90%	Quality of treatment	PRODUCT	2
65%	Diagnostic facilities		3
54%	Specialist service		3
55%	Service 24/24 hr	PLACE	4
75%	Consultation without appointment		7
67%	Distance to clinic		5
55%	Internet site	PROMOTION	6
30%	Articles in a journal		
80%	Experiences of others		9
96%	Communication qualities	PERSONNEL	2
95%	Animal handling		1
80%	Telephone approach		8
	Intervention costs	PRICE	10
	Costs of products		10
95%	Reason to choose this clinic	PRESENTATION	
62%	1* visit for orientation/ familiarisation		

Communication within a team

After Stephan van der Vat, Dier-en-Arts, No. 10, 2008, pp 418—419

People's personalities influence what they take away from communications. This perception is affected by certain filters: **the profiles**. There are six types of profiles that contribute to a **personal profile**.

YELLOW	The one who wants to understand everything; the analyser
GREEN	The one who puts human issues and social interactions first; he seeks harmony
ORANGE	The one who wants to show progress, results, performances
BLUE	The one who wants to create certainty, clarity, concrete issues
RED	The one who claims speed, power, courage, action
VIOLET	The one who seeks security (**home**)

Within each personal profile, there are dominant and recessive profiles; each personality has an individual mix of the profiles. People also develop **aversions** towards certain aspects to their personality. Additionally, the profiles do not tend to change with age.

Cooperation

Cooperation concerns (1) the different profiles, hence the personal profiles, and (2) sharing information.

Profiles.

In a team, differences between profiles and personal profiles can be observed. The motivation for, and relevance of, cooperation are deduced from these personality profiles.

For example:

The YELLOW´s in a team will be stimulated by new ideas; for them a meeting must be stimulating and inspiring.

The GREEN´s will try to avoid conflict in a team; they prefer positive interactions between people.

The ORANGE´s want to see results that are relevant; if they do not occur rapidly they will drop out.

The BLUE´s want to have clear agreements and they drop out when these agreements are not followed or tracked, or when promises are not kept.

The RED´s only want action . . .

while for the VIOLET´s, the mutual trust and confidence in a team are paramount features

In a team, one should have a balanced mixture of (all) colours

Communication within a team

Sharing information

The BLUE's want to know what will be done with the information and whether private or personal matters are taken into consideration and protected. The BLUE's focus on the process, the procedure and the "how".

The ORANGE's will evaluate the information from a point of view of competition and want to see that no one else has access to that information.

Management of personality profiles

YELLOW	He will not easily pronounce to be pro or con cooperation. He wants to know what such cooperation is about and what the rationale behind it is.
GREEN	Sharing is okay; everyone can choose to contribute but must not be forced into it. However, everyone should feel comfortable enough to cooperate.
ORANGE	What's in it for me? What do I gain by cooperation? What could I lose? Is it going to benefit me? Will I be judged?
BLUE	What do they want from me? And what do they want from the others? How will the process be run? Is my private life and quality assured? Who monitor this? What will happen when agreements or promises are not kept?
VIOLET	In the context of trust and confidence everyone does for the other what he needs to do

When you are able to identify the different (most dominant) profiles in a team, you will be much more able to handle and manage that team.

The different profiles require each a different communication approach.

Questions & complaints in veterinary practices

L.van Veen, NL, Nov. 2008 - professional comm.

When dealing with services, it is often not the quality of service or the interventions that determine success but rather the way in which we respond to questions and handle complaints from clients. The latter determines the level of satisfaction.

The number of complaints increases each year, so either the performance of the vets is decreasing in quality or the vets have lost the skills to properly manage client complaints. Maybe the expectations of the client have increased more rapidly than the performance of the veterinary clinics. Interestingly, only20% of clients with a complaint will actually express their dissatisfaction.

Correct management of a complaint results in

- Clients who remain loyal
- Positive marketing effects
- Less sensitivity to costs of treatment/service
- The client wish to buy more than before.

A complaint = each question, related to a certain performance, that the client rises

Complaints must be

- Expressed
- Recorded
- Analysed

Success factors of correct complaint management are:

1 Rapid reaction

2 A good attitude, spoken language, body language

3 Compassion

4 Analysing the complaints in-depth

5 Conduct a follow-up inventory of satisfaction of the client

What does the client with a complaint want?

- To be taken seriously
- To feel they are being heard
- To feel that we understand their problems
- To feel that we understand that they are truly upset
- Sometimes compensation
- A rapid solution for their problems
- A solution tailor-made for them
- That we prevent their problems from occurring in the future
- That we assure them that it will not happen again
- To hear that their complaints lead to a structural change

Some tips for complaint management

- Stay calm and friendly; no emotions try to see it from the client's point of view.
- Try to get the client to say 'yes' to your proposals. Listen carefully; sum-up.
- Show respect for the client's opinion. Do not defend yourself!
- Do not criticize. If mistakes have been made, say so.
- Be honest in saying that client was right to come to you with his complaints.
- Tell him he is valued as a client.
- Look for win–win solutions.

Professional network

(L. van Veen, Dier-en-Arts, no. 3, 2009)

Creating a professional network means creating opportunities. This network serves to obtain additional information from other specialists.

It is not easy for **young veterinarians** to create such a network because, initially, they do not have many contacts. Below are some tips to facilitate a professional network.

Before attending a function, **prepare** with some answers to the following questions: who would I like to meet with and about what do I wish to speak? Is there anyone going to be there who may want to speak with me? Make a **list of the 10 most important subjects** for discussion - they can be used to start a conversation.

It is always useful to **carry business** cards but be selective about who you give them to.

Smile as you enter the room and say hello to people.

If you do not immediately see someone you know, go to the refreshment table , a place most often frequented so provides a good opportunity to talk to someone about trivial things. **Trivial things** are important to start a conversation and should not be viewed as a waste of time at all! It is a good way to create rapport and obtain confidence to move the discussion forward.

The DOs:

1. Listen carefully to what people say to you and respond
2. Raise open questions that cannot be answered with yes or no (questions starting with 'How' or 'How much' or 'Where is... ' etc.)
3. Be curious
4. Look for similarities between you and the person you are talking to
5. Do not talk about irrelevant subjects (with the exception of the trivia mentioned above to open a conversation)
6. Be interested, look people in the eye, do not look away.
7. Without trivial issues, a good relation is impossible
8. Participate in the conversation if you are with several people; be sure to address everyone
9. Do not be too serious and certainly do not play the teacher
10. Focus your attention on the conversation, do not become distracted by other things.

The DON'Ts:

1. Try to sell yourself everywhere (a network is not a market place)
2. Try to put the spotlight on yourself
3. Fail to keep promises
4. Making insincere, superficial remarks
5. Treat people as if they are idiots
6. Distribute business cards everywhere and to everyone
7. Talk too much
8. Becoming over-familiar with someone you don't know very well or have only just met
9. Spread gossip

Finishing a conversation often seems difficult. Chose phrases, such as 'Nice to have met you' before saying goodbye.

You must always follow up as soon as possible! The follow-up must be that you send the information that you promised.

Create a **database** (key words), simple but effective, of people you meet that may be of importance.

Conflicts, coaching & motivation

Adaptation after Ravensberger *et al.*, 1999

1. Restrictive

This is based on fear or punishment.

- **Coercive** I have to do this because ....'
  This does not arrive consciously, and one can observe this in daily routine activities or when there are persistent delays. It is the farmer who would like to participate in a HHPM programme but is not able to execute the tasks he is supposed to do. Without purpose, his actions turn against him.
- Inhibitive 'I cannot do that because something terrible will happen '
  If a farmer says 'if next month the SCC is too high, my vet will criticize me'

  If the HHPM programme encounters such things, you should never expand the HHPM to other domains because that farmer would not cope with more, what he perceives as, criticism . He has to learn to adapt first.

Implementation strategies

There are **approximately** three important strategies for the implementation of programmes, such as HHPM. The choice of each depends on the goal of implementation. Characteristics of strategies are given below.

1. Goal= dissemination & adoption

Credible interventions with clear efficacy and provided with evidence

Not too complex, clear-cut interventions.

Interactive training of people.

Look for opinion leaders in the region and intermediates for meetings and discussions or instructions.

2. Goal = implementation

Present the measures that affect daily activities

Measures that support the execution of interventions (written reports; newsletters)

3. Goal= a certain compensation (satisfaction; honour; money)

Create a good organisation and structure. Visits carried out by others (for example, other specialists) for disseminating information, support and knowledge in the field

NOTE: when a standard HHPM programme is to be implemented, others will try to adapt the programme and design interventions or protocols in their own personal style.

It is futile to fight this unless these people abuse or completely change the concept and principles of the HHPM programme.

(HHPM= veterinary Herd Health & Production Management Advisory Programme

Conflicts, coaching & motivation

Adapted after Ravensberger *et al.*, 1999

During the execution of a HHPM programme, the veterinarian is often confronted by problems at the level of the herd, in areas like animal health, animal welfare, milk production and reproduction. By following appropriate protocols, it is easy to go from one step to the next, tackling the individual problems.

Examples of such steps are: definition of the problem; diagnostic herd evaluation, additional examinations; work hypothesis, interventions and advice. All these steps are fully **technical in nature**.

Advisory work, however, is not always adopted by the farmer. Why would that be so?

In principle, one could speak of a **"conflict"**: the vet wants the farmer to do or change something and the farmer does not want to do it.

In the process of decision-making there are several elements, which play a substantial role: emotions, perception, the current situation as compared with a future situation that is not clear, psychology.

To avoid a "conflict situation", it is best that the vet shows some empathy to the farmer and their problem, that he makes the farmer feel that their problem is his problem too, that it may not be an easy problem to deal with but that he (the vet) will work with the farmer to solve the problem. It is here that the feature of "**coaching**" enters the picture.

In a HHPM programme, the two approaches (technical and coaching) must be implemented at the same time but in an effective way (Figure 12.2).

This applies mainly at the level of farm visits, because it is then that farmer and vet have the most intensive contact and will discuss certain issues or findings.

Definition of problem
↓
Diagnosis
↓
Formulate Options
↓
Evalution of options
↓
Take a decision
↓
Implementation
↓
Evaluation & follow-up
↓
WHAT are we going to do?

Technical content
STATIC APPROACH

Take initiatives
↓
Collect data
↓
Convictions
↓
Manage a conflict
↓
Take a decision
↓
Criticize the decision
↓
HOW are we going to do it?

How to proceed
DYNAMIC APPROACH

Figure 12.2 Questioning techniques

Conflicts, coaching & motivation

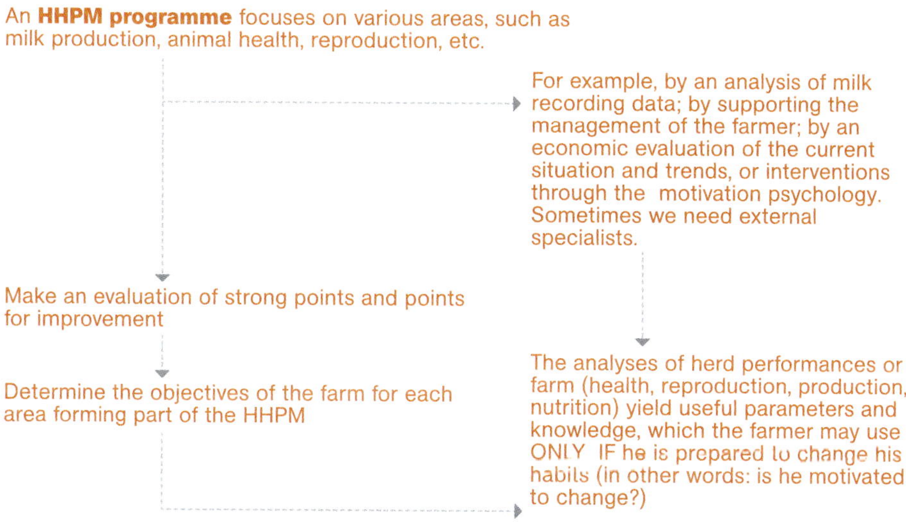

Figure 12.3 Conflicts encountered in herd health programmes: the need for change

To manage this type of discussion (mixture of technical advice + coaching) the vet must know what questions to raise, how to phrase them and when to address them.

An HHPM programme focuses on various areas, such as milk production, animal health, reproduction, etc.

> Motivation = reasoning that is sufficiently strong to provoke certain actions, to provoke a change in management, or to provoke a certain change in behaviour.

Psychology → ***people always try to keep their affairs in order.**

So, motivation means that someone is prepared to change habits. This new equilibrium can be reached through mental pictures.
Two important pictures are:

1 Constructive

2 Restrictive.

1. Constructive

This means 'what would be the benefits of change?' (in money or other compensation)

How does it feel when you achieve something? The difference between the current situation and the objective may be perceived as « hard to achieve »!

- **Incentive** - a motivation based on tangible compensations (money or other)
  \+ Disadvantages: if there is no money to be gained, motivation decreases.
- **Attitude** - a motivation based on self-esteem and approbation.
  +++ This is the best motivation and is sustainable. You often see it among (young) farmers who start on their farm and among entrepreneur-like dairy farmers who know exactly what they want.

Conflicts, coaching & motivation

After Ravensberger *et al.*, 1999

1. Restrictive

This is based on fear or punishment.

- **Coercive** I have to do this because ....'
  This does not arrive consciously, and one can observe this in daily routine activities or when there are persistent delays. It is the farmer who would like to participate in a HHPM programme but is not able to execute the tasks he is supposed to do. Without purpose, his actions turn against him.
- Inhibitive 'I cannot do that because something terrible will happen...'
  If a farmer says 'if next month the SCC is too high, my vet will criticize me'

  If the HHPM programme encounters such things, you should never expand the HHPM to other domains because that farmer would not cope with more, what he perceives as, criticism . He has to learn to adapt first.

Implementation strategies

There are **approximately** three important strategies for the implementation of programmes, such as HHPM. The choice of each depends on the goal of implementation. Characteristics of strategies are given below.

1. Goal= dissemination & adoption

Credible interventions with clear efficacy and provided with evidence

Not too complex, clear-cut interventions.

Interactive training of people.

Look for opinion leaders in the region and intermediates for meetings and discussions or instructions.

2. Goal = implementation

Present the measures that affect daily activities

Measures that support the execution of interventions (written reports; newsletters)

3. Goal= a certain compensation (satisfaction; honour; money)

Create a good organisation and structure. Visits carried out by others (for example, other specialists) for disseminating information, support and knowledge in the field

NOTE: when a standard HHPM programme is to be implemented, others will try to adapt the programme and design interventions or protocols in their own personal style.

It is futile to fight this unless these people abuse or completely change the concept and principles of the HHPM programme.

(HHPM= veterinary Herd Health & Production Management Advisory Programme

The five major conditions for successful certification

(WIAS, Wageningen Agricultural University, The Netherlands, 1996)

1. A high degree of confidence between the parties involved (e.g. the market)

2. A high level of participation from farmers (degree of participation)

3. Liability regulations (insurance)

4. Willingness of the livestock production sector to accept their responsibility within the food production chain (networking)

5. Protection against risk-prone farmers (legislation)

7 Golden management rules

1	Be attentive, visible and conscious;
2	Be intelligent;
3	Be reasonable in your judgments;
4	Be responsible;
5	Be devoted and engaged;
6	Do not forget to laugh and have fun.
7	Show authority if necessary

The management of personnel has two components:

A Technical
B Coaching

The **technical management** is relatively simple because the manager is technically skilled and experienced. However, the employee might be (much) less skilled and in that case he/she has to be approached differently or sent on a technical course. It takes time to acquire skills and the manager has to allow this time!

The **coaching** is (often) more important than the technical component to get the employee to execute things in the way the manager prefers. It has everything to do with creating an appropriate atmosphere and team spirit, next to a personal approach to encourage motivation, engagement and cooperation from the employee. This personal approach implies issues, such as perceptions, emotions, behaviour and spirit of the employee.

An optimal mix between these two components is the key to a successful cooperation. Most often, these two components should be handled in an elastic manner: some days or moments require more emphasis on technical issues, others more on coaching issues.

Perception

Perception is something odd and not absolute. Everyone has their own interpretation and image of the situation! The perception of the environment in which we live imposes a certain perception, which, when we discuss with someone else, is often in contrast to their own perception!

Table 12.4 Perception of personnel

Perception of the manager about personnel	Issue of concern	Perception of the personnel about their own priorities
8	being appreciated	1
10	feels like being part of	2
9	understand job	3
1	salary	5

Ranking order of important issues as given by personnel and as perceived by the manager.

In order to achieve understanding in discussions , the parties need an idea of what and how to other thinks: that is, they must try to find out the image of perception of the other.

A good demonstration of people's different perceptions can be found in the following pictures:

Different perceptions

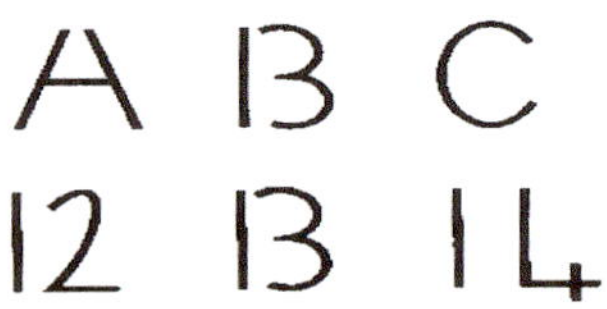

Perception 1

Do you see the figure '13' or the letter 'B'?

The context dictates what we observe, perceive or interpret.

Perception 2

Do you see a white vase or two face profiles

Perception 3

How does person 1 perceive situation S? And how does person 2 perceives it? They probably have different points of view.

Therefore, both persons have to invest in such a way that they will find a common (solid central area in the figure) area or common way of communication about the same issues. If not, they will fail in understanding each other.

The only way for a manager to create a true team spirit, a proper motivation and cooperation among

Management issues

J.van der Kamp, Veeteelt, April 2010, p 40

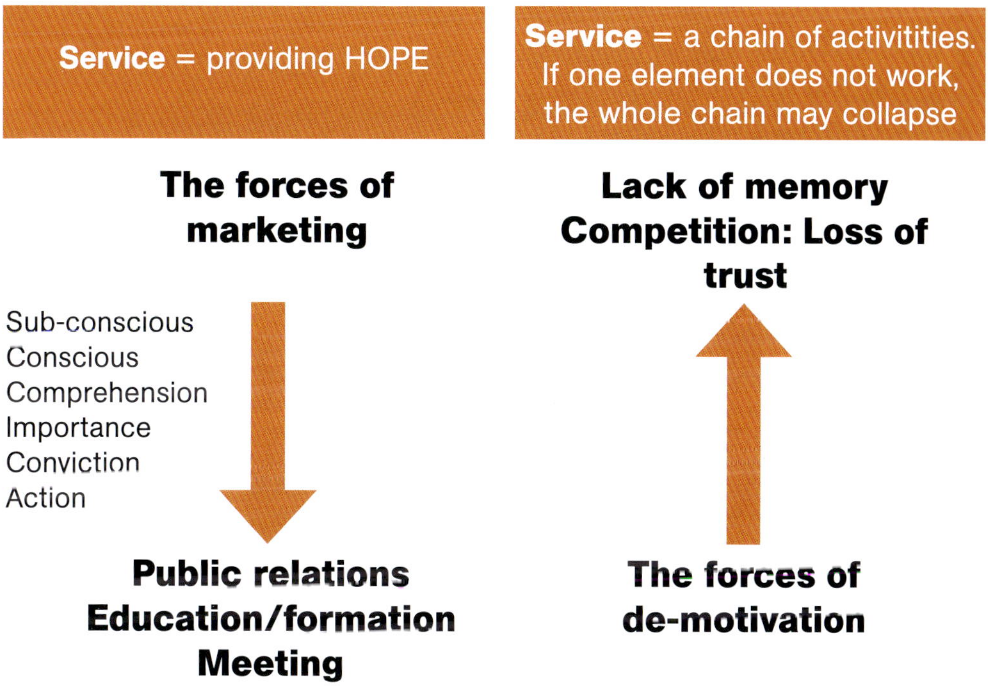

Figure 12.4 Marketing of veterinary services

Marketing in veterinary practises

Mitchell, 1978

Marketing = What does the client want? → Segmentation of wishes!!

Package of activities:

- Which products /services do we propose?
- Who are our clients?
- What methods do we have to contact our clients?
- How can we understand their true needs and their behaviour?
- What are our unique advantages and strong points?
- How do we plan to develop the practice over the next three, five, ten years?
- How do we carry out a comparison between products/services provided, those which seem obsolete and those which seem important for the future, to aid performance development?
- procedure:
  - objectives; material & methods;
  - evaluation criteria; results;
  - our means of control.

Problems in the marketing of services:

- You cannot have tangible ′displays′ with services, but still they can be marketed.
- You cannot show what the service is without carrying it out or give a demonstration.
- You cannot guarantee the results.
- You cannot provide samples of the services.
- You cannot put the service in a nice bag.
- Protection of diplomas and skills (e.g. pharmacy; legal right to exercise veterinary medicine) does not exist for advisory service.
- Quality of the services can be variable.

Clients

- Buy benefits, not characteristics
- Want to be without stress and worries
- Want a quick recovery for their animal
- Focus on their own comfort
- Desire convenience
- Want to be assured
- Want to become more confident
- Want to see good results.

Types of clients:

1. survivors = old people; immigrants; those with few choices; closed
2. companions = materialists; average class people; educated labourers
3. competitors = young people; double income; extroverts; yuppies
4. arrived = enterprising people; stylish; with money 22%
5. academics= pioneer; mentally rich; engaged 0%

The five major conditions for successful certification

1 A high degree of **confidence** between the parties involved (e.g. the market)

2 A high level of **participation** from farmers (degree of participation)

3 **Liability** regulations (insurance)

4 Willingness of the livestock production sector to accept their **responsibility** within the food production chain (networking)

5 **Protection** against risk-prone farmers (legislation)

Scoring management on large-scale dairy farms

After Goodger, Ruppaner, Slenning & Kushman, 1984, J.Dairy Sci. 67: 675-685

Table 12.5 An approach to scoring management on large-scale dairy farms

Management domain	Indicators	Score:	Action:
1. Dairy farm sanitation	Retains sanitary equipment		
	Maintains clean farm areas		
	Buildings are in good repair		
	Standard milk handling procedures		
2. Identification method	Records and authenticates identification		
	ID system is updated routinely		
	Cow numbers are visible from distance		
	Replacement cattle identified on arrival		
3. Heat detection ability	Has knowledge of oestrus signs		
	Heat observation time is scheduled		
	Accuracy of heat detection is evaluated		
	Knows about first post-partum heat		
4. Use of farm records	Current barn sheets present		
	Participates in Milk Recording scheme		
	Utilizes breeding records and sire data		
	Detailed medical records present		
5. Financial practices	Current financial records present		
	Financial management skills present		
	Has knowledge of operating costs		
	Pays attention to budget		
6. Feeding procedures	Utilizes group feeding		
	Weighs feed (components) accurately		
	Consults with nutritionists		
	Checks condition of feed bunk contents		
7. Herdsmen's attitudes	Have a developed 'cow sense'		
	Show responsible job performance		
	Have incentive to learn		
	Have the ability to manage personnel		

Scoring management

Management domain	Indicators	Score:	Action:
8. Drug use	Shows responsible drug & hormone use		
	Drug source is reputable		
	Good condition of drug storage		
	Has a Herd Treatment Advisory Plan		
9. Milking parlour	Plans for expansion exist		
	System cleaning capabilities		
	Efficient use of labour		
	Degree of automation		
10. Herd health management	Raises own replacement heifers		
	Uses scheduled veterinary visits		
	Mastitis is under control		
	Applies Diagnostic Herd Evaluation protocol*		
	Biosecurity Plans are operational		
11. Calf management	Is a good calf manager		
	Calf housing is adequate		
	Shows optimal care in first two weeks after birth		
	Mortality rate is acceptable		
12. Housing design & use	Efficient traffic flow		
	Animals can be segregated		
	Design is focused on individual animal		
	Barn climate and housing suits cow comfort		
13. Heat stress management	Prevention & Control Plan is operational		

This scoring list has been developed following the failure of many studies that tried to explain differences between farms in milk production level or herd health status to include management issues in their analyses, or they have used more qualitative parameters (such as education level; farming experience years; sociological indexes; psychological parameters) to estimate the impact of management.

The current scoring list is estimated to be more objective, comprehensive and specific than others in use. The indicators are all 'observable characteristics of a dairy farm'. The 13 management areas together form the **"management index".**

Scoring management on large-scale dairy farms

Care must be taken when interviewing the farmer or herdsmen, because there could be a considerable bias between what is said and what is actually carried out. A series of two or three scoring exercises over time could (at least partly) overcome this problem. Ideally, observations should be carried out on the farm without the farmer to get an idea of what is being done. Several areas and/or indicators are suitable for this purpose, at least as a control.

In order to calculate the management index, each indicator is scored between 0 and 20. The potential, maximum **score** is 1000 points; the maximum **score** per management area is 80 (except for # 10 with a maximum of 100; and # 13 with a maximum of 20).

In a study reported by Goodger *et al.* (1984) the mean overall score was at 74% (with a range from 60 to 90%).

The **weakest management domains** were:

1. Housing
2. Calf rearing
3. Herd health management.

Given the fact that this fact sheet is called “Approach to….” one should be careful with conclusions drawn from this scoring system, because the goal of this scoring is important: is it for developing future farming strategies or for getting a better insight in stronger and weaker points in management on large-scale dairy farms? There is not enough reliance on the outcome for the former, while the latter provides indications for management changes.

See also at www.vacqa-international.com for a practical scoring method on strong points and weak points on a dairy farm)

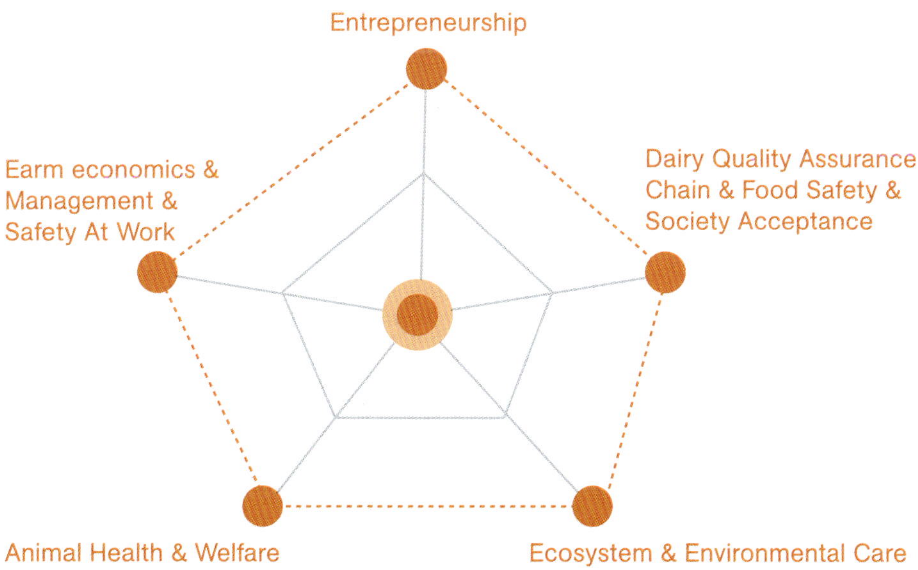

Figure 12.5 Scoring Pentagram

Managing the dry period

Adapted after Ruth Bouwstra & Tine vanWerven, UGCN, The Netherlands in Veeteelt, February 2011, 46-47

Optimal management of the cows entering the dry period and during the dry period is the best investment for safeguarding an optimal start and udder health in the subsequent lactation.

Table 12.6 Ten golden rules for the successful management of the dry period

1	Check the somatic cell count of cows due for drying off. SCC levels > 250.000/ml indicate a likely udder infection; take a milk sample for bacteriological culturing and treat the cow with the most appropriate drug
2	The optimal drying off level of milk production is around 10 kg milk. Above this level, milk leaking often occurs with the risk of infection. To achieve the 10 kg level, an "end of lactation group" should be considered: low-energy ration
3	An excessive BCS (> 3.0 on a 5-point scale) at drying off and a loss of BCS during the dry period by > ½ point will likely enhance a negative energy balance prior to and post-calving. This negative balance will be more severe after calving of those cows. BCS scores should be taken at around 210 days of lactation, at the start of the dry period and about a week prior to calving. Conduct ration-adjustments when indicated by the BCS pattern.
4	Dirty udders increase the mastitis risk in both lactation and the dry period. Conduct a routine Hygiene Score in the herd, on both lactating and dry cows.
5	Prevent, as far as possible, a severe and/or long-lasting negative energy balance, ketosis and milk fever because they induce a loss of appetite, distress, reduced disease resistance. Milk fever cows have a 7-times higher risk of retained afterbirth, a 9-times higher risk of endometritis and an 8-times higher risk of mastitis
6	Prevent, as far as possible, the occurrence of teat-end callosity as it increases the risk of udder infections (namely Staph. aureus). Teat-End Callosity scoring should be done every three months, including the dry cows. Cows with TEC scores ≥ 3 on a 4 point scale at one week before calving have a 6-times higher risk of clinical mastitis. At rates > 15%, a full check-up of milking machine function should be done.
7	About 20 to 25% of cows with mastitis at calving have a poor keratin plug at the teat end. Cows dried off at > 10 kg milk often show a poor function of the keratin plug
8	Cows/heifers with udder oedema at calving have a 65% higher risk of clinical mastitis. This oedema is most often associated with the dry period ration composition.
9	It is advised to dry cows off at seven to eight weeks before their expected calving date (HF breed) in order to achieve the best possible activity of the applied dry-off therapy
10	The best parameter with which to evaluate the dry period management is an optimal start to the lactation. The latter can be assessed by < 12% clinical mastitis in the first 100 days in milk; < 5% retained afterbirth; < 10% high cell count cows in the lactation

Quick scan - General issues of the enterprise

Adapted after Van Eenige, 2007

Table 12.7 Quick scan - general issues of the enterprise

This score sheet can be used for subsequent discussion between farmer and veterinarian. It comprises both the stronger and the weaker points on the dairy farm as highlighted by the previous quick scan.

Topic of concern	Details	Current Score Date:	Weak	Strong	Would you like to know more?	Would you like fact sheets?
Leadership	Clear vision					
	Knowledge & competencies					
	Structural training					
	Attitude toward environment					
	Pro-active in nature conservation					
Strategy & Policy	Farming targets beyond legal ones					
	Business plan					
	Financial (part) plans					
	Availability of risk capital					
	Robustness of cattle, fauna and flora					
Employees	Focused on results in the chain					
	Engaged in the farm business					
	Structural training programs					
	Presence of a safety plan					
	Safe farm processes and working methods					
	Compliance to environmental concern measures					
Means	Cash flow					
	Structural reporting					
	Realization versus Planning					
	Feed & water supply					
	Hygiene in housing					
	Landscape, culture-historical elements					
Processes	Investing in the chain					
	Landscape and farm open to the public					
	Animal Health status					

Quick scan - General issues of the enterprise

Topic of concern	Details	Current Score Date:	Weak	Strong	Would you like to know more?	Would you like fact sheets?
Processes	GMP for feedmill & GMP for transport					
	Disease incidence					
	Responsible use of antibiotics					
	Daily feed and water supply					
	Hygienic and preventive work					
	Contamination (bacterial/chemical/ physical)					
	Intensity of grazing					
	Presentation and appearance					
	Habitats for fauna					
	Buffer zones					
	Ecological linkage areas					
	Responsible use of pesticides					
Improvement & Renewal	Quality (risk) management					
	Pro-active attitude for nature conservation					
	Hygienic and preventive work					
	Financial (part) plans					
	Farm hygiene plan and integral improvement actions					

Physical problems in milking of dairy cows & ergonometry

Alice Booij, Veeteelt 2, March 2011

Milking 100 cows in 90 min = 400 teats to be cleaned + 400 teat cups attached + 400 post-milking disinfection conducted = 13 equal handlings per min. This occurs, on average, twice daily. Reason enough to pay attention to the ergonometry of the milking parlour.

Body parts most frequently affected are:

- Shoulders
- Neck
- Lower back

Some tips:

1 Milking should last no longer than 60 to 90 min.

2 Milking should be a pleasure instead of an obligation; if needed, change the milking conditions.

3 Carry a rubber milking sleeve: with this, the farmer will be more at ease than with an overall.

4 Carry milkers gloves (not least in order to avoid a latex allergy); otherwise use silicone teat cup liners.

5 Adapt the depth of the milking parlour: the elbow should be 20 to 25 cm higher than the platform ridge where the cows stand. Too deep = shoulder + neck pain; too shallow = back pain. A flexible solution (€6000) to parlour depth is an inflatable cushion (such as those found in slaughterhouses), topped with plastic swine stall grids; this can change height by 10-20 cm.

6 Adapt the weight of the milking claw, which should not exceed 2 kg (a blacksmith can alleviate each cup by about 110 g).

7 Install arms to hold the milking cluster; then the farmer can use both hands for attachment.

8 Install proper lighting in the parlour: 10 Watt per m2. With a neon-light of 36 Watt this covers 3 ½ m2. Install them at a height of 250 cm above parlour depth. Keep walls in light colours.

9 If appropriate, distribute the milking routine over more, well-trained persons.

10 Do weekly exercises in a fitness-club or at home to keep body muscle condition in good shape.

General issues of the enterprise - Quick scan

Adapted after Van Eenige, 2007

	Weak →			Strong
1. Entrepreneurship transparent and planned				
I have a clear vision for my farm's future and position				
I have a clear, updated business plan				
I have several detailed sub-plans available				
2. Pro-active entrepreneurial behaviour				
I have insight into my strong and weak points and work on them				
I attend focused training regularly				
I make clear agreements with vets, feed mill, hired labour, others				
I free money and labour easily and invest to improve farm processes				
3. Justified economic decision-making				
I focus on disposable cash (= profit minus depreciation costs)				
I often use financial and non-financial reports for decision-making				
I invest (money) in nature conservation, animal health, environment				
My real benefits and costs approach those that were planned				
4. Progressive entrepreneurship justified for society				
I work regularly to improve (integral) quality				
I have regular contact with third parties around me and make agreements				
I often conduct things beyond legal demands or standards				
5. Dairy Quality Assurance Chain partnership				
I pay attention to, and invest in, achieving and retaining a high animal health status				
I work with protocols toward improvement of feed and water quality				
My feed mill and feed transporters comply with GMP+ regulations				
6. Responsible and food safety associated production				
My performance figures for cattle health continually improve				
My daily feed and water (facilities) quality for cows in milk continually improve				
My daily feed and water (facilities) quality for dry cows continually improve				
My daily feed and water (facilities) quality for young stock continually improve				
Chemical, microbial and physical contamination of feed/water/milk diminishes				
7. Cattle welfare & cow comfort				
My cattle look healthy and clean and are quiet				
My housing facilities contribute to good welfare, comfort and hygiene				

General issues of the enterprise - Quick scan

Adapted after Van Eenige, 2007

	Weak → Strong			
Hygiene on my farm is well-organized and regularly I conduct improvements				
I apply a long grazing period (or: invest in high quality housing facilities)				
8. Improving the ecosystem				
Part of my nature plans and improvement actions are directed to reinforce robustness of cattle, fauna and flora				
I have an updated landscape development plan				
I pay attention to habitats for flora and fauna on/around my farm and carry out improvement measures				
I reduce diseases and plagues by the creation and expansion of buffer zones				
I work towards creating and retaining ecological passage zones				
I apply a restricted use and dosage of fertilizers and pesticides/herbicides				
9. Accessible and landscape harmonious farming				
Farm strategy toward nature conservation				
I try to retain landscape cultural elements on/around farm in shape				
My landscape and farm are (at least partially) open to the general public				
10. Pro-active behaviour as employer				
I include my family members and employees in improving the farm management				
My employees attend regularly focused training courses				
11. Healthy and safe working environment				
I have a safety plan and update it every six months				
My farm processes and working practices have been organized such that they are safe for employees and children				
I apply clear and demonstrable measures for the protection of the environment				
My family members, employees and service people comply to the farm measures for the environmental protection				

This quick scan comprises 11 chapters and totals 41 scoring items. These scoring values can be summarized into a pentagram or hexagram. The larger the surface covered in the graph, the better the overall farm score will be (each axis runs from 0% in the center to 100% in the corners). An acceptable score would be at 60% (see in the graph) or higher at each domain defined. At the same time, points for discussion and for improvement are detected. They can be listed separately and dated. See figure 12.5 for a scoring pentogram.

Quantifying safety risk indicators

Adapted after Risk assess. of safety on animal prod. farms. LEI report 5.02.08; 2002; the .Netherlands

Risk indicators are defined as those factors that assist in detecting (potential) hazards on cattle farms, for example, animal health and welfare issues and food safety-related disorders.

The management of these hazards and risks contribute to the overall safety status of the farm.

Animal health		Specifications	Scoring (1-5 scale*)
	Diseases		
1		IBR status	
2		BVD status	
3		Paratuberculosis status (Johne's)	
4		Salmonella status	
5		Curative medicine use	
6		Somatic Cell Counts	
7		% diseased cattle observed	
	Hygiene		
8		Closed farm features	
9		Young stock housed separated	
10		Overall + boots available	
11		Use of hygiene barrier + tub	
12		Traffic routing on the farm	
13		Rodent control measures	
14		Separate calving pen	
15		Separate sick cattle pen	
16		Quarantine house used	
17		Log book for visitors	
18		Total bacteria count	
	Veterinary medicines		
19		Log book treatment recording	
Food safety 20		Total use of drugs	
21		Withdrawal periods respected	

Quantifying safety risk indicators

`Animal health		Specifications	Scoring (1-5 scale*)
	Bedding, feed and water		
22		GMP for concentrates	
23		Cleaning of silos	
24		Cleaning of forage storage	
25		Covers of the silage clamps	
26		Forage quality (fungi)	
27		Water quality tested	
28		Hygiene of drinking troughs	
29		Dry storage of bedding material	
30		Quality of bedding (fungi)	
	Culled cattle		
31		Culled young stock	
32		Culled adults cattle	
Welfare 33	Age distribution in herd		
34	Grazing > 120 days		
35	Absence slippery floors		
36	Farm typology		
37	Housing system		
38	Cubicle sizes		
39	Lame cattle		
40	Participates in HHPM		
Total Score			……..points

(*1 – good; 3 – moderate; 5 – poor)

Quantifying safety risk indicators on cattle farms

Note that in the scoring sheet the factors for animal health, animal welfare and food safety are often interrelated. Therefore, there is no differentiation between these three aspects.

Ideal score (ideal situation)
= 40 x 1 = 40 points

Moderate score (moderate situation)
= 40 x 3 = 120 points

Poor score (worst situation
= 40 x 5 = 200 points

Graphically represented, traffic light colours, associated with a certain level of points, can be used to highlight problem farms and/or problem areas. The problem areas can also be identified using a bar chart graph.

40 — At the lowest total score (<40) the smallest safety hazard is achieved

120

200

Scoring can be conducted once every three, six or 12 months, in order to detect improvement and/or deterioration of the situation.

For those who want to apply weighting factors to each of the 40 indicators, it is stipulated that they must be robust enough to balance out chance variation or subjectivity.

Food safety is predominantly influenced by animal disease scores, veterinary medicines scores and, to a lesser extent, feed/water/bedding, culling and hygiene.

Animal welfare is predominantly influenced by welfare scores, culling, feed/water and animal diseases.

Animal health is largely influenced by diseases, medicines, feed/water and culling.

Those who desire a more specific and detailed scoring of the above are referred to the website www.vacqa-international.com

© www.lely.com

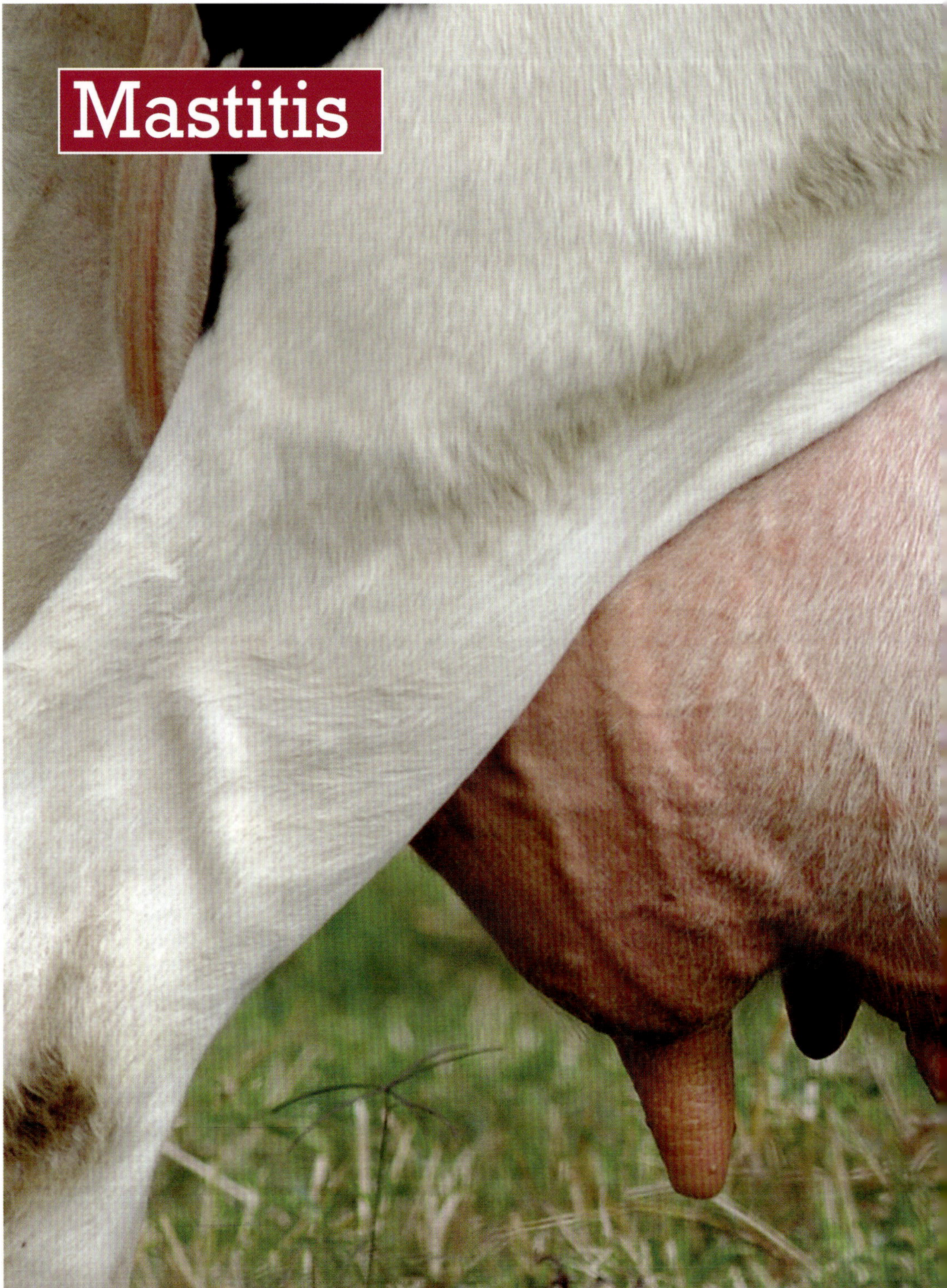
Mastitis

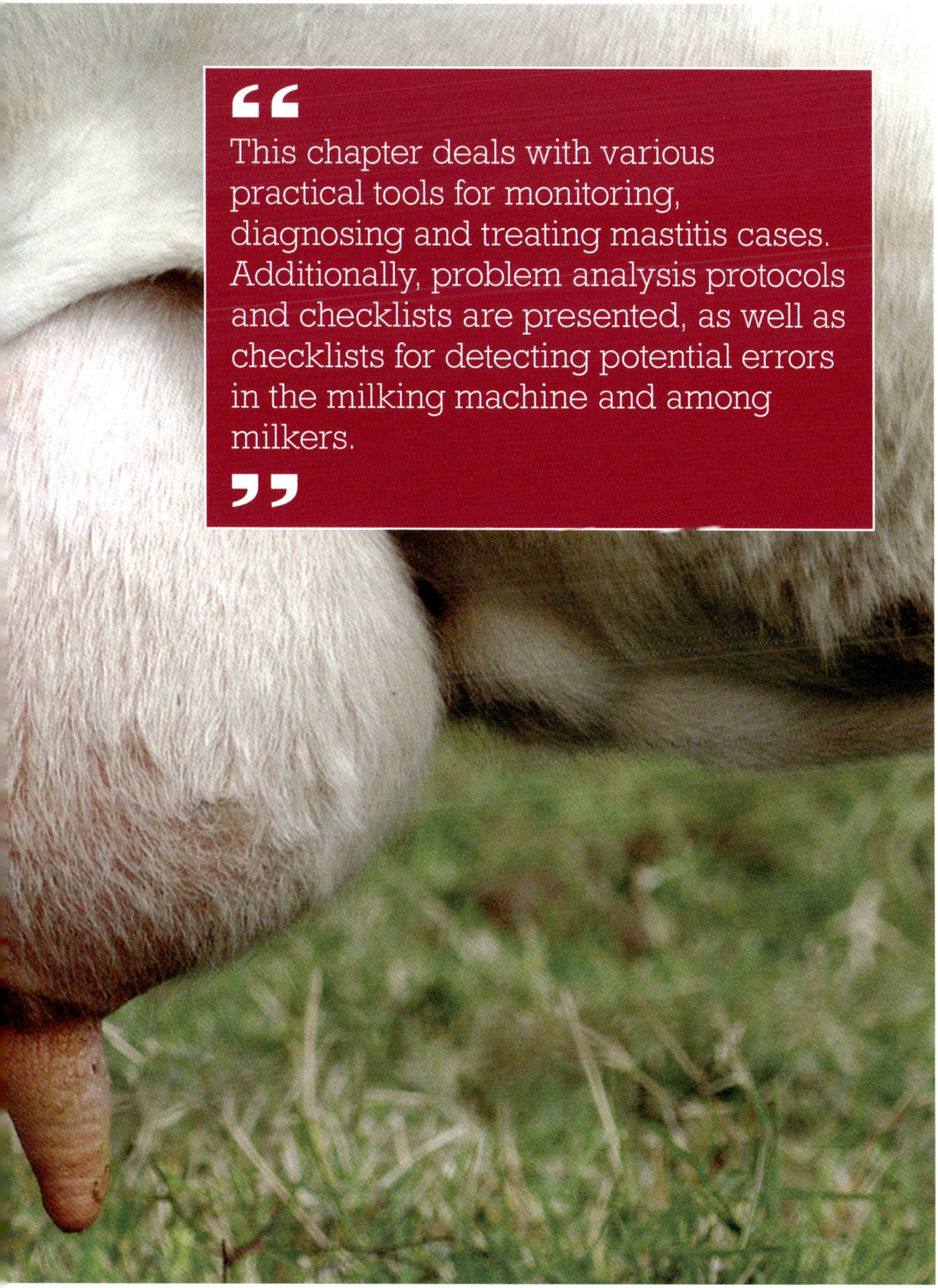

> "This chapter deals with various practical tools for monitoring, diagnosing and treating mastitis cases. Additionally, problem analysis protocols and checklists are presented, as well as checklists for detecting potential errors in the milking machine and among milkers."

Teat end callosity scoring

Teat-end callosity can be scored in a very detailed manner (Neijenhuijs *et al.*, 2001 & 2005), a less detailed practical manner (with a 1-5 scale) and a simple practical manner (with a 1-3 scale).

The latter is sufficient for a Diagnostic Herd Evaluation. It is always possible to go into more detail once a problem has been detected.

Teat-end callosity is associated with an **increased risk of mastitis**, in particular *Staph. aureus* mastitis.

There are two criteria for determining teat-end callosity: a ring and thickening of tissue:

- The ring can be smooth or frayed
- The thickening can be present or absent

The scoring method has been described above, with lowest scores (1-2) being preferred and highest scores (3-4) being undesirable.

See also the paper by F. Neijenhuijs *et al.* (2005) in the proceedings of the 4th IDF congress in Maastricht, Holland, 11-16 June 2005, pp 376-382 ; or their paper in the Journal of dairy Science 84 : 2664-2672 (2001)

Some remarks

After milking the teat must leave the teat liner and cluster in a nearly dry condition.

If the teat is wet, there is a problem with the speed at which the milk is transported to the milk container. The teats end up being "washed" in their own milk. In this situation, bacteria can easily travel from one quarter to another.

Score 1 **Score 2** **Score 3** **Score 4** **Score 5**

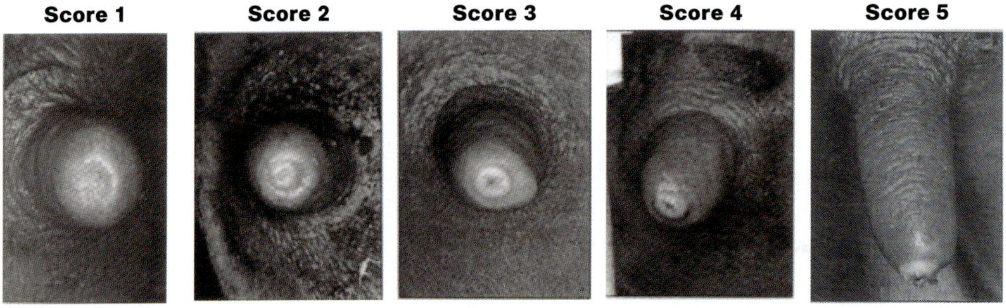

Score 1: This is the perfect score. No edema after milking; a round smooth ring. An indication for the least lesions caused by the milking machine.

Score 2: The first signs of callosity are the round but thick rings around the teat base. The teat end is a little swollen.

Score 3: The teat end is swollen ; the ring is round but irregular and thick.

Score 4: At the teat end one can observe the round thick ring, very edematous and eversed. There are not yet looking like warts.

Score 5: This score is given to cases which look and feel like warts. The teat end is eversed and frayed.

Score 1 **Score 2** **Score 3**

Figure 13.1 Teat end callosity scoring

Teat end callosity

Solution:

- Install milk reception containers in the milking parlour at each post
- Install claws with a milk volume that is much larger (eg 100 or 200 ml)
- Use other installation measures for rapid milk transportation in the system.

A (sometimes blue) ring is visible at the teat base after milking.

During milking, the teat liner can crawl up the teat towards the base, for example when wet udder prepration has been carreid out and teats are wet when the cluster is attached. Another causal factor may be teat liners, which are too wide, long or short, or poor functioning of pulsators, too long a duration of milking, or a too high a vacuum. Cows with blue rings are often poorly milked. First lactation heifers and cows in early lactation are most frequently affected.

Flat teats.

This teat has been blocked in the teat liner, which can be painful for the cow. The cause is a too long D phase (empty phase) of the pulsator, worn out teat liners, or teat liners, which are too hard or wide.

Petechia on the teat skin.

These little haemorrhages are caused by too high a vacuum under the teats or teat liners that are too large.

and finally...

The skin of the teat must feel flexible, without cracks or lesions. This can be achieved by applying good cosmetic teat-care products and good housing (cow comfort).

Taking milk samples

Only bacteriology, carried out on milk samples taken from cows with udder infections, can give information about causative pathogens causative of mastitis in dairy farms. A bacteriological profile at herd level is very useful for a more effective mastitis treatment and, therefore, more economical. This applies to both new and recurrent infections.

A milk sample must be taken before any antibiotic treatment.

1. Write the ID or name of the cow to be sampled, the quarter and the date on the label with a waterproof pencil
2. Take a clean towel and clean the teats and teat ends of one (1) cow
3. Eliminate the first milk streams
4. Disinfect the teat orifice (teat end) with 80% alcohol
5. Again, eliminate some milk streams; take the cap off the sampling container in using an aseptic technique
6. Milk a few milk streams into the sampling container but do not fill above ¾; tilt the sampling container while milking to avoid contamination with dirt
7. Put the cap of the sampling container back on using an aseptic technique and seal it
8. Write the ID or name of the cow, the quarter and the date on the label with a waterproof pencil
9. Put the sampling container in the freezer until the next farm visit from the vet. Or send it to the laboratory in the appropriate manner

California mastitis test

The CMT can give you an indication about the health status of the udder quarters.

Moreover, the test is cheap and easy to perform. A T-pol solution is mixed with an equal volume of milk sample and gently shaken. The solution will become immediately viscous if the somatic cell count (SCC) of the milk sample is higher than 400.000/ml; the more cells are present, the thicker the mix will be. The CMT can be performed at the individual quarter level and at the cow level when a composite milk sample is taken. Note that *Staphylococcus aureus* does not always yield increased somatic cell count levels and, hence, not necessarily a positive CMT result despite the fact that these bacteria are present in the milk.

Procedure:

1. Clean the 4-quarter milk plate thoroughly prior to use.
2. Clean and dry the udder; discard the first milk streams.
3. Position the milk plate well under the cow (respective quarters are indicated on the plate) and take two streams of milk from suspect quarters onto the plate in the corresponding hole.
4. Add the same volume of T-pol solution to that hole and start shaking the plate for 10 sec.
5. Fully evaluate the CMT result for that quarter (see below). If more than one quarter is suspected, fill the other holes in the plate accordingly.

Interpretation:

- Normal milk = the mix of milk and T-pol remains fluid.
- Score 1 = the mix is less fluid and more mucous SCC > 400.000 cells/ml.
- Score 2 = the mix is quite creamy SCC > 750.000 cells/ml.
- Score 3 = the mix is very viscous SCC > 1 million cells/ml.

Picture of a 4-quarter milk plate for CMT:

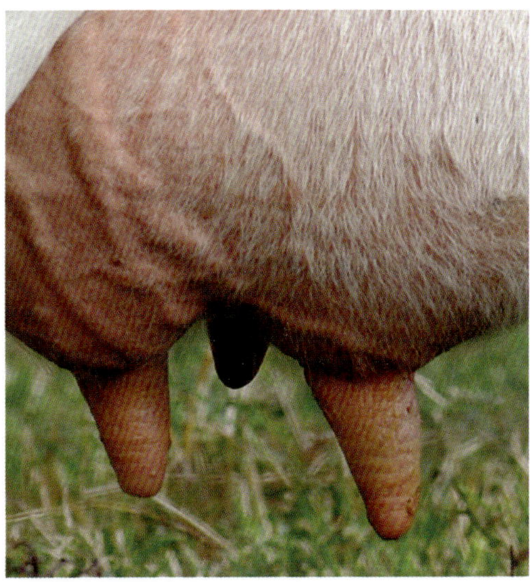

Herd treatment advisory plan

Farm code = **Date =**

Description of the mastitis	Suggested treatment (injectors – injections)	Withdrawal period milk	Withdrawal periodmeat
Mild case - clots in the milk, cow is not sick, cow eats normally; possible slight decrease in milk yield.			
Moderate case - cow is sick (fever), the affected quarter may be hard, swollen and warm; cow eats less. The cow produces less milk.			
Severe case - very sick cow, high fever (cold extremities), cow is dull. Cow does not eat. Milk contains clots and/or is yellowish. Milk yield is reduced.	Call your veterinarian!!		

Note:

Always complete any treatment as prescribed (e.g. three or five days)

Herd treatment advisory plan 1

Remember that optimal farm **management in the dry period** is crucial for subsequent udder health. On average, only 67% of cows that have been dried off with a cell count of > 400.000/ml will recover.

At the same time, 45% of cows dried off with a cell count < 400.000/ml enter the next lactation with an already high cell count.

It is crucial to know the bacteriological profile of udder pathogens in the herd.

Not all cows need treatment; culling of chronic cases is a good alternative.

Subclinical cases may warrant early treatment (Table 13.1).

One of the critical issues in udder health is the **detection of cows with new infections** (or with increased somatic cell count > 300.000/ml) for the first time.

Strict monitoring of such cows can lead to early detection and rapid treatment.

Table 13.1 Guide to when to treat lactating and/or dry cows suffering from mastitis (after Tischer, 2010)

Treat lactating cows when the following pathogens are implicated....	Treat dry cows when the following pathogens are implicated....
Str. agalactiae *Str. canis* *St. aureus* (new infection; start of lactation; heifers)	*Str. uberis* *St. aureus* (cows with > 1 quarter infected)
Why? Good recovery rates Infections are contagious No further spread of infection induced	Why? Good recovery rates Sole effective therapy in most St. aureus affected cows

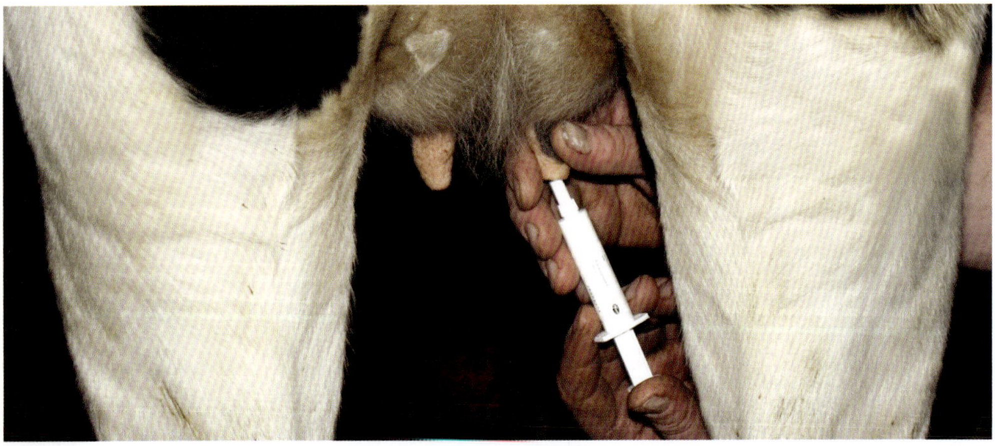

Herd treatment advisory plan 2

Major pathogen present : Staph. Aureus ß-lactamases, penicillin resistant, spectinomycin, +/- ampicillin.

Table 13.2 Clinical Mastitis (3 choices in order of importance)

	AB Intramammary	WT mi	WT be	+ AB injectable	WT mi	WT be
1.	Ampiclox® :1 tube, morning and evening, at least 3 times	4 m	7 d	Tylan® : 40 ml IM first, and 30 ml next 2 days	6 m	14 d
2.	Cobactan LC®: 1 tube, morning and evening, for at least 3 milkings (up to 6)	10 m	4 d	Cobactan® : 20 ml per 500 kg, IM, for 3 to 5 days	1 m	5 d
3.	Synulox intramammary® : idem	4 m	7 d	Amphoprim® : 35 ml, IM, for 3 days	4 m	5 d

WT= withdrawal time for milk (mi) or beef (be); m= milkings; d= days; SCC= somatic cell count
Subclinical Mastitis: SCC > 250 000/ml for several milk recording dates
Treatment in lactation or at an earlier drying off if close to end lactation
Treatment:
1 Ampiclox® +/- Tylan® injectable
2 Rilexine® treatment : 1 tube, after 4 subsequent milkings (WT : 7d, 6 m)
3 Pirsue® : 1 tube, 1x/d, for 8 days (WT : 23 days, 10 milkings)

Table 13.3 Drying off (systematic treatment of all cows!)

	Treatment	WT mi	WT be	In which case ?
1	Rilexine HL® intramammary	0 / 14 d	0	Cows with low SCC
2	Cobactan DC® or Fatrox® 18 t intramammary	36 d ap ttt 0 /18 m	2 d 0	Cows with higher SCC but still < 300
3	Idem + Tylan® injectable (50 ml, IM, 2x at 3 days interval)	6 m	14 d	Cows with high SCC (>300 at both last milk recording dates)

SCC= somatic cell count
Culling:
Cows that are a major source of infection in the herd and assessed as « incurable » are the cows that:
1 The oldest
2 Have been affected by several infected quarters
3 Are mid- or end-lactation, and have shown:

SCC > 600.000 or 800.000/ml (sometimes repeatedly) these cows are to be culled as a priority!
The same is valid for cows which are persistently high in SCC, in spite of treatment, and for cows that calve with high SCC again in spite of dry off treatments.

Sub-clinical mastitis treatment

Pharmacia Animal Health 2003

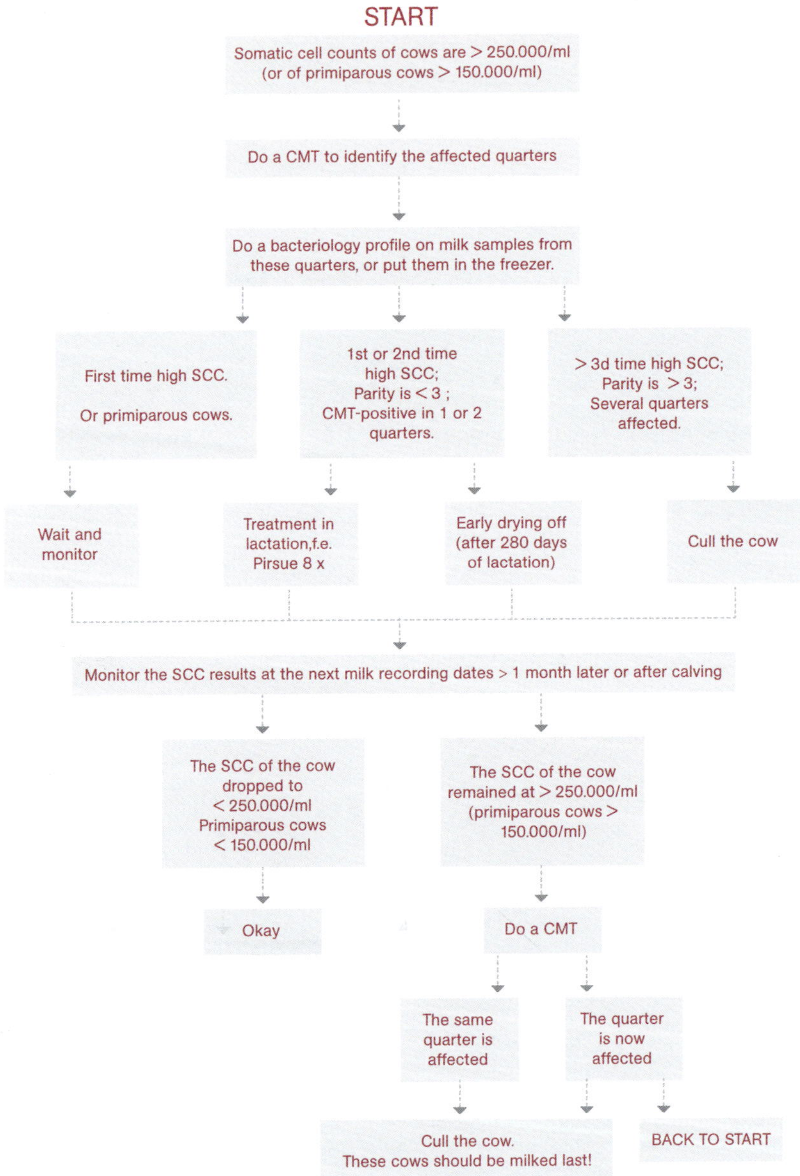

Figure 13.2 Treatment decision scheme for sub-clinical mastitis.

Mastitis treatment

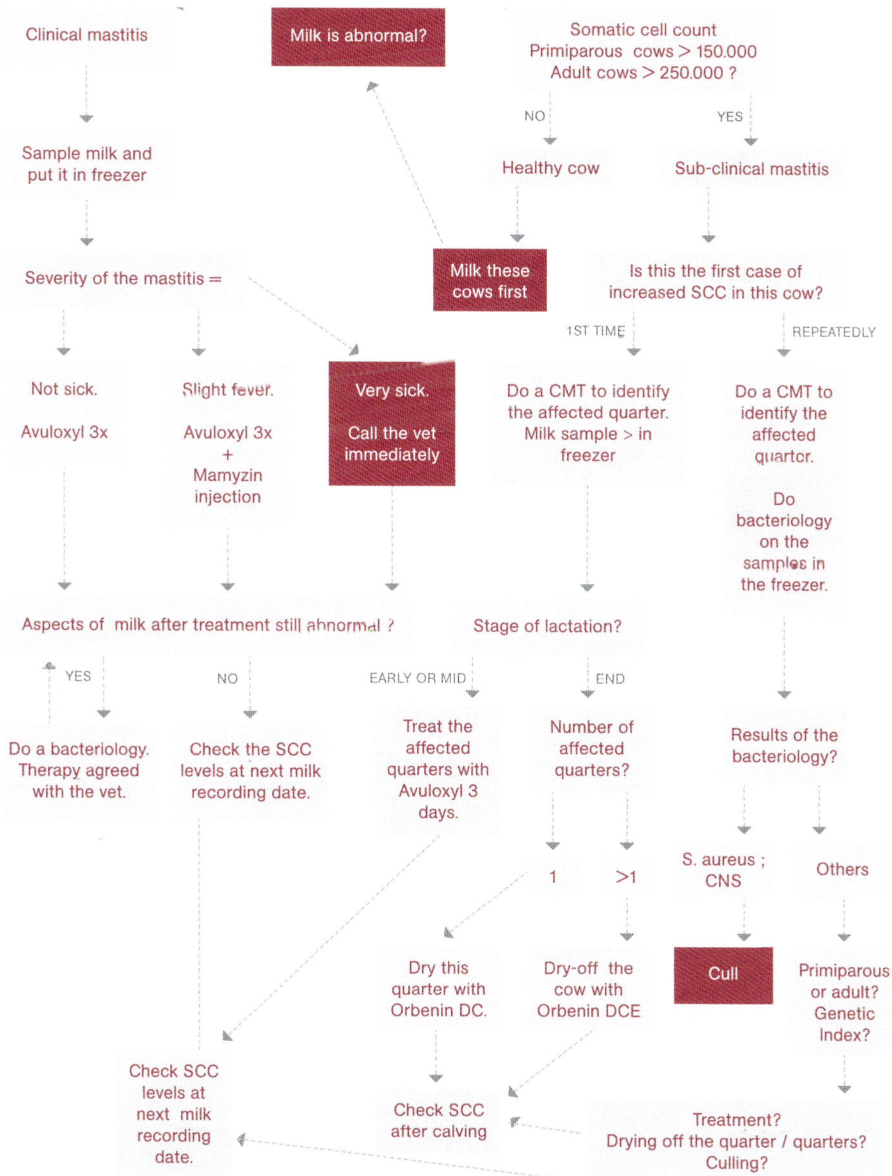

Figure 13.3 Advisory plan for the treatment of mastitis (+high SCC) at herd level

Formularium - dairy cows

Formularium 1

During lactation and at drying off →

Table 13.4 *Staph. aureus* clinical and sub-clinical mastitis:

	Injectables		Intramammary
1	Penethamate HJ	1	Cloxacilline/ampicilline Benzylpenicilline/ nafcilline/ dihydrostreptomycine
2	Erythromycine or Tylosine **2	2	Amoxycilline/clavulane Lincomycine/ neomycine
3	Trimethoprim-sulpha	3	Cefoperazone Cefquinone

The injectables and intramammary therapies are ranked in order of preference – 1 being the preferred choice, followed by 2 etc.
** = IM injections may provoke tissue lesions!!

Formularium 2 antibiotics of choice for *Staph. aureus* mastitis (2008)

Clinical mastitis (sensitive to Penicillin)

- Cobactan udder injector; to be combined with Cobactan or Mamyzin as injection fluid.
- Pirsue udder injector; to be combined with Trimethoprim-sulpha or Erythromycin injection fluid.
- Pathozone udder injector; to be combined with Cobactan or Mamyzin injection fluid.

Clinical mastitis (Penicillin-resistant)

- Cobactan udder injector; to be combined with Cobactan injections.
- Pirsue udder injector; to be combined with Trimethoprim-sulpha or Erythromycin injection.
- Pathozone udder injector; to be combined with Cobactan injection.

Sub-clinical mastitis and/or at drying off

- Pirsue udder injector, combined with Trimethoprim-sulpha or Erythromycin injection.
- Dihydrostreptomycin, combined with Cobactan or Mamyzin.

Biosecurity protocol for contagious mastitis pathogens

After htpp://nmconline.org/transl/NMCchecklistNA_fr.pdf & www.mastitisnetwork.org

General prevention measures dealing with transmission	Tick if applicable
Cows with a confirmed infection and cows with a persistently high SCC (> 200.000/ml) are segregated from herd mates for few months in order to reduce transmission probabilities	
Infected cows are segregated from herd mates for milking (separate units are used, or infected cows are milked last)	
Cows with chronic *Staph. aureus* infection and cows with other pathogens, which do not respond to treatment *(Pseudomonas, Arcanobacterium, Nocardia, mycoplasma)* must be culled or permanently isolated	
Check udder health status of first lactation cows (CMT; bacteriology)	
Implement a management practice to prevent maiden heifers raised together to suckle one another	
Do not feed infected or potentially infected (high SCC) milk to calves	

Prevention measures for the sale or purchase of cows	Tick if applicable
Check individual cows' SCC data and bulk tank SCC data at the seller's office. Check thoroughly (CMT; bacteriology) the individual quarters before purchase.	
Get aseptically collected milk samples from all quarters for bacteriological culture before purchase.	
Quarantine recently purchased cows upon their arrival and milk them separately until confirmation of absence of udder infection	
When purchasing lactating cows, first conduct a test for antibiotic residues. In the case of dry cows and late pregnant heifers, make sure that the seller writes down all treatments given to the cows (product; date; dosage; withdrawal period; etc) and test the milk after calving for presence of antibiotic residues.	

Note: testing for antibiotic residues is not necessary when the seller writes and signs a certificate that the animal has not been treated with antibiotics for at least the last 60 days prior to this sale, or is free from antibiotic residues, or, if treated, the appropriate withdrawal period has been observed.

Analysis of mastitis & high cell counts

Protocol for a structured analysis or high cell count problems

It is essential to establish the bacteriological cause(s). For example, *Staph. aureus*.

Cows

Establish the severity of the situation:

- How many cows are affected?
- For how long have they been affected?
- What are the persistent infections?
- What are the new infections?
- How many quarters are involved?
- What is the incidence of clinical mastitis?
- What is the incidence of sub-clinical mastitis?
- What is the incidence of latent infections?
- What effects have been noted (production; fertility; health; other)?

Carry out a diagnostic herd evaluation (see elsewhere in this series) to include:

- General condition: body condition score (BCS), rumen fill (RF), faecal consistency (FC), undigested fibre in the faeces (UFF), coat, skin
- Health of claws and legs (hocks); lameness prevalence.
- Hygiene & cleanliness of cows - thighs + udder.
- Teats: callosity scores; blue or red at teat baserings; depigmentation; lesions.

Milking Machine

Carry out a dynamic (= wet) test twice a year to ensure correct functioning

Look for risk indicators.

Practice & Hygiene of milking

Conduct an interval study during milking on a sample of cows. Inspect teats and udder immediately after milking (lesions, deformations, decolourisations, etc). Wet udder preparation must be followed by drying with absorbent paper; check the number of towels used. Look for risk indicators.

Housing, barn climate and other issues

Look for risk indicators (insufficient manure scraping; straw yard is dirty; relative air humidity in barn is too high; ventilation is poor; etc)

Antibiotic Treatments

- Lactation
  - * short duration (3-5 days; normal treatment duration)
  - * long duration (8 days)
- Drying off

Efficacy of treatments:

- Find out the cell counts before and after treatment
- Determine the level of production before and after treatment
- Estimate the volume of milk lost (litres)
- Calculate costs minus benefits.

Strategy for culling

- What criteria are used?
- Which cows appear on the farmer's list for culling and why?
- Which cows should be on that list for udder health reasons?

Mastitis, cell counts & environmental bacteria

How does the approach to mastitis and or cell count problems differ when environmental bacteria are suspected?

There are several different types of environmental pathogens:

- *E. coli*
- *Klebsiella spp*
- *Strep. uberis*
- Certain sub-types of *Staph. aureus*! (no practical test for these)

Mastitis cases caused by environmental bacteria are most often related to a **decrease in immune-responsiveness** of the cows. This decrease can be associated with a negative energy balance (NEB), of short or long duration and varying in severity, or ketosis in the herd.

Other stress factors can also contribute to this lowered immune-responsiveness:

- Social stress
- Stress related to poor housing
- Stress related to poor barn climate
- Stress caused by the behaviour of farmer or employees
- Infectious diseases.

In short, all factors that negatively affect cow comfort may cause decreased immune-responsiveness in cows. Risk factors associated with udder infections by *Streptococci* or *Staphylococci* (and associated with transmission and contamination) are less important for infections by environmental pathogens. This means that one should look for factors, which can contribute to this decrease in immune-responsiveness, that is, factors at cow level and at environment level and less at contamination and transmission, such as the milking machine.

> ***General hygiene standards, hygiene in the barn and at milking, in the exercise area and personal hygiene must be respected.***

On farms where *Staph. aureus* is an issue, problems can be of an infectious and/or environmental nature, making treatment more complicated.

Mastitis problems protocol for analysis

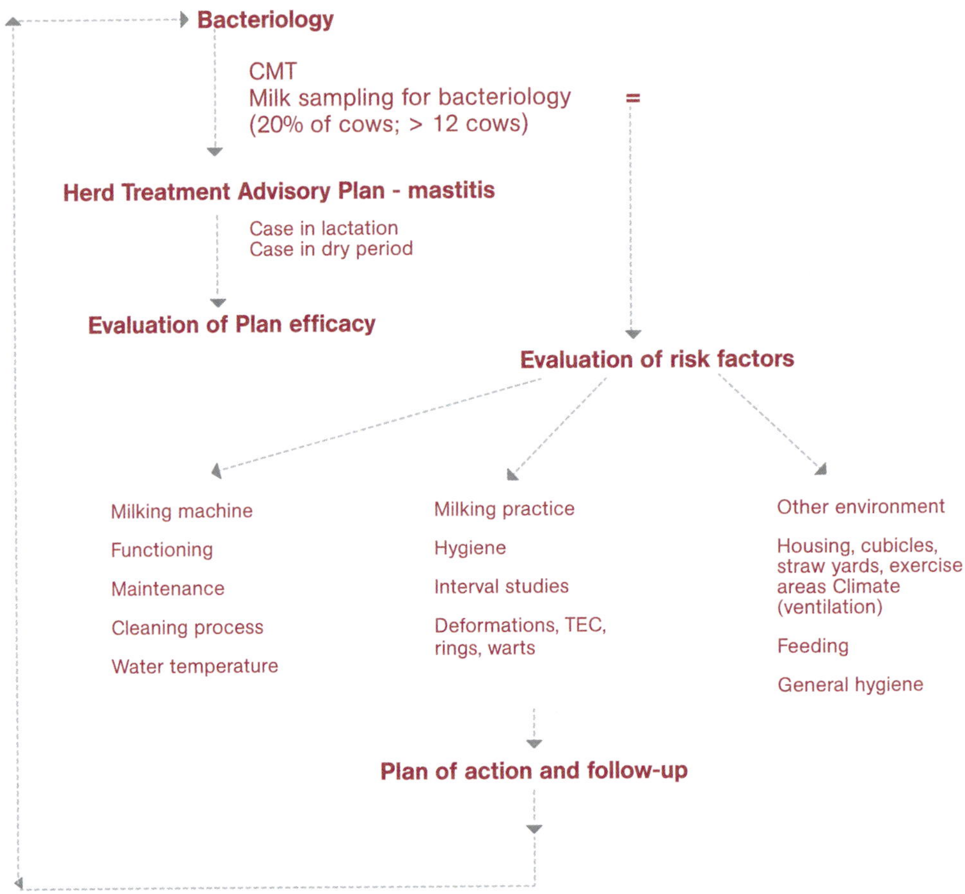

Figure 13.4 Protocol for analysis of mastitis problems

Staph. aureus mastitis & recommendations

Table 13.5 Risk factors for *Staph. aureus* mastitis and recommendations

The herd (principal reservoir)	Practice of and hygiene at milking	Milking machine	Others
Culling Treatment in lactation (early choice) Routine treatment at drying off → cf see Protocol, for reducing infection pressure Detect carrier cows and/ or affected quarters (latent cases) Teat health (attention to cracks, lesions, warts)	Wear gloves during milking; clean them between cows; avoid touching the quarters unnecessarily (contamination by hands of milker) Dry teats with paper towels after proper cleaning with individual towel before the cluster attachment Milk first streams (on CMT plate) to look for mastitis signs (clots) Increase the duration of udder preparation to 60–90 sec: cluster attachment at 60–90 sec after stimulation Avoid air entrance during cluster attachment (phenomenon of air impact: contamination of other quarters) Post-milking teat dip with disinfectant, cleaning of cup after infected cow; do not dip up to teat base Cleaning/disinfection of clusters after an infected cows has passed milking, or determine a milking order (mastitic cows to be milked last) Timely detachment: avoid over-milking (lesion of sphincter); set threshold at 200 or 300 or even 400 gr/min	Change teat liners every year or after 2500 milkings (otherwise (accumulation of pathogens in the cracks; loss of flexibility) Regular maintenance and proper setting of vacuum for a less traumatic milking. Dynamic evaluation of machine function twice yearly (subscription with dealer) Cleaning procedure of the machine: check a whole cycle; check of hot water temperature at the start and end of the cycle	Feeding Barn climate Cow environment Housing Cleanliness Cow comfort Avoid all types of stress →to achieve a maximum immunity

Coagulase-negative *Staphylococci*

After Otlis Sampimon, 22 December 2009, Doctoral thesis at University Utrecht

Coagulase-negative *Staphylococci* (CNS) are increasingly cultured from milk of cows with clinical and sub-clinical mastitis and are associated with a high somatic cell count (SCC) in milk. These bacteria are particularly found on farms with an average bulk milk tank SCC of less than 250.000/ml. In a study in The Netherlands comprising 50 dairy herds, a CNS prevalence of 11 percent over all quarters was found and no herd was without CNS. The rate of CNS was higher in lactating heifers than in adult cows (see also CNS in heifers).

A mastitis control programme based on the classic five-point scheme is ineffective for CNS. It is probably that the environment plays a more substantial role in the epidemiology of CNS than previously thought, although the exact mechanisms have not been elucidated thus far.

Primary risk factors for CNS are:

- Drinking water in troughs not linked to the mains water system
- Dry cows housed in one group instead of two groups
- Individual cell count tests each month (this factor is possibly hiding something else)
- Check of udder quarters by the veterinarian (Irritation of tissue? Contamination?)
- Grazing during summer season and/or the presence of flies (Haematobia irritans)
- The proportion of contaminated cubicles
- Bulk milk tank SCC higher than 250.000/ml.

Secondary risk factors are:

- Poor hygiene in the houses
- Poor fly control measures
- Heifers suckling each other
- Milk leaking in the cubicles

For a proper CNS diagnosis, the phenotypic API Staph. ID32 or the StaphZym test may be used but these are generally unsatisfactory due to their low sensitivity and poor identification capacities.

The SCC for quarters without bacteria is lower than the SCC for quarters with CNS and the latter is usually lower than the SCC for quarters with mastitis. In most cases, the quarters with CNS should be treated with antibiotics (*Cloxacillin 600 mg*). Coagulase-negative *Staphylococci* show greater resistance against antibiotics than *Staph. aureus.*

Many heifers are already infected with CNS or with pathogens at the time of calving. Treatment of heifers at 8 to 10 weeks before expected calving date resulted in a substantially reduced load of minor pathogens and true pathogens compared with non-treated heifers (the authors). The heifers treated during their dry period showed a lowered risk for mastitis in lactation than non-treated heifers (9 versus 18 percent), as well as a lower SCC (55.000 versus 71.000/ml for treated vs. non-treated, respectively). The milk yield in treated heifers was around 1 L/day more than in non-treated heifers. Although pre-calving treatment of heifers demonstrated positive results, it must be emphasized that one should look for the causes of these CNS and design an appropriate control programme.

Coagulase-negative *Staphylococci* in heifers

Study of the Department of Reproduction, Obstetrics & Herd Health, University of Gent, Belgium

In around 80% of heifers, bacteria can be found in one or more udder quarters around their first calving and approximately 4% of heifers showed clinical mastitis at calving. The CNS bacteria represent ~85% of the bacteria found in sub-clinical mastitis cases, the other infections include *Staph. aureus* and *Streptococci (agalactiae; dysgalactiae)*. *Staph. aureus* and the *Streptococci* spp. named above are the most pathogenic, while the CNS are less pathogenic.

Somatic cell counts of heifers with pathogenic udder bacteria tend to be higher than those of CNS heifers during lactation.

On day 100 of lactation, there were ~18% clinically infected heifers of the 20% originally non-infected heifers, while there were 32% with clinical mastitis from the previously infected group. Interestingly, the CNS heifers contributed only 2% of the infections. By day 200, these percentages were 18%, 34% and 4%, respectively.

CNS heifers had a lower culling rate (1%) than heifers with a case of clinical mastitis (5%), or healthy heifers (3%) in the first lactation.

The milk production level did not significantly differ between non-infected and CNS heifers.

About 10% of heifers make a poor start to lactation with a case of clinical mastitis. The CNS infections are frequent but there appears little cause for great concern in these cases.

The average somatic cell count of ≥ 150.000/ml in these heifers should be considered as an indication of udder health problems. Ideally, milk samples should be taken from these heifers immediately after calving to determine the bacteria profile and design a more accurate Herd Treatment Advisory Plan for mastitis, as well as a Herd Mastitis Prevention Plan.

See also www.m-team.ugent.be

Action when *Klebsiella spp.* are found in milk

Otlis Sampimon, GD Deventer, The Netherlands, 2010

Course of action when Klebsiella spp are found in the bulk milk tank.

The *Klebsiella* bacteria are classed under the Gram-negative bacteria, like *E. coli, Serratia, Proteus.* These bacteria can be found on the skin, in the intestines and in the environment (soil; bedding material). The most important *Klebsiella* spp. are oxytoca and *pneumonia*. These bacteria can be the cause of a very severe clinical mastitis but can also cause sub-clinical mastitis.

The mode of infection

The teats can be contaminated by soil, manure, contaminated sawdust, especially under warm, humid weather conditions. Even with good udder preparation and cleaning practices, these bacteria may still be present at the teat orifice. Another route of infection is during milking when one quarter transfers infection to another through the teat liner and milk flow (teat washing). *Klebsiella* can also be passed on via the cubicles or straw yard.

If *Klebsiella* spp are found in the bulk milk tank, then action must be taken:

- Make an inventory of the severe cases of clinical mastitis in the herd
- Take milk samples from quarters with severe mastitis for culturing
- Sample cows with elevated SCC for the first time from the DHIA records
- Check the soil and bedding of cubicles and the conditions of the straw yard; ask whether changes have been made. Check the calving pen and sick cow pen.
- Check the hygiene standards in the cubicles, straw yard, exercise area, milking parlour
- If in doubt, take samples from soil and or areas for bacteriological culturing
- If in doubt, take samples from the bedding material for culturing; if positive cease using that material and change storage location.
- Conduct a strict examination of recently purchased cows and heifers
- If necessary, adjust the Herd Mastitis Treatment Advisory Plan (see elsewhere in this series).

Once the **source** of infection/contamination has been found, it is not sufficient to simply eliminate that source: all areas the material has been in contact with must be cleaned and disinfected (cubicles; straw yard; sawdust stocking; milk parlour; passageways and areas).

Moreover, the *Klebsiella* **carrier cows** must be identified within the next month by sampling and culturing the high cell count cows in the herd. These carrier cows must be culled because they usually do not respond to treatment.

For treatment of cows at **dry-off** in a *Klebsiella* herd, a product active against Gram-negative bacteria should be used. Often, after treatment, affected cows retain high somatic cell counts.

Milking machine and/or personnel problems

Table 13.6 Checklist to detect milking machine and/or personnel problems

Problem	Checkpoints for milking machine	Checkpoints for personnel	Other potential causes
Teats are not in good condition		Water to clean udder and teats is too hot. Teat disinfectant is too aggressive.	Sun Weather Photosensitivity skin Formalin
Teats are blue and/or hard	Vacuum too high. Ratio Pulsation to Rest phase is poor. Malfunction of pulsators. Teat liners are worn.	Duration of milking is too long.	teats are too long First lactation heifers
Callosity of teat-ends is severe	Vacuum too high Malfunction of pulsators Teat liners are worn or too wide Ratio Pulsation to Rest phase is poor.	Duration of milking is too long.	
Infections in teats (bacteriological or viral)		Vector Hygiene Disinfectant	New cows have been purchased
Teat lesions			Age, udder conformation ; housing, claw health.
Mastitis	Substantial vacuum fluctuation Tissue lesions due to vacuum or pulsation or worn teat liners. Poor cleanliness during milking	Duration of milking is too long. Personnel is brutal. Poor udder/teat preparation (1 towel for all ; wet ; no first stream checks ; interval until attachment is too long >90 sec ; interval until attachment is too short <60 sec.) Intervals during milking too irregular.	Other known risk factors

Milking machine and/or personnel problems

Problem	Checkpoints for milking machine	Checkpoints for personnel	Other potential causes
Somatic cell count is too high	See known risk factors	Adaptation of milking practice ?	Other known risk factors
Too many bacteria in the milk	Cleaning of the machine after milking is poor. The water temperature is too low at start (<80°C) and/or at end of cycle (<40 °C). Machine parts are worn.	A non-hygienic milking.	Temperature of bulk tank is too high (>4°C). Cleaning of tank is insufficient.
Loss of milk production	Insufficient milking: Vacuum; ratio; pulsation; teat liners.	Poor udder/teat preparation (variable; disturbed routines; too many different people).	Feeding Weather Genetics Housing Climate etc.

Odour-based diagnosis

Kasper HETTINGA, 14th January 2009

The diagnosis of mastitis based on the odour profile of the milk

Quality control of raw cows´milk by headspace analysis – a new approach to mastitis diagnosis

- In normal cow's milk of high quality, seven different odours may be detected.
- An increased level of bacteria in the milk is associated with a specific milk odour profile.
- Chemical residues originating from cleaning products for the milk tank also cause a specific milk odour profile.
- In case of mastitis, milk odour profiles that are specific to the different pathogen species can be identified.
- The origin of these odours is the bacteria and not the milk.
- On the basis of these profiles, it is possible to predict which bacteria are causing the mastitis.
- Incubation of milk samples for 8 hrs at 37°C is sufficient to achieve a reliable diagnosis using odour profiling. This process is obviously much more rapid than bacterial culturing.

(« patent pending » for this diagnostic method)
PhD thesis, University of Wageningen, Holland, Department of Sciences of Milk products.

Efficiency & profitability in the milking parlour

(Neil Birkett, Int. Dairy Topics, vol 9 nr 3, 2010)

Improving efficiency: & profitability of time spent in the milking parlour

The milking parlour is where milk quality management takes place. An important part of that management is the availability of protocols and the execution of appropriate routines. The veterinarian/coach/consultant should not only assess what is happening in the parlour but also what happens before and after milking.

Mastitis management forms part of managing milking.

This management comprises, among other factors, the reduction of new udder infections:

- Reduce the contamination of teats between milkings → straw yards and/or cubicles must be clean and comfortable; exercise areas must be clean and dry; lighting in the barn must be optimal; the feed bunk must have space for every cow.
- Avoid bacteria entering the teat canal before milking → disinfection of teats before milking; good udder preparation practice and subsequent drying; sufficient time between preparation and attachment of clusters (60-90 sec; see Figures 13.5 and 13.6).
- Minimise the time spent milking per cow → improve the udder preparation; avoid milking on a empty quarter; check the threshold of detachment for automatic cluster removal (400-600 ml/min).
- Make efforts to keep teats clean and their skin healthy → disinfection of teats after milking and application of a product, which protects the skin.

The practice of milking must always be the same (same order; same activities; same person; same treatment method) because cows are creatures of habit. They should be collected in the waiting area without any stress, irrespective of the methods used to move them.

Evaluation of the milking practice and management of cows during milking must be part of a herd health management programme.

No udder preparation or poor preparation appear to cause few problems initially as the udder cisterns are full with milk and that milk enters the container. However, after that phase, there is very little milk flow. The full oxytocin response takes around 60 to 90 sec after the first stimulation to occur, so good udder preparation pays off here:

- duration of milking is reduced by 0.6 min
- milk yield increases by 0.32 L (milk flows more rapidly)
- there are fewer udder tissue lesions.

Cleaning the teats must be carried out in a **circular motion**, finishing up at the bottom of the teat. Cleaning the teat orifice is a good way to stimulate oxytocin production and, moreover, contributes to the reduction in hyperkeratosis (**teat-end callosity**).

Pre-dipping followed by drying reduces the number of environmental bacteria by 85% compared with a wet udder preparation or a spray. Pre-dipping alone can reduce bacterial load by 46% compared with using cleaning towels only.

Gloves should ALWAYS be worn during milking.

Post-milking teat dipping must be done immediately after cluster detachment as the teat skin will become too dry.

The cows should leave the parlour, again, without

Good udder preparation will reduce milking time!!

any stress and move to exercise areas that are clean and dry.

Additional advice for achieving good milk quality:

- Make sure that cows are always clean and in a dry, clean and comfortable environment.
- Make sure that teats are always clean and dry and well-stimulated before attachment.
- Use only authorised germicide products for post-milking teat dipping.
- Check the functioning of the milking machine several times per year; check operational machine issues daily.
- Use appropriate products for drying off as indicated by the herd bacterial profile.
- Chronic mastitic cows must be kept separated from healthy herd mates and preferably culled.
- High somatic cell count cows must be identified and kept under surveillance and, if necessary, treated or culled.

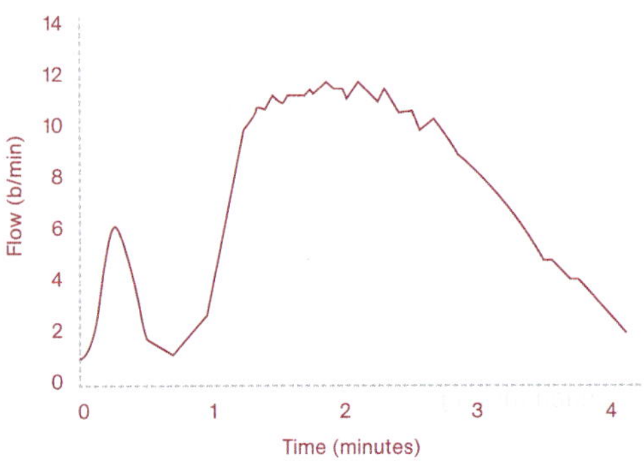

Figure 13.5 Example of a bimodal graph of milk flow caused by no pre-milking preparation.

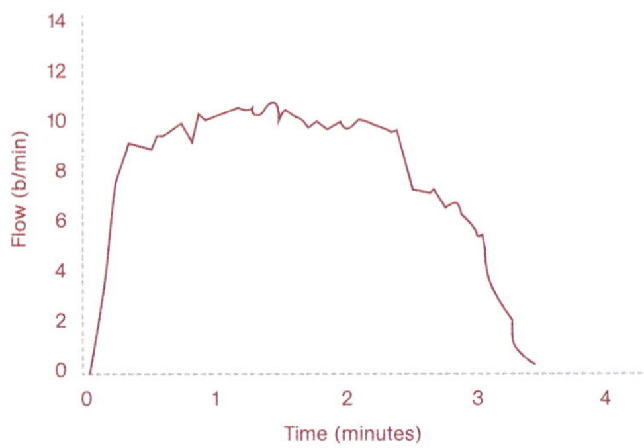

Figure 13.6 Example of an uni-modal graph of milk flow caused by a good pre-milking preparation (interval between first stimulation and attachment is between entre 60 and 90 sec) ; note that the milking duration is reduced.

Mastitis in first-lactation heifers & associated economic losses

(Huijps, de Vlieger & Hogeveen. 2007)

A clinical case of mastitis costs ~€209 on average, or €14per average heifer present (ranging from €0 to €180) based on average herd size and calving patterns.

- Milk production loss = €68
- Non-deliverable milk = €17
- Premature culling = €118

A sub-clinical case of mastitis costs ~€25 per average heifer present (ranging from €0 to €250)

- Premature culling = €20
- Milk production loss = €3
- Other costs €2

Total economic losses per average heifer are estimated at €55 with a range from €1 – €321 in an average dairy herd
(Proceedings Heifer Mastitis Conference, Gent, Belgium, June 24-26 (de Vlieger, editor)

Recommended mastitis control programme

Source: www.nmconline.org

1. Establishment of Goals for Udder Health

- Set realistic targets for average herd somatic cell count(SCC) or linear score and clinical mastitis rate.
- Review goals on a timely basis, with input from theHerd Udder Health Advisory Team (veterinarian,producer, herd manager, milking personnel and advisors).
- Prioritise management changes to achieve stated goals.

2. Maintenance of a Clean, Dry, Comfortable Environment

- Ensure proper stall usage by ensuring adequacy of stall size and design.
- Maintain clean, dry, and comfortable stalls through appropriate bedding management.
- Keep cow lots or housing and traffic areas clean and dry.
- Ensure ventilation system is functioning properly
- Ensure proper stocking density in facilities.
- Control detrimental environmental influences (heat stress, frostbite, stray voltage, insects etc.).
- Ensure that cows remain standing after milking (provide fresh feed and water).

3. Proper Milking Procedures

- Examine foremilk to facilitate early detection of clinical mastitis and proper milk letdown.
- Apply pre-milking teat disinfectant that completely covers the teat skin and allow it to remain on teats for at least 30 seconds.
- Dry teats using a properly washed and disinfected cloth towel for use on one cow, or a single service paper towel.
- Wear clean gloves during the milking process to limit spread of contagious pathogens.
- Attach teat cups squarely and level with the udder within 90 seconds of udder preparation
- Adjust cluster during milking to prevent liner slips and squawks.
- With manual removal, avoid machine stripping and shut off vacuum to the claw before removing cluster.
- Apply teat disinfectant immediately following teat cup removal, and assure complete coverage of teats.
- Pre- and post-milking teat disinfectants should be selected based on documented efficacy data which can be found on the NMC website (www.nmconline.org)
- To optimize mastitis control and reduce costs, teat dipping is preferred to spraying as the method of disinfectant application.
- Milk cows with confirmed contagious intramammary infections last.

4. Proper Maintenance and Use of Milking Equipment

- Install or update equipment to ASABE S518 (American Society of Agricultural and Biological Engineers, "Milking Machine Installations–Construction and Performance").
- Service, maintain, and regularly evaluate equipment function according to manufacturer's guidelines, using dynamic evaluation methods and an appropriate record form.
- Replace inflations and other rubber and plastic parts regularly, according to manufacturer's guidelines.
- Replace broken or cracked inflations and short milk tubes immediately.
- Sanitize equipment prior to each milking and thoroughly wash and sanitize equipment after each milking.

5. Good Record Keeping

- For each case of clinical mastitis, record cow identification, date detected, days in milk, quarter(s) affected, number and type of treatments, outcome of treatments (i.e. return to normal milk, time to discard milk) and the causative bacterial pathogen if a sample was cultured on-farm or in a laboratory.

Recommended mastitis control programme

- Use a computerized or manual record system to manage information, such as individual cow SCC data, on the prevalence and incidence of subclinical mastitis.

6. Appropriate Management of Clinical Mastitis During Lactation

- Develop and implement a herd clinical mastitis treatment protocol with the Herd Udder Health Advisory team. Carefully consider the economic ramifications of therapy
- Collect a pre-treatment milk sample aseptically for microbiological culture so that antimicrobial susceptibility tests can be used when appropriate.
- Use an appropriate therapeutic regimen; use drugs according to the protocol, or as recommended by the health advisors.
- Prior to infusion, disinfect the teat with a germicide and scrub the teat-end with an alcohol swab.
- For infusion of intramammary antibiotics, use a single-dose, regulatory approved product by the partial insertion method.
- Do not treat chronic non-responsive infections.
- Observe the correct withdrawal period for the antibioticused, as stated on the label. If extra-label drug use is necessary, follow regulatory guidelines under the supervision of a veterinarian (i.e. in the systemic treatment of coliform mastitis).
- Always follow recommended drug storage guidelines and observe expiration dates.
- Clearly identify all treated cows, and record all treatments in a permanent record.

7. Effective Dry Cow Management

- Decrease the energy density of the ration during late lactation to reduce milk production before dry-off.
- Dry cows off abruptly and dry treat each quarter immediately following the last milking of lactation.
- Disinfect teats and scrub the teat-end with an alcohol swab before infusing.
- Treat all quarters of all cows with a commercially available approved [long-acting] dry-cow antibiotic and/or an approved internal teat sealant.
- Use the partial insertion method of dry treatment infusion. Disinfect teats immediately following infusion using any approved post milking disinfectant teat dip.
- Provide adequate dry cow nutrition to enhance immune system function.
- Maintain a clean, dry, comfortable environment for dry cows. Dry cow environmental management is important to minimize exposure to pathogens.
- In situations of high environmental pathogen exposure, use an internal or external teat sealant for dry cows in addition to any antimicrobial product.
- In herds with coliform mastitis problems, vaccinate with a core antigen endotoxin vaccine following manufacturer's directions.
- Clip flanks and udders to remove excess body hair. Udder singeing may be useful to ensure hair removal.

8. Maintenance of Biosecurity for Contagious Pathogens and Marketing of Chronically Infected Cows

- Request bulk tank and individual cow SCC data.
- For suspect animals, further diagnostic efforts maybe indicated to identify cases of subclinical mastitis prior to purchasing cows.
- If possible, obtain aseptically collected milk samples for bacteriological culture from cows prior to purchase.
- Isolate recently purchased cows, and milk separately, until there is assurance of the absence of intramammary infection.
- Segregate cows with a persistently high SCC or linear score (i.e. SCC greater than 200,000

8. Maintenance of Biosecurity for Contagious Pathogens and Marketing of Chronically

Recommended mastitis control programme

Infected Cows

- Request bulk tank and individual cow SCC data.
- For suspect animals, further diagnostic efforts maybe indicated to identify cases of subclinical mastitis prior to purchasing cows.
- If possible, obtain aseptically collected milk samples for bacteriological culture from cows prior to purchase.
- Isolate recently purchased cows, and milk separately, until there is assurance of the absence of intramammary infection.
- Segregate cows with a persistently high SCC or linear score (i.e. SCC greater than 200,000 or linear score greater than or equal to 4.0 for several months) and observe response to dry treatment or other recommended therapy.
- Market or permanently segregate cows persistently infected with Staphylococcus aureus or other non-responsive microbial agents (Mycoplasma, Nocardia, Pseudomonas, or Arcanobacterium pyogenes).
- Consider udder health status of first calf heifers as this can impinge on herd biosecurity.

9. Regular Monitoring of Udder Health Status

- Enroll in an individual cow SCC program or use some other monitor of subclinical infections.
- Use a sensitive cow-side monitor of inflammation in cows suspected of infection and in high risk periods (i.e. early lactation).
- Monitor distributions of high SCC cows, and rates of change to elevated SCC.
- Conduct milk bacteriological culture of clinical cases and high SCC cows regularly.
- Monitor udder health for the herd using reports from the regional regulatory agency or milk marketing organization and DHI.
- Calculate clinical mastitis rates and distributions regularly, paying particular attention to infections in heifers.
- Use SCC and clinical mastitis records to evaluate protocols,and to make treatment and marketing decisions.

10. Periodic Review of Mastitis Control Program

- Obtain objective evaluations from veterinarian, industry field person or extension representative.
- A step-by-step approach to the review, and a standard evaluation form are useful.
- Make use of the entire Herd Udder Health Advisory Team: veterinarian, producer, herd manager, milking personnel, and advisors.

Milking practice

"Protocols for evaluating milking procedures are provided in this chapter and the analysis of milking machine problems is addressed. Various material dealing with good milking practices are given, which can be considered part of Good Dairy Farming Codes of Practice."

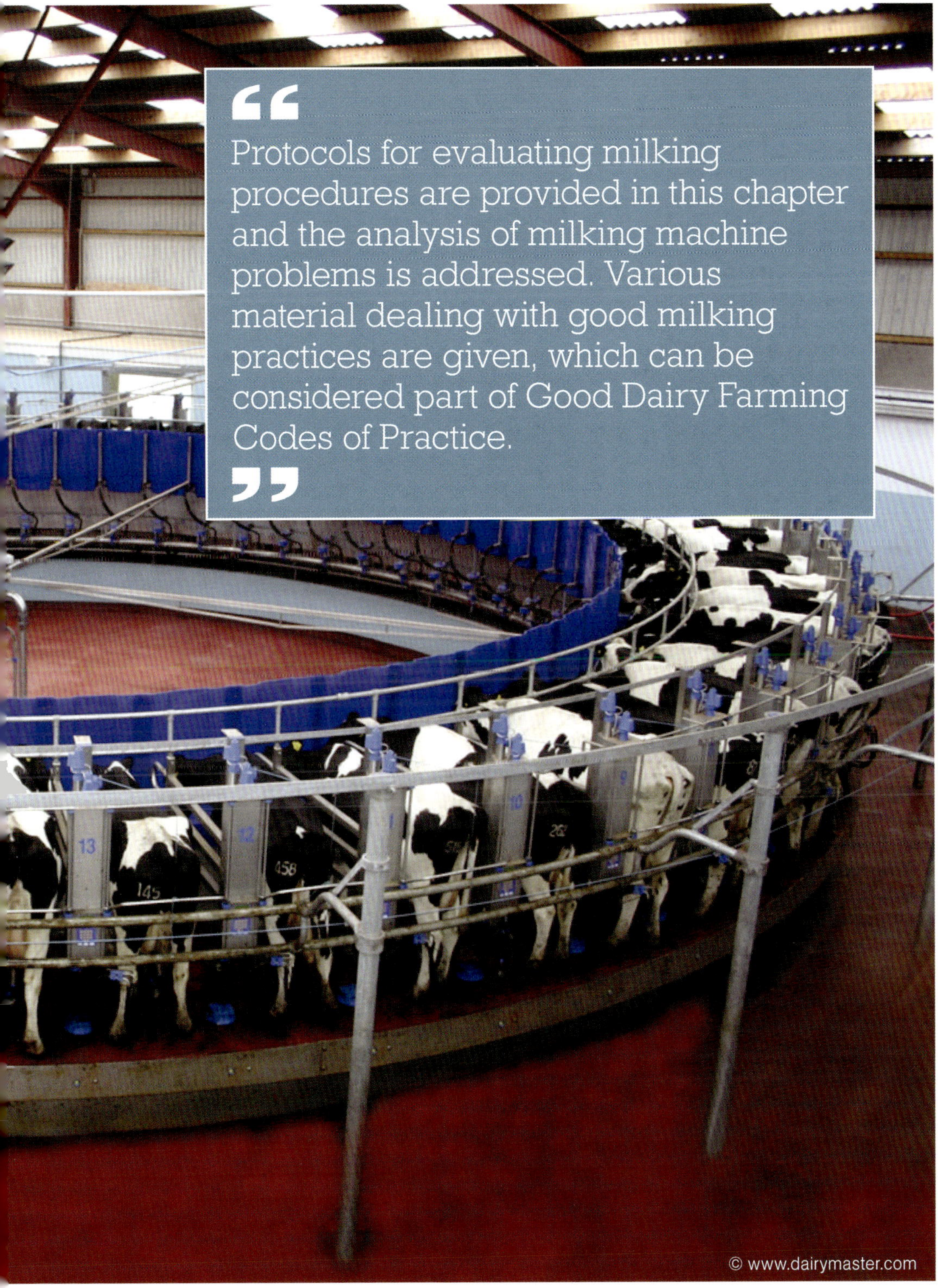

© www.dairymaster.com

Evaluation of milking practice (1)

Dairy Herd Health & Production Management Advisory Programme **MILKING INTERVALS**

Farm = **Veterinarian =** **Number of cows in the herd =** **Date =**

	Action.........................Cow ID →	....	....	......	......	......	......	......	......	......
1.a	Concentrates supplied									
1.b	Cleaning of udder and teats									
1.c	Drying of teats									
2.a	Attachment of milking claw									
2.b	End of milk flow									
3.a	Continuation of milking (but no milk)									
3.b	Detachment of milking claw									
4	Disinfection of teats									
5	Release of the cow from the parlour									

Remarks & conclusions:

Enter for each cow in the parlour the respective time (min + sec) spent on a given action.
Afterwards, calculate the intervals and evaluate them.

Evaluation of milking practice (2)

Intervals during milking

It is not necessary to check all cows; a representative sample is sufficient (20% of cows).

However, attention should be paid to the fact that dominant cows will enter the milking parlour first, while mastitic cows (or sick cows) will enter later. Evaluation can be carried out in two shifts.

The five steps in evaluation:

1. Record the identification number of each cow in the sample.
2. Record the time (for example, 1832) that the first cow is prepared.
3. Then the time at which the milking claw is attached
4. Record the time of detachment of the cluster.
5. Intervals can be calculated and evaluated at the end of the entire cycle

Remarks:

Preparation - dry ☐ wet ☐

One towel ☐ several ☐

Cows dirty ☐ clean ☐

Automatic detachment - yes / no

Threshold for detatchment (200 or 300 or even 400 ml/min).

Disinfection of teats after milking - yes / no

Disinfection is appropriate - yes / no

Residual milk must be less than < 5%.

Deviations / lesions:

Refusal of cows to enter the parlour - yes / no; if yes, number of cows =

Air leaking during milking - yes / no

Defaecation during milking - yes / no

Teats blue after milking - yes / no

Teats with ring at the base - yes / no

Eversions of teat end visible - yes / no

Teats flat after milking - yes / no

Other anomalies:

Cow ID	Start of the udder preparation	The duration of the preparation (norm 30-60 sec)	Interval from preparation to cluster attachment (norm is 60 sec)	Time of cluster attachment	Time of cluster detachment	Duration of milking (norm < 6 min)	L of milk	L per min.

Improving teat health

Derek Davies, Avon Rubber, Internatl Dairy Topics, vol 8 no 3, 2009

The teat liners must be replaced before 2500 milkings (silicone liners last longer).

Even if you cannot visibly detect abnormalities, the rubber tissue will have degraded and its performance reduced. By the time cell counts increase, it is too late because the problems of teat and/or udder health already exist.

Teat liners are exposed to different chemical products, milk (depositions, sometimes absorbed by the rubber) and water and can open/close around 2 million times over a year, depending on the number of cows milked.

The depositions in the teat liner form a good environment for bacterial growth. Moreover, they may cause cracks in the material, which is traumatic for teat tissue. Thum (1975) found that the number of bacteria significantly increased after 2500 milkings.

Between 2000 and 3000 milkings, the number of teat lesions increases, as well as red discolouration (should remain below 15 % of the herd). Secondly, the teat loses its natural defense (barrier of the skin) and teat end callosity may appear. The latter increases the cell count and the risk of mastitis. The volume of residual milk also increases, which is yet another risk factor for mastitis.

The eight points of teat health

1 Milk with healthy and clean teats. Provide good cow comfort (housing; feeding; water; health; specific behaviour). Clip the hair on the back of the cow for winter; clean the teats with a separate dry towel for each cow. Udder preparation should last between 60 and 90 sec per cow.

2 Check each quarter and teat for mastitis before attachment of the cluster. Remember that total udder preparation for milking is between 60 and 90 sec.

3 If necessary, apply pre-milking teat disinfection.

4 Dry teats well with a paper towel.

5 Attach the clusters (90 sec after stimulation) and ensure they are well-positioned. Correct the cluster position during milking if necessary.

6 Detach the clusters when the milk flow drops substantially (for automatic detachment devices, set the threshold for detachment to not lower than 300 or 400 ml milk per min). Shut down the vacuum before detachment!

7 Disinfect the teats immediately after detachment and ensure the whole circumference teat is disinfected.

8 The exit from the milking parlour must be clean and an easy passage. Leave the cows first at the feed rack for 20 min, to allow teat-ends to close. Thereafter, the animals should lie down.

Milking machine problems

Problems are caused either by the machine, the operator, the cow or combinations. As a starting point, all personnel must be well-trained and qualified and carry out their duties in a hygienic way. The following highlights some of the problems that may transpire in the cows or the machine itself:

Cow problems

1 There is insufficient `*let down*` after stimulation of the udder
- The preparation of udder and teats was insufficient
- The cow did not receive concentrates in the parlour (if this is the practice)
- The cow is in heat, sick or frightened
- The weather conditions are (very) poor

2 The cow is disturbed
- A young animal, not yet used to milking
- Teats have cracks and lesions
- A mastitis case
- The personnel are brutal and cruel
- Too many flies in the milking parlour
- The pulsator function is poor
- The milking process continues without milk flow

3 Milking is slow
- Milking speed is low
- The conformation of teats is poor, too long
- The tension in the teat liners is insufficient
- The vacuum is too low
- The position of the milking cluster is wrong

4 Too much residual milk and the cow has to be milked manually

5 Milking from some cows varies between quarters
- The udder is not well-formed or symmetrical
- There is a lesion on one teat
- One teat liner is in torsion in a cluster
- A milk tube or a pulsation tube is trapped
- Obstructions are present in a tube
- The pulsator function is poor

6 The teats are hard and slightly blue immediately after milking
- Teats are too small or too fine
- Swollen udder
- Milking process continued without milk flow
- The vacuum was too high and wet udder preparation was carried out
- The small air inlet of the pulsator was obstructed

7 Teat ends show callosities
- Vacuum level is too high
- The pulsator was not functioning correctly
- Milking continued without milk flow
- Teats are too long for the teat liners
- Milking speed was too low

Machine problems

1 After having started the motor, the vacuum is insufficient or absent
- The pump does not work as it should (too slow)
- The valves are obstructed
- The interceptor is not closed
- There is an opening somewhere
- There is a tube open somewhere

2 The vacuum is too high
- Valves are stuck in the regulator
- Regulator not correct (after cleaning!!)
- Regulator is closed (after inspection!!)

3 The vacuum diminishes slowly during milking
- A milking cluster had fallen down
- A tube is detached
- The security valve has opened

4 The pulsator function is poor
- Pulsator is not working
- Parts are stuck
- There is no vacuum in the pulsator
- There is no electricity (electric pulsator): burnt; short-circuit; safety fuse blown

5 The number of pulsations is too high or too low
- The pulsator is poorly set
- The pulsator is dirty or worn

6 The milk clusters constantly fall down
- The vacuum is too low
- A mastitis case
- Air is entering the system
- The weight of the cluster is not well-balanced

Good milking practice

Adapted by Jos Noordhuizen, May 2008 (ENVL) from
Source: De Koning ; Bos & Neijenhuis (Wageningen University & Research, Lelystad, NL, 2002

The process of milking a cow may be graphically represented by plotting **kg of milk expelled per minute**.

These data usually form one of four profiles:

1	square
2	rectangular
3	triangular
4	trapezoid

Square and rectangular profiles are considered the normal, desirable situation as milking speed increases gradually after cluster attachment, up to a peak where it plateaus and then decreases steadily until detachment.

Triangular (3) and trapezoid (4) profiles are considered undesirable (abnormal) profiles that lead to a less effective milking and elevated cell counts in affected cows. These profiles are formed due to a decrease during the first minute after attachment, which give these profiles a bi-modal image (Figure 1) and a plateau of short duration.

Over the course of the lactation, time spent at the milking machine will continually decrease, while the rate of bi-modality increases. The higher the parity of a cow, the more she will show triangular profiles and the number of square profiles diminishes.

The **milking profile of a cow can be influenced** by management:

- Adequate udder preparation (cleaning + first streams + massage = 1 minute) (or a short preparation + waiting time up to one minute)
- Adapted pulsation (shortening the phases a and c result in longer machine-on times, more agitated cows and more blue rings at the teat base)
- Earlier detachment of the milking cluster (at 300 gram/min or 400 gram/min, without cell count problems or mastitis or milk leaking in cubicles)

This way, we are able to change profiles 3 and 4 into profiles 1 or 2.

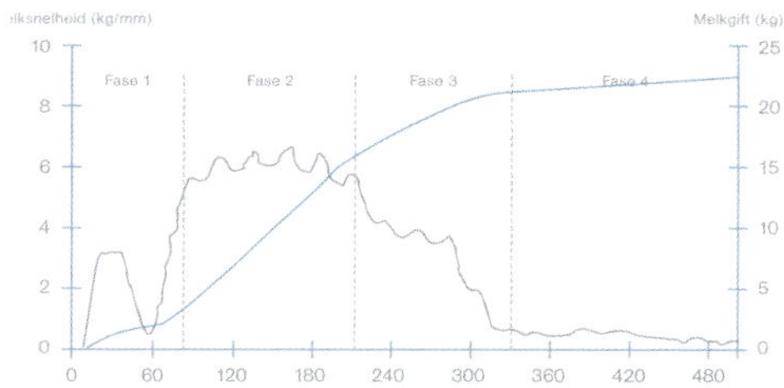

Figure 14.1. A typical bi-modal milking profile

Analysis of milking profiles (in four phases)

Phase 1. Ascendance

The milking cluster is attached, which takes about 10-12 sec. After attachment, the milk flow starts and increases quickly (Figure 1). During the first minute, there is a drop in milk, which then recovers. This drop is related to milk from the udder cistern being expelled rather than milk from the alveoli (or true let-down). This phenomenon causes a bi-modality in the milking process and is mainly caused by insufficient stimulation (not long or intensive enough) resulting in a relatively low blood concentration of oxytocin. A bi-modal milking curve means milking empty quarters and an excess force on teat-ends (monitor teat-end callosity scores) together with all the risks of infection and contusion of tissue (mastitis; high cell counts).

Bi-modality can be more frequently observed in low producing cows (because there is less milk in the cystern). Due to the stimulation by the milking machine, the level of oxytocin then increases causing the second peak, evident in Figure 1.

Good milking practice

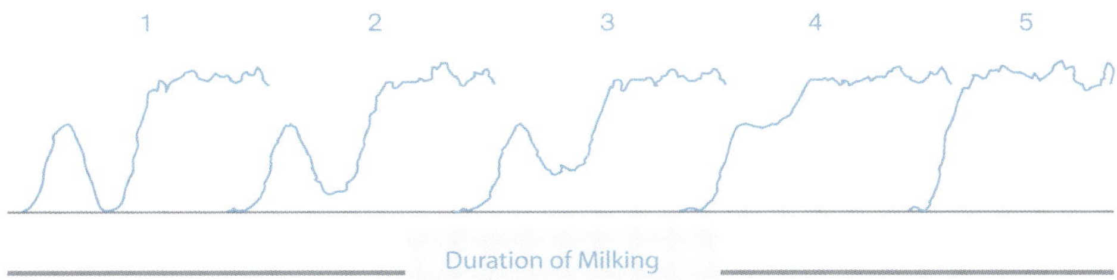

Figure 14.2. Examples of milking profiles and bi-modalities. The horizontal axis represents the time of milking (seconds) and the vertical axis the milking speed (L/minute)

Figure 14.2 gives examples of milking profiles and bi-modality classes and also figure 14.2 shows that Phase 1 becomes increasingly shorter in profiles 4 and 5. Profiles 1 and 2 are considered normal, while the last three are considered undesirable (abnormal).

This means that at the start of milking, the speed must be a minimum of 300 g milk/min. If the speed drops below this for six sec or more, the profile becomes bi-modal.

After around 90 sec, the bi-modality will be over, the volume of milk per unit of time increases and a plateau is reached (Figure 14.1).

Phase 2. Plateau

During this phase, the maximum milking speed is reached. The harvest in Phase 2 can be calculated by the production of milk (l) in this phase divided by the total milk production (l) for the duration of that particular milking. This can be used as a parameter of efficacy.

Table 14.1 Classification of bi-modalities and their qualifications.

Bi-modality class	Milk flow during a bi-modal phase	Qualification
1	0-25%	Highly bi-modal
2	26-50%	
3	51-75%	Bi-modality moderate
4	76-100%	
5	no drop	Non-bi-modal

Phase 3. Descendance

The descent will be progressive or in small steps. The small steps may be observed when there is, for example, one quarter or two quarters empty. Phase 3 finishes when the volume of milk reaches a threshold of 500 g milk/ min. The inclination of the descent will depend on individual cows: ease of milking; milking speed; symmetry of the quarters; genetics. Often a bi-modality can be seen when Phase 3 is prolonged.

Good milking practice

Adapted from
De Koning ; Bos & Neijenhuis, Wageningen University & Research, Lelystad, NL, 2002

Phase 4. End phase

Phase 4 finishes with the detachment of the cluster. This phase may be prolonged when the teat liners 'climb' the teats (leading to rings at the teat base; mostly caused by wet udder preparation), milking is not well-balanced, quarters are hard to milk, or insufficient udder preparation resulting in a relatively low oxytocin blood level.

The shape of milking profiles

Milking profiles can take one of four shapes: triangular, trapezoid, square or rectangular.

In the square and rectangular profiles, Phase 2 (the plateau phase) occurs at a slightly reduced milking speed but the duration of this phase is longer. This means that the volume of milk harvested is larger and the milking more efficient.

A non-bi-modal milking cow will show a shorter Phase 4 and will be milked more efficiently (less milk remaining; fewer risks). A prolonged Phase 4 causes a longer "machine-on time" but without the increase in volume of milk.

For example, because one quarter is emptied later than the other three, this causes blind milking in those three quarters increasing the risk of mastitis risk and a higher cell count.

The triangular and trapezoid profiles often concern the front quarters, which are already empty while the hind quarters are still being milked.

The aim is to milk all cows with square or rectangular profiles as these represent more efficient milking.

Table 14.2 shows some parameters of the different milking profiles.

Relationship between milking profiles and cell counts

Figure 14.3 shows the relationship between milking profile and level of cell counts. The cows with the triangular profile (prolonged phase 4) show a cell count level twice that of cows with a square profile.

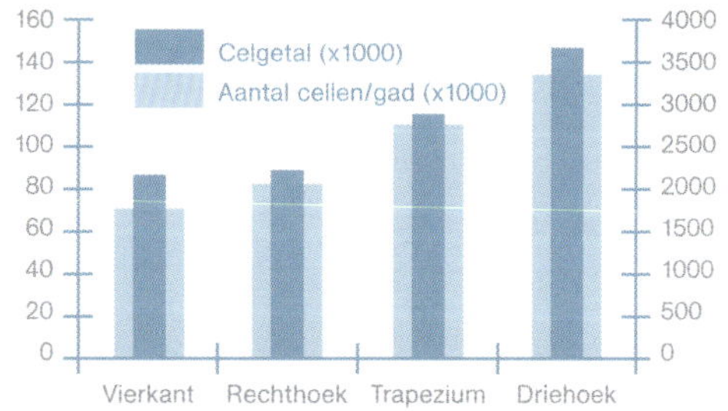

Figure 14.3 Effect of milking profile on somatice cell count (x 1000 cells/ml)

Table 14.1 Parameter values for the different milking profiles

Profile	Production of milk (L)	Duration (min)	Speed of milking	Phase 1 (in %)	Phase 2	Phase 3	Phase 4	Milk harvest plateau (%)
triang	14.0	8.3	4.5 L/min max	17	8	51	24	20
trapez	13.5	7.4	4.0	16	24	37	24	45
square	13.8	7.0	3.4	16	43	23	19	68
rectang	12.3	9.4	2.1	12	53	16	20	75

Good milking practice

Udder preparation and stimulation

Udder preparation has three goals:

1. Clean teats and teat-ends
2. Stimulation of oxytocin
3. Check of first milk

A quick preparation will take 20 to 30 sec but will induce bi-modality !

It is better to **spend 60 to 90 sec on a complete preparation**.

This ensures a more efficient milking (more milk harvested during the plateau phase; phase 1 is shorter; milking speed is higher; total duration of "machine-on time" is shorter). The latter is also important to minimise force on the teats and teat-ends and cell count increase helping to reduce mastitis incidence.

A preparation of 60 to 90 sec will have the same effect as a preparation of 30 or of 60 sec combined with a waiting period of 40 or 30 sec.

A **post-preparation-waiting-period of 1 min** can be achieved when four cows are prepared consecutively and the clusters then attached to the same four cows in the same order!

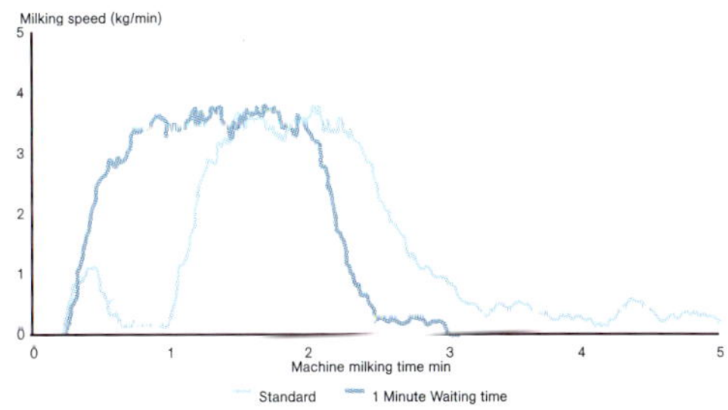

Figure 14.4 Effect of preparation and waiting time on milking profile

Table 14.2 Effect on milking profiles of different udder preparation methods

Activities →	Clean the teats Attachment	Cleaning + First streams check Attachment	Cleaning + Stimulation First streams check Attachment
Time :	10 sec	16 sec	31 sec
Duration of machine-on	5.20 min	5.12 min	4.91 min
L of milk	11.5	11.5	11.4
Milking speed (min)	2.28 L/min	2.33	2.43
Idem (max)	4.02	4.08	4.17
Type of profile :			
Phase 1 (sec)	75	70	56
Phase 2	98	93	101
Phase 3	113	119	113
Phase 4	25	26	25
Milk harvest in phase 2	31%	30%	33%
Bi-modality degree	80%	66%	40%

Milking machine and/or personnel problems

Table 14.3 Checklist to detect milking machine and or personnell problems

Problem	Checkpoints for milking machine	Checkpoints for personnel	Other potential causes
Teats are not in good condition		Water to clean udder and teats is too hot Teat disinfectant is too aggressive	Sun Weather Photosensitivity skin Formalin
Teats are blue and/or hard	Vacuum too high Ratio pulsation to rest phase is poor Malfunction of pulsators Teat liners are worn	Duration of milking is too long	Teats are too long First lactation heifers
Callosity of teat-ends is severe	Vacuum too high Malfunction of pulsators Teat liners are worn or too wide Ratio pulsation to rest phase is poor.	Duration of milking is too long	
Infections in teats (bacteriological or viral).		Vector Hygiene Disinfectant	New cows have been purchased
Teat lesions			Age, udder conformation housing, claw health
Mastitis	Substantial vacuum fluctuation Tissue lesions due to vacuum or pulsation or worn teat liners Poor cleanliness during milking	Duration of milking is too long Personnel is brutal Poor udder/teat preparation (1 towel for all ; wet ; no first stream checks ; interval until attachment is too long >90 sec ; interval until attachment is too short <60 sec.) Intervals during milking too irregular	Other known risk factors

Milking machine and/or personnel problems

Problem	Checkpoints for milking machine	Checkpoints for personnel	Other potential causes
Somatic cell count is too high	See known risk factors	Adaptation of milking practice ?	Other known risk factors
Too many bacteria in the milk	Cleaning of the machine after milking is poor. The water temperature is too low at start (<80°C) and/or at end of cycle (<40 °C). Machine parts are worn.	A non-hygienic milking.	Temperature of bulk tank is too high (>4°C). Cleaning of tank is insufficient.
Loss of milk production	Insufficient milking: Vacuum; ratio; pulsation; teat liners.	Poor udder/teat preparation (variable; disturbed routines; too many different people).	Feeding Weather Genetics Housing Climate etc.

Machine milking - what can go wrong

Table 14.4 Machine milking: what can go wrong and what can be done

Cow level	Complaint	Potential approaches for solution of the problem
	Cows have no milk let-down	Insufficient udder preparation (too short; none)
		Cow has not been given the usual concentrates
		Cow is in heat, sick, nervous, or frightened
		Unusual bad weather conditions
	Cow is troublesome	Young animal (heifer) and not used to machine
		Sore teats (see also teat end callosity scoring)
		Udder infection
		Rough treatment by milker
		Flies are troubling the cows
		Cows have not received the usual concentrates
		Pulsator does not function or not properly
		Milking continues when it should have stopped
	Cows milk slowly	Low milking speed
		Teats do not have an appropriate form (too long)
		Liner barrels of teat cups have insufficient tension
		Vacuum is too low
		Position of the milking claw is not correct
	Too much rest milk; cow stripped	
	Milking is not square (milk flow from the quarters is uneven)	A kink in the short milk tube or in the pulsation tube
		One teat is injured
		A liner barrel is twisted in the teat cup
		Cow's udder is not square
		Obstruction in the short milk tube
		Pulsator does not function "squarely"
	Teats are bluish and hard after detachment	Teats are too small or too thin
		Udder inflation
		Milking continued after it should have stopped
		Tension in liner barrels is too low
		Vacuum level is too high
		Air inlet of pulsator is obstructed

Machine milking - what can go wrong

Cow level	Complaint	Potential approaches for solution of the problem
	Hardened and bulging teat openings	Vacuum level is too high
		Pulsator is incorrectly set
		Milking continued after it should have stopped
		Teats are too long with regard to liner barrels
		Milking speed is too low
Milking machine		
	After starting the motor, there is no vacuum or insufficient vacuum	Pump does not turn rapidly enough
		Valves in the pump are blocked
		Lid of the interceptor is not air-tight
		Somewhere there is an open tap
		Somewhere a tube has got loose
	Vacuum is too high	Sticky valve in the regulator
		Regulator is incorrectly set
		Regulator does not function
	Vacuum fades away during milking	Safety overflow has started
		Another milking unit dropped off
		A loose vacuum tube
	Pulsator does not function	Pulsator is out of order
		Some parts stick together
		There is no vacuum in the pulsator
		There is no electricity (blown fuse; insufficient contact between tap and magnet head; coil of electromagnet has burnt out)
	Number of pulsations too high	Pulsator is not correctly set
		Pulsator is dirty or worn out
	Teat cups drop off	Vacuum level is too low
		Udder inflation
		False air penetrates round mouth piece
		Weight of the milking unit is not balanced

Of course there could be many other problems confronting the milker. A regular (daily and weekly) and adequate maintenance of the milking machine can prevent a lot of problems. A technical assessment of the milking machine function during milking at least twice yearly also contributes to the prevention of problems.
Finally, the proper '*reading*' of the signals that cows may give may contribute to an early detection of problems related to machine milking.

The purpose of pulsation during milking

After Jerry Slattery, Boumatic www.boumatic.com in Internatl Dairy Topics, vol. 10, nr.4, 201

Proper pulsation is critical for a good performance of the milking machine and for a good udder health. However, there are farmers who do not bother too much once the pulsators have been installed. The major goal of pulsation is to limit congestion and oedema in the teat tissue during milking. Additionally, pulsation assist in: (1) maintaining a high milk flow rate from the teat within a pulsation cycle; (2) counteract the possible negative effects of teat congestion , oedema or other discomfort during milking; (3) reduce the new infection rate of mastitis; (4) stimulate milk ejection.

The pulsation system can be characterized by two parameters: pulsation rate and pulsation ratio. Pulsation rate is defined as the number of complete pulsation cycles per minute. Common pulsation rates of milking machines are between 45 and 60 cycles per minute. Pulsation ratio is defined as the percentage of time that the inflation is in the milking phase or the rest phase.

There are 4 phases in a pulsation cycle:

- Increasing vacuum phase: when the vacuum level in the teat-cup pulsation chamber increases from atmospheric pressure to machine vacuum level. The teat liner begins to open and milk starts flowing.
- Maximum vacuum phase: when the vacuum in the pulsation chamber is similar to the vacuum level inside the liner barrel. Teat liner is open, milk flows.
- Decreasing vacuum phase: when the pulsator connects the pulsation chamber to atmospheric air pressure. When air enters the pulsation chamber, the teat liner collapses and milk stops flowing.
- Minimum vacuum phase: when the pulsation chamber is at atmospheric pressure level. The teat liner is closed and exerts maximum force on the teat.

A vacuum recorder can be used to monitor and evaluate any changes in pulsation chamber vacuum graphs. The graph produced can indicate the different phases in percentages. For example, a ratio of 15:45:15:25. The latter means that in one pulsation cycle the liner takes 15% of the cycle duration to open (increasing vacuum phase), is 45% of the time fully open (maximum vacuum phase), 15% of the time it is closing (decreasing vacuum phase) and 25% it is closed (minimum vacuum phase). This is the only proper way to evaluate whether pulsators are functioning best.

Pulsation should always be represented with units attached and teat cups plugged or attached to cows. In many milking parlours the duration of milking has increased. To guarantee that the system is working at peak performance, regularly scheduled maintenance evaluation visits should be made by the machine dealer (e.g. at least twice yearly). Daily maintenance recommendations should be followed strictly, including cleaning and disinfection. It is rare that pulsators suddenly drop out; they rather degrade gradually due to dust and dirt.

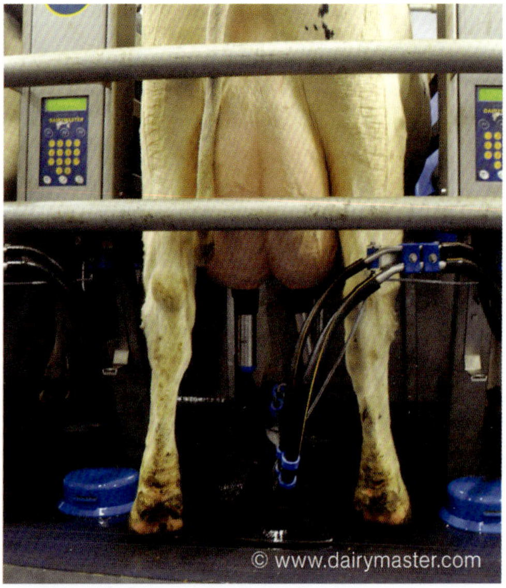

Some issues to improve ease of milking

After DA Reid, Boumatic www.boumatic.com at International Dairy Topics vol.8, nr.3, 2009

Ease of milking is the process of a proper stimulation of cows before milking, leading to teats filled with milk which can flow once the cluster is attached to the cow. Good ease of milking is when milk flows as soon as the cluster is attached; the cow milks out rapidly, completely and evenly. Next, milk flow will stop rather suddenly and soon after the clusters can be detached. During such a milking the cow will not move, kick or left a foot.

Many factors affect the ease of milking: several are in the hand of the milkers (Table 1), others are associated with the milking machine (Table 2).

Closely monitor the milk flow from the milking claw to the milk line. That path should be properly downhill over the whole distance. Are there any uphill sections in the milk line? Are milk hoses too long and hence hindering easy transport by forcing milk into loops? The 'yes' answers point to potential behavior problems in the parlour and affect ease of milking because it impacts on peak milk flow vacuum. When peak milk flow claw vacuum levels decrease, the duration of milking per cow will increase. It also means longer periods of low flow since cows run out of their oxytocin stimulatory hormone levels. Such cows are at higher risk of new udder infections too. High peak milk flow claw vacuum levels are the best way to increase the average milk flow. It could be indicated to shorten milk hoses to milk high yielding cows faster, without exposing the others to a too high vacuum towards the end of milking. Worn out or cracked hoses can be another problem, disturbing vacuum and hence milking. If the bottom hoses on milk jars are restricted, then vacuum drops of up to 6 kPa may occur each time the jar tries to empty its milk. Such drops will increase the duration of milking, while cows endure longer low flow milking. Pinch valves may also affect peak milk flow claw vacuum negatively. The time that milking clusters are actually attached to the cows is not more than 25 to 35% of the total milking time between first unit on till last unit off. This means that the pinch valve is compressing the hose for at least 65 to 75% of the time. Pinch valves should be moved to another position on the hose once every 7 days.

Table 14.5. Milker related factors affecting ease of milking

Factor	Impact
Cow handling Cow preparation	Affect milking time and milk production
Cluster attachment procedure Cluster alignment	Affect complete, even milking and liner slips

Table 14.6. Milking machine related factors affecting ease of milking

Factor	Impact
System vacuum Pulsation features Milk hose length & diameter Restrictions in milk flow from teat ends	Affect peak milk flow
Cluster alignment & support Teat liner type & condition Cluster weight Automatic detachment settings Regulator response	Affect complete, even milking
Cluster alignment & support Teat liner type & condition Cluster weight & weight distribution Teat liner vacuum & restrictions Size & condition of cluster air vents	Affects slipping or falling of teat cups

Production

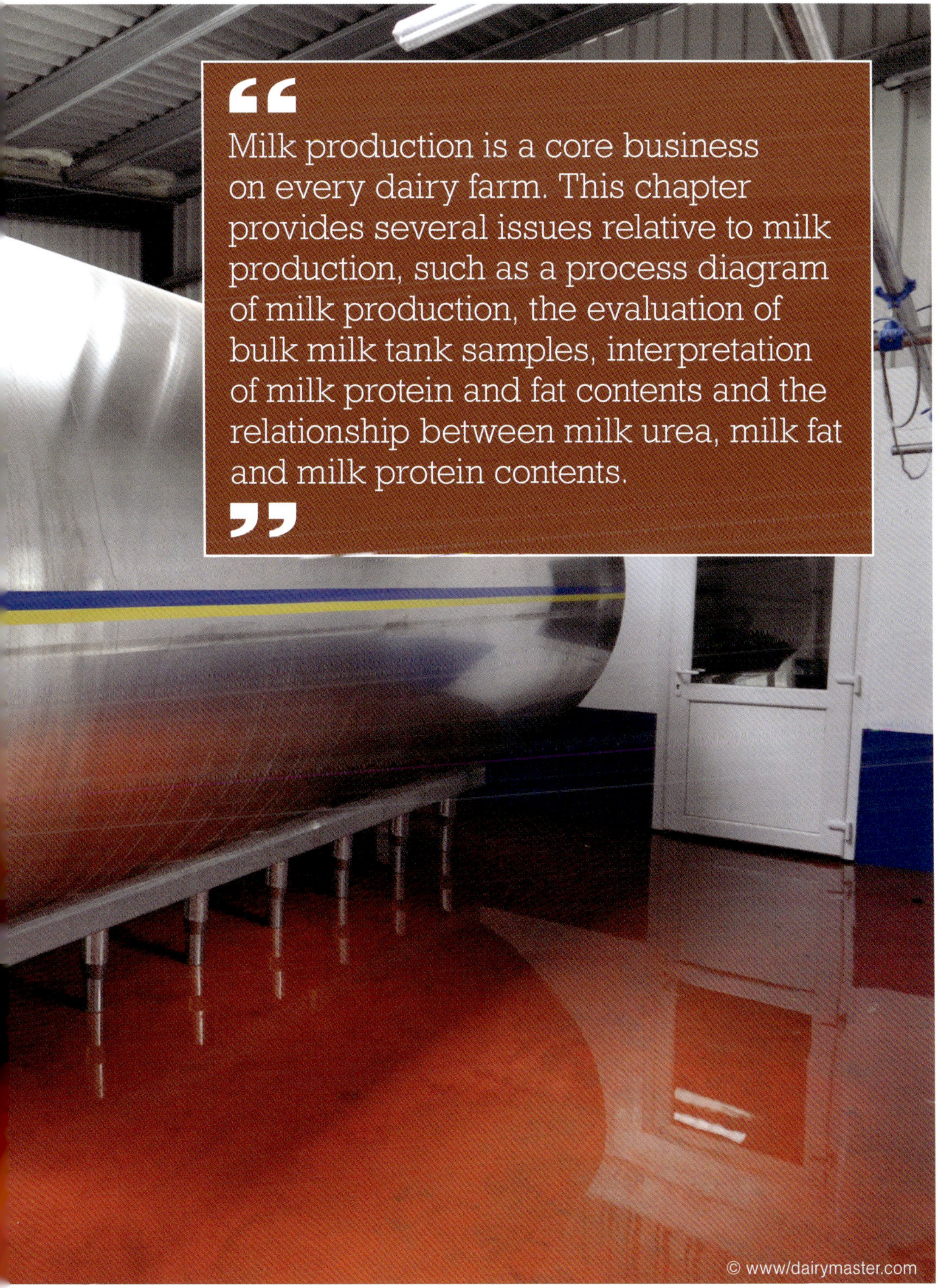

> "Milk production is a core business on every dairy farm. This chapter provides several issues relative to milk production, such as a process diagram of milk production, the evaluation of bulk milk tank samples, interpretation of milk protein and fat contents and the relationship between milk urea, milk fat and milk protein contents."

Milk production at farm level

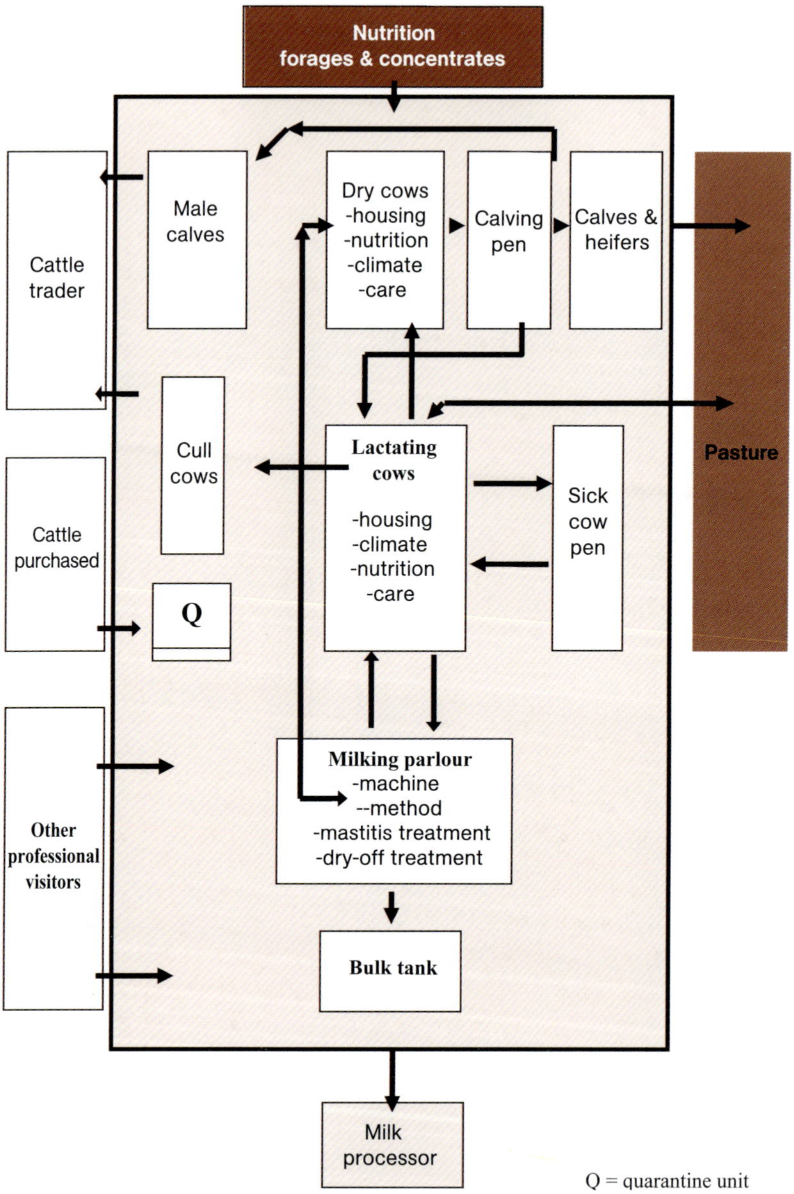

Figure 15.1 Milk production at farm level: different production units

Evaluating bulk milk tank samples

Veerle Fievez, University of Gent & DLO, in Veeteelt, June 2006 (80) & January 2007 (1/2), The Netherlands

Table 15.1 Evaluating bulk milk tank samples for acidosis based on the fatty acid profile

Fatty acid	Indicator for	Reference value*	Problem area
Iso C 14:0	Fibre digesters	2.53	< 2.53
	Acidosis		
Iso C 15:0	Fibre digesters	5.88	< 5 or > 6
	Acidosis		
C 18:1-10	Acidosis	0.38	>> 0.40
Total UFA	Lowered rumen pH and or rumen dysfunction (acidosis)	50.5	> 50.5

*expressed as a percentage of total unsaturated fatty acids (UFA)

Table 15.2 Negative energy balance in cows

Negative energy balance & ketosis occurs when:		
Long-chain fatty acids	> 40%	
Medium-chain fatty acids	< 40%	in % of total fatty acid methyl-esters, 2-6 wks pp (g/100 g)
Oleic acid	> 30%	

Managing two groups of dry cows

R. Knook, De Heus Voeders, NL, Veeteelt, Holland October 2008.

Objective:	To achieve optimal DMI before and after calving
Duration:	2 months (50 to 60 days)

The cation-anion balance should only be addressed during the last two weeks of the dry period. The addition of concentrates may be used to reduce the balance by 200 mEq per kg of DM.

Benefits of managing dry cows as two separate groups include:

- 42% fewer dull calves,
- 17% fewer endometritis cases,
- 20% fewer mastitis cases,
- 32% fewer acidosis cases.

However, there can be an increase (~31%) in abomasal displacements if too high a level of concentrates is fed post-calving.

Milk fat & protein content: Relationship to health

www.vacqa-international.com

To evaluate milk fat and milk protein content, 20 % of cows in early lactation should be monitored according to the the milk recording data (minimum of 12 cows).

To evaluate the ratio milk fat to milk protein, at least 20 % of cows in early lactation should be monitored (12 cows minimum).

Remarks:

Madouasse *et al.* (2009) carried out an analysis of more than 8 million lactations and reproductive performance in the UK. They discovered that a **delay in the interval between calving and conception** is largely influenced by parameters, such as the ratio of fat:protein, milk production at the second recording, milk fat at the first recording, SCC at first and second recordings.

Table 15.3 Milk fat and protein contents & their relationship with animal health

Milk fat good	Milk fat poor	Milk protein good	Milk protein poor	Ratio Fat to Protein good	Ratio Fat to Protein poor	NEB problems suspected if >20% of cows between 40 and 70 days in milk show ….
More than 0.3 above herd mean	More than 0.5 below herd mean	More than 0.3 above herd mean	More than 0.3 below herd mean	> 1.2	> 1.5 NEB Ketosis	Milk fat > 4.8%
						Milk fat < 3.5%
					< 1.2 = SARA	Milk protein < 2.9%

Table 15.3 'mean' = the mean of the herd according to the date of the milk recording.
NEB= negative energy balance
SARA= sub-acute ruminal acidosis

Other parameters:

- **ratio milk fat$_1$ / milk protein$_1$** < 3.0% (at first recording date)
- **milk fat minus milk protein** < 3.0% (Holstein) → > 20% deviations / month = problem
- **milk fat minus milk protein** < 0 (Montbéliarde)

Milk fat:protein ratio is lowest in April–May.

The probability of conception decreased with increasing interval between calving and conception:

- 0-60 days pp 8% probability of conception
- 61-80 days 13%
- 81-100 days 17%
- After 100 days, there was a linear decrease towards a 4% pregnancy probability.

Interactions between milk urea, fat & protein

After J. Bedouet. L'étude des composants du lait. Proceedings des Journées Nationales GTV, Nantes, France, 2005, pp 515-521

Table 15.4 Interactions between milk fat and protein

	Low milk fat	High milk fat
Low milk protein	Under-feeding	Fat mobilisation → loss of BCS Lack of energy in ration
High milk protein	Ration hyper-energetic Acidosis if milk fat < milk protein Too much by-pass starch	Ration rich in energy (maize; sugarbeet)

Table 15.5 Interactions between urea, fat and protein

Urea	Milk fat	Milk protein	Interpretation
High	High	Low	Lack of energy
High	Normal	Normal	Excess degradable nitrogen
Low	Normal	Low	Lack of degradable nitrogen
Low	Low	High	Lack of degradable nitrogen + acidosis

Table 15.6 Urea level in milk

A normal urea level = 2 to 5 mmol / L milk	220 - 290 mg/dL milk (winter e.g. maize)
	260 - 340 mg/dL milk (summer e.g. grass)
An abnormal urea level=	< 2 mmol / L milk
	> 5 mmol / L milk

Quality

"Quality often refers to dairy products but is also relevant to the production methods on farm. Therefore, this chapter addresses several product quality issues (milk, air, water, soil, wildlife, society), as well as quality risk management, for example based on the HACCP principles. Checklists are provided to allow assessment of the quality of management."

Flow diagram - detecting antibiotic residues in milk

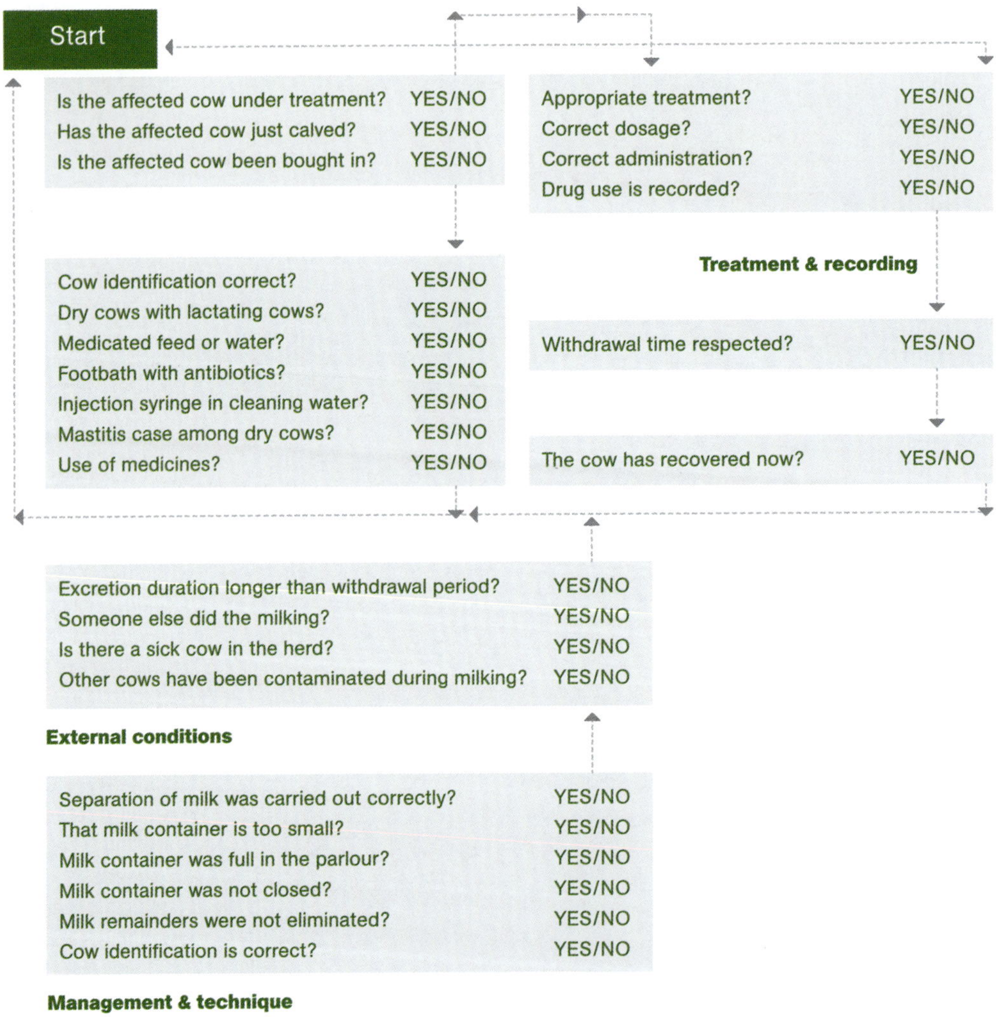

Figure 16.1 Flow diagram for the detection of possible causes of antibiotic residues in milk

Milk quality - milking procedures & practices

Adapted after N. Schuring, IDT, Vol 8, No 6, November 2009

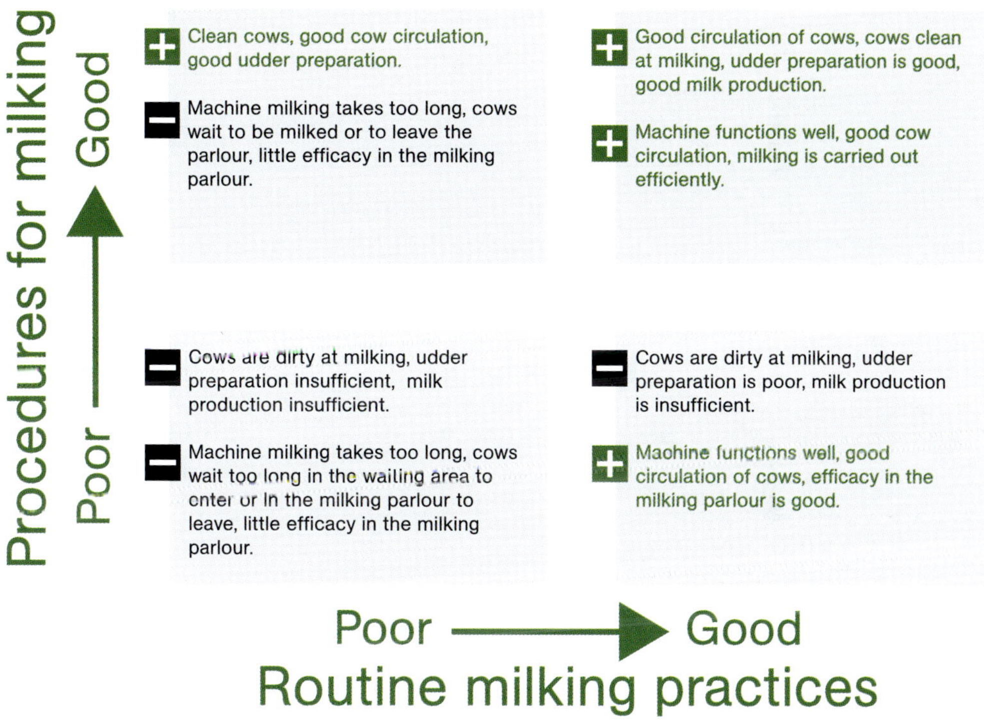

Figure 16.2 Milk quality: where milking procedures and routine practices meet

Procedures = how to milk cows for optimal production?

- Visual inspection
- Pre-milking teat dip
- Detection in the first milk streams
- Preparation of the udder
- Attachment of the milking claw; check alignment
- Detachment of the milking claw (automatic removal threshold at 300 – 500 ml/min)
- Post-milking teat dip

Routines = how to carry out milking in the most efficient, productive way?

- Cows should enter the milking parlour immediately, rapidly and quietly
- The first cow should be subjected the above procedure immediately (60 – 90 sec)
- Organise the udder preparation and attachment in groups of four-six cows.

Quality risk management

In order to create a quality risk management plan that addresses environmental issues, the farmer or his vet needs to identify the fields they will be working within

These fields include:

- Best management practices for environmental issues
- Quality of air and control of odour
- Water quality
- Soil quality
- Wildlife conservation
- Aesthetics and societal/community relationships

Dairy farmers have *moral and societal obligations* regarding the environment, towards their neighbours, the community and society. The following checklist is meant as a self-evaluation exercise, to find out where a dairy farmer stands.

If the farmer identifies areas for improvement, he can ask his veterinarian to help draw up a *quality risk management (QRM) plan* for the environment. This QRM should suitably address the balance between environmental needs and economic farm profits. Below is presented a "Farmer's Checklist" with six different sections.

(based on the Quality Environmental Stewardship – consultation guide 2000, KR Carlson author, by Milk & Dairy Beef Quality Assurance Center, Stratford IA, USA –DQA; www.DQACenter.org/university)

Quality risk management

Table 16.1 Farmer's checklist

Best management practices		YES	NO
1	I am aware of, and comply with, legal quality standards for the environment (land–air–water)		
2	I believe it is more profitable to voluntarily be in charge of my own environmental efforts than to be forced to make specific changes to meet standards		
3	Records readily track and validate to others that a QRM plan for the environment is in place		
4	I am familiar with the major provisions of the law regarding surface water		
5	I know the water management system for the area my farm is located in		
6	I have attended at least one environmental meeting with other farmers in my area		
7	I can explain the difference between a point and non-point source pollution type		
8	I have made efforts to obtain support or funds to reduce environmental risks		
9	My manure storage and treatment facilities are designed for runoff and rainfall 24 hrs		
10	I conduct inspections of my waste management facilities, wildlife provisions and water quality programme once a week		
11	I have employee training plan in place, including pollution prevention techniques		

Air quality and control of odour		YES	NO
1	Animal and waste handling facilities are isolated or screened from roads and neighbours		
2	Windbreaks and other plants are used to mix air around my dairy farm		
3	Manure storage facilities have sufficient capacity relative to the number/size of cattle		
4	Manure storage facilities are screened or otherwise out of view of the public		
5	I use an above-ground manure storage facility		
6	I avoid spreading manure late in the afternoon during warm weather or the summer season		
7	I use local weather conditions to decide when to spread manure		
8	I have explored using deodorants, masking agents etc to improve air quality		
9	I produce electricity and hot water and improve odour control via a methane digester		
10	I have installed solar panels for electricity		

Quality risk management

Manure or nutrients that are properly managed, do not contaminate ground or surface water. A nutrient management system does not simply happen, it is implemented.

11	I use bottom filling methods in my manure storage facilities to reduce surface disturbances		
12	Liquid manure is injected into the soil or incorporated immediately after spreading		
13	I have attended a meeting, seminar or other class to learn more about air quality		
14	I know that the best way to control odours is through facilities and management measures		
15	Manure odour, which might damage air quality is controlled on my dairy farm		
16	Building interiors are equipped with monitored ventilation to exchange fresh air		

Water quality as relevant part of environmental quality management		YES	NO
1	My manure management does not allow effluents into groundwater or surface water		
2	My well is located and constructed so as to prevent pollution of groundwater by nitrogen and microorganism runoff from lots, lagoons, or other sources of manure		
3	I have a series of test wells to monitor groundwater quality near my dairy farm		
4	I have a system to monitor surface water that my be leaving my dairy farm		
5	The location of my animal waste and/or waste handling facilities is down-slope and more than 3 m away from wells or surface water		
6	I manage phosphorous, nitrates, nitrites and potassium in manure to avoid danger to humans or wildlife		
7	I understand that dairy cattle, deer, raccoons, rabbits and other species may contaminate water with pathogens		
8	I have all the proper legal permits to operate my dairy farm		
9	My dairy farm has a management plan to prevent manure from entering lakes, streams or other surface water		
10	I know the depths to groundwater under * confinement lots, * discharge areas for wash water, * row crop fields, * forage areas, non-crop pastures		
11	I test my well water for nitrates, chemicals and bacteria at least once a year		
12	I have an earthen waste storage pit, pond or lagoon that was designed or installed according to legal specifications		
13	I favour an above-ground storage facility as it eliminates groundwater seepage concerns and any leaks are more easily detected		
14	My storage area is pumped down regularly so that overflows never occur after heavy rainfall		
15	All surface water is diverted away from animal waste storage facilities		

Quality risk management

Management personnel must be involved, feel responsible and make efforts to make it happen

16	I have a regular (daily or weekly) schedule to inspect animal waste storage facilities for leaks, spills or overflow		
17	Any manure stacked on my farm is on concrete and protected from surface water; any leakage is diverted into a treatment pond or filter area		
18	All stream/ditches are protected by a strip of vegetation of at least 3 m width		
19	Dairy cattle have no access to streams or ditches		
20	Silage is harvested and stored between 65-70% moisture and storage is located at least 3 m down-slope from any well or surface water		
21	All silage leakage from my farm is collected and not allowed to enter low areas or water ways		
22	Any trench silos include an impervious base and are covered; surface water is directed away from these trenches on my farm		
23	Milking parlour waste water is directed to a liquid manure storage or septic tank on my farm		
24	I understand how manure-digestion to make electricity and hot water improves the quality of water discharged from my farm		
25	I have a written emergency action plan (a protocol or working instruction), which includes plans for containment, an emergency response team, location of emergency equipment and emergency phone numbers		

Quality of soil as part of environmental quality management		YES	NO
1	I am aware that imported nutrients in purchased feed and fertilizer must be balanced with milk and meat sales, as well as crop nutrient needs		
2	I have a phosphorous-based nutrient management programme		
3	I have records of all feed analysis reports from the laboratory		
4	I manage dairy manure so that it is neither toxic nor hazardous		
5	I take advantage of nutrient value and soil organic matter benefits of dairy manure by spreading the manure to cropland at rates to meet crop nutrient requirements, especially for phosphorous		
6	I am aware that storage of manure needs to be such that nutrients are best matched to soil needs		
7	I utilise facilities to produce methane fuel and bio-solids to be used to improve soil tilt		
8	I inject, or immediately incorporate into the soil, manure produced by my farm to capture the value of nitrogen		
9	I have a carefully designed nutrient management programme that includes feed testing, annual manure analysis and soil testing for P, N and K		
10	I consider soil texture, soil pH, climate and soil water-holding capacity before spreading manure on pasture plots or cropland		

Quality risk management

11	I understand the sustainable process involved in the nutrient cycle (= feed nutrients and manure nutrients applied to soil, which is producing maize, grass, or hay fed to cattle)		
12	Since compounds found in cattle manure are naturally occurring and not applied in excessive amounts, my farms fits well into the natural ecosystem		
13	My farm is large enough (> 300 head) to consider installing digesters to produce methane and bio-solids		
14	My farm prevents excessive sedimentation from wind erosion or water runoff by using practices to control the volume of wind or water → * filter strips; * contour farming; * no-till farming; * windbreaks; * other		
15	I have facilities in place to store and manage manure and a runoff system that limits discharges		
16	I carefully manage irrigation activities to prevent salty soil and other soil problems		
17	I use integrated pest management techniques to reduce any chance of soil contamination by pesticides		
18	I always keep records of all manure, pesticide and fertilizer applications to each field or plot		
19	I manage cattle grazing to prevent overgrazing, which exposes soils, increases erosion, endangers fish habitat and reduces natural sediment filtration		
20	I recognise the danger of exceeding crop nutrient needs of fields near buildings and/or lots		

Wildlife as a relevant part of environmental quality management		YES	NO
1	I use at least 3 m buffer zones of trees, grasses and/or wetland habitats by streams or ditches		
2	My farm encourages owls, hawks and other carnivorous species, which assist in controlling mice and various rodents in pastures, hay fields, grain fields and other areas		
3	I consider and avoid nesting areas when harvesting forages		
4	I encourage partnership with hunters, trappers and other groups, which may seek use of my properties for recreation		
5	I maintain odd areas in fields as habitat for wildlife		
6	I leave some standing maize for turkeys, deer and other wildlife to feed upon		
7	I belong to an organisation that supports wildlife management practices (e.g. for quail, duck, pheasant)		
8	I fence all open water to protect banks of streams, ponds etc		

Quality risk management

Aesthetics and community/society relations		YES	NO
1	I am aware that protecting the environment is in my long-term business interest		
2	I am committed to protecting my local environment for my family, neighbours and community		
3	I use windbreaks, buffer strips and other plantings to improve the appearance of my dairy farm		
4	I use above-ground storage facilities to promote the formation of a manure crust to help prevent odours and reduce flies, and create a visibly pleasant manure storage facility		
5	Building exteriors and fences are painted, clean and free of debris; buildings are landscaped, grass is normal. All of these are part of my efforts to be a good neighbour		
6	I avoid hauling manure on weekends and during evenings		
7	I remove dead cattle from sight and dispose of carcasses according to legal regulations		
8	I take advantage of every opportunity possible to help consumers understand how a dairy farm actually works		
9	I am perceived as a good neighbour		
10	I communicate with the community: * whenever I plan expansion or other changes; * by conducting tours, open days etc; * to determine sensitive times or dates to spread manure; * when mistakes happen; * so they come to me with their complaints rather than to regulatory people first		
11	I purchase my supplies locally so that local businesses are my spokesperson		
12	My dairy farm donates to community activities just like a town business		
13	I treat community relations as very important		
14	I volunteer for boards and strive to improve my communication skills		
15	I encourage my employees to be active in community organisations		
16	I know that all dairy farms, regardless of size, need to have an Environmental Quality Management plan		
17	I understand why the public wants "green" electricity and why they encourage on-farm manufacturing of electricity		
18	I use good sanitation to control flies		

There are six sections, each detailing several items to be scored. The end-score can be calculated for the farm as a whole and/or for each section. At least **70% of the total score** (per farm level or per chapter level) must be achieved to enter the category of "**good environmental quality status**".

Evaluating high quality management

After The Unigate Business Deal, Superior Stockmanship, version 8, January 1997, UK

The following document comprises three sections, each of those comprises different scoring questions.

If a given situation is NOT applicable to the farm, please answer YES to that question. At the end, a section score and a total score can be calculated to classify the farm.

Section A-key areas

Nutrition		YES	NO
1	Do you have a written nutritional plan?		
2	Is this plan updated at least three times per year?		
3	Does this plan cover the three lactation stages?		
4	Do you have feed labels of all feedstuffs?		
5	Are all records of feed deliveries, mixtures etc present?		
6	Are feed bunkers and storage sites free from birds and rodents and clean?		
Herd health			
7	Are all cattle movement records complete and up-to-date?		
8	Are all AI certificates or sire details available?		
9	Are all medicinal drugs in a locked storage and correctly recorded?		
10	Do you have proper segregation and calving facilities?		
11	Do you have appropriate handling facilities, holding and cattle loading facilities?		
12	Do you have a written herd health plan, regularly updated by the vet?		
13	Does the herd health record clearly identify trends for disease or lesions?		
Staff training			
14	Is the person responsible for the herd qualified (eg three years experience)?		
15	Has the person carrying out this self-assessment attended relevant courses?		
16	Have you read, and do you have, the Good Dairy Farming codes of practice?		
Milking machine			
17	Has your milking machine been inspected and tested by the dealer twice a year?		
18	Have the teat liners been inspected and changed according to the recommendations?		
Action plans			
19	Do you have regularly updated, written action plans?		
20	Do you have a list of telephone numbers for emergencies in the office?		
21	Do you have evacuation and emergency plans in the office?		

Evaluating high quality management

Stockhousing - For cubicles housing only			
22	Are your milking cow cubicles large and comfortable enough for the cows?		
23	Can you prove that you have at least one cubicle per cow?		
24	Is the bedding kept clean and dry?		
25	Can the cows get up and lie down without restriction or affecting other cows?		
26	Can the cows lunge forward to get up without striking their heads?		
27	Is the heel stone height and cubicle lay-out of an appropriate design?		
28	Is the loafing area 1.5 times the lying area?		
29	Are the yards frequently scraped?		
30	Is the building adequately lit (> 200 lux) and ventilated?		
31	Has your vet given you the written commentary on your accommodation?		
Stock housing - For loose housing only			
32	Can you demonstrate that your loose housing provides the required space?		
33	Is the bedding kept clean and dry?		
34	Is the loose housing cleaned out regularly?		
35	Is the loafing area at least 35% of the bedding area?		
36	Is the building adequately ventilated and lit (> 200 lux)?		
37	Has your vet given you the written comments on your loose housing?		
All stock housing			
38	Can you prove that watering facilities conform to recommendations?		
39	Are the watering facilities clean?		
40	Do the feeding facilities conform to standards? feed is clean and fresh		
41	Has your vet given you the written comments on your housing?		

Section A -- score	Scores YES	Scores NO
Nutrition	...	
Herd health	...	
Staff training	...	
Milking machine	...	
Action plans	...	
Stock housing	...	
TOTAL A	...	

Evaluating high quality management

Section B - The Five Freedoms

Hunger & thirst		YES	NO
1	Do newborn calves receive the recommended amount of colostrums?		
2	Are purchased calves treated and fed according to recommendations?		
3	Is the milk replacer prepared according to recommendations?		
4	Do all livestock have free access to clean, fresh water?		
5	Is the water tested daily for manure contamination?		
6	Is the continuity of water supply ensured in emergencies?		
Discomfort			
7	Is the calf accommodation well-ventilated, dry and clean?		
8	Are cattle protected from extreme temperatures?		
9	Are the cattle sheltered from wind, rain and snow?		
10	Are all buildings well-ventilated; airborne load low?		
11	Are cattle protected from unnecessary discomfort/stress?		
12	Are cattle protected from flies using indicated products?		
13	Are the water troughs of specified design and location?		
14	Are the yard surfaces kept clean and dry; scraped daily?		
15	Does the height of the accommodation allow heat behaviour?		
Pain, injury, disease			
16	Are cattle protected from disease as far as possible and treated promptly in the case of disease?		
17	Are stock inspected at least twice per day?		
18	Are all drugs administered by a competent person or vet?		
19	Are records for treatment kept three years and present for inspection?		
20	Are all veterinary investigations/procedures recorded?		
21	Is there a monitoring system available for cattle health & environment?		
22	Is there competent, experienced or knowledgeable staff present?		
23	Are adverse trends in herd performance/health investigated?		
24	Is body condition score regularly monitored and action taken when necessary?		
25	Are barns, parlour, and yards regularly hosed, cleaned and disinfected?		
26	Does housing have adequate ventilation to remove dust and bacteria?		

Evaluating high quality management

		YES	NO
27	Are teats always clean before milking?		
28	Are cows monitored daily for mastitis; infected cows milked last?		
29	Is the milking machine and associated equipment maintained properly?		
30	Is accommodation constructed to minimize the risk of trauma?		
31	Is the floor non-slip and not causing claw damage?		
32	Are exterior cow tracks maintained with the aim of minimizing claw lesions?		
33	Are cattle handling facilities designed for holding securely?		
34	Are the claws of all cows inspected for growth /infection; trimmed?		
35	Are foot baths present and used properly?		
36	Are isolation boxes present for sick or injured cattle?		
37	Are the impermeable walls of these boxes high enough?		
38	Are (down-)calving cows routinely inspected?		
39	Are surgical procedures performed by trained competent persons?		
40	Is cattle marking carried out by trained competent persons only?		
41	Are paints, preservatives, chemical products out of cattle reach?		
42	Is all electrical equipment often tested and inaccessible to cattle?		
43	Are regular checks made to ensure absence of toxic plants/weeds?		
Normal behaviour			
44	Are all cattle provided with the required space to lie down correctly?		
45	Is space allowance made based on age, sex, weight, horns?		
46	Do all passages and pens allow two animals to pass freely or turn?		
47	Is all lighting for cattle barns to the required standard?		
48	Do cattle have sufficient freedom of movement, grooming, stretching?		
49	Are cattle only confined under specified conditions?		
50	Are cattle provided with opportunities to perform normal behaviour; is remedial action taken when necessary?		
51	In loose housing are polled and fully horned cattle kept separately?		
52	When cattle are segregated (except for sickness or injury), are they still within sight and sound of herd mates?		
Fear & distress			
53	Are cattle handled quietly but firmly at all times?		

Evaluating high quality management

54	Are cattle kept such that they are not continually or often frightened?		
55	Do buildings and alleys avoid dead ends, sharp corners, light contrasts?		
56	Do stockpersons have a caring and responsible attitude towards cattle?		
57	Are cattle treated such that fear of human contact is minimized?		
58	Are facilities used such that slipping and panic are prevented?		
General issues			
59	Are all stockpersons informed of the prevailing cattle welfare codes?		
60	Are cattle markets, abattoirs, haulers selected on welfare issues?		
61	Are contingency plans for fire, flood, loss of service etc in place?		
62	Is there an updated telephone list for emergencies present?		
63	Are records of all events occurring in the cows breeding cycle kept?		
64	Are the sires selected for breeding chosen for minimizing calving problems, congenital disorders?		
65	Are all stockpersons familiar with the normal behaviour of cattle, the signs of health disorders and contentment?		
66	Are all records required by law maintained + accessible for inspection?		

Section B scores	Scores YES	Scores NO
Hunger & Thirst	...	
Discomfort	...	
Pain, injury, disease	...	
Normal behaviour	...	
Fear & Stress	...	
General issues	...	
TOTAL B	...	

Evaluating high quality management

Section C - Overall farm inspection

Milk storage area		YES	NO
Milk truck access 1	Is there a concrete area for the milk truck?		
2	Is this concrete in good condition?		
3	Is the concrete area frequently cleaned?		
4	Is the access for the milk truck unobstructed?		
5	Are there preparations for poor weather conditions?		
Doors 6	Is the outer door complete, undamaged and weather sealed?		
7	Can the outer door be closed?		
8	Can the milking parlour and barn doors be closed?		
9	Can all doors be locked?		
10	Are the doors locked for the night?		
Floors, walls 11	Is the floor complete and undamaged?		
12	Is the floor clean?		
13	Is the floor well-drained?		
14	Are the walls complete and undamaged, with an impervious finish?		
15	Are the walls clean?		
Ceiling, windows 16	Is the ceiling/roof complete and undamaged, with a sealed finish?		
17	Is the ceiling clean?		
18	Are the beams, girders and ducting clean?		
19	Are the windows unbroken and clean?		
20	Are the lights clean; no exposed bulbs or tubes?		
Environment 21	Is the lighting bright?		
22	Is the ventilation good?		
23	Is there an absence of dust and foul odour?		
24	Is the area clear of accumulated waste, junk or articles (chemicals)?		
25	Are all surfaces clean?		
Pest control 26	Are birds, including swallows, excluded from the area?		
27	Have control measures been taken against flies and vermin?		
28	Have control measures been taken for dealing with waste?		
29	Is the area clear of animal feed?		

Evaluating high quality management

Bulk tank 30	Are the emergency milk storage/collection arrangements in place?		
31	Is the exterior clean?		
32	Is the parlour white board clean, complete and up-to-date?		
33	Is the temperature of the tank cooling set for 4°C?		
34	Is the outlet clean and capped?		
Chemicals 35	Do cleaning agents used comply with regulations?		
36	Is the rinse water supply clean and suitable?		
37	Is the area clear of pesticides and medicinal drugs?		
38	Is the area free from aromatic disinfectants?		
39	Are oil, paint and preservatives absent?		
Hygiene 40	Is there a hand sink with hot and cold water?		
41	Is there unscented soap and a plastic nail brush present?		
42	Are there paper towels present?		
43	Is there a frequently emptied waste bin ?		
44	Is there a hosepipe and hard broom for exclusive dairy use?		
Milking parlour			
Floor, walls 45	Are the floors undamaged?		
46	Are the floors clean?		
47	Are the floors well-drained?		
48	Are the walls undamaged?		
49	Are the walls clean?		
Ceiling, window 50	Is the ceiling/roof complete and undamaged, with a sealed finish?		
51	Is the ceiling/roof clean?		
52	Are the beams, girders and rails clean?		
53	Are the windows unbroken and clean?		
54	Are the lights clean; no exposed bulbs or tubes?		
Environment 55	Is the lighting bright?		
56	Is the ventilation good?		
57	Is there an absence of dust and foul odour?		
58	Is the area clear of accumulated waste, junk and stored articles?		
59	Are all the surfaces clean, incl. pipe work supports, cable ducting?		

Evaluating high quality management

Exterior of milking equipment 60	Is the pipe work clean?		
61	Are the jars clean?		
62	Are the clusters and jetters clean?		
63	Are the hoses clean?		
64	Is the ancillary equipment clean?		
Interior of milking equipment 65	Is the pipe work clean?		
66	Are the jars clean?		
67	Are the clusters and jetters clean?		
68	Are the hoses clean?		
69	Is the ancillary equipment clean?		
Hygiene aids 70	Is udder wash available?		
71	Is teat dip or spray available?		
72	Are paper towels available?		
73	Is a water hose and water point available?		
74	Is a manure scraper or squeegee available?		
Cleaning system 75	Are the systems regularly checked and maintained?		
76	Are chemicals used, measured or dispensed correctly?		
77	Is there sufficient hot water available (80-82°C)?		
78	Are there dedicated plastic brushes available?		
Health & Hygiene			
79	Are the milkers clothes and caps clean?		
80	Are the milkers hands and arms clean?		
81	Are skin cuts and wounds well-covered?		
82	Are staff free from any notifiable disease?		
83	Is the area a “no smoking” area?		
Safety			
Electric systems 84	Is the electric system according to regulations?		
85	Are the safety devices effective?		
86	Are all switches, plugs, sockets etc unbroken, protected and secured?		
87	Are the appliances and instruments working correctly?		

Evaluating high quality management

88	Are the security lights and alarms functioning?		
Chemicals 89	Are all pesticides/chemicals in a locked store?		
90	Are all containers sealed and labeled?		
91	Are all pesticides and chemicals measured and used as recommended?		
Training & Instr. 92	Has pesticides training been received; approved contractors used?		
93	Has recognised training been undertaken?		
94	Has first aid training been received?		
95	Are instruction and working sheets been displayed?		
96	Are safety instructions given?		
Precautions 97	Is protective clothing clean and ready, including respirators?		
98	Is there a first aid box, fire extinguishers and instructions?		
99	Are chemical data sheets available?		
100	Is safety information available?		
101	Are warning signs displayed for bulls, slurry, etc?		
Documentation			
102	Are records kept of pesticides and herbicides used?		
103	Are any health and safety inspector's reports satisfactory?		
104	Are any environmental reports satisfactory?		
105	Is the official Ministerial registration identification valid?		
106	Are notifiable disease tests negative for cattle?		
107	Are service contracts up-to-date?		
108	Is any (quality) handbook available?		
Farm surroundings			
Entrance 109	Is the entrance tidy?		
110	Is the farm sign clean and visibly displayed?		
111	Is the drive well-maintained?		
112	Are the gates in good working order?		
113	Is there adequate unobstructed access?		
Buildings, yards 114	Are the buildings well-maintained?		
115	Are the guttering, down pipes and drains well-maintained?		
116	Are the yards hard surfaced and maintained?		
117	Are the yards drained and scraped clean?		

Evaluating high quality management

118	Are buildings/yards free from accumulated waste, rubble, scrap?		
Waste 119	Is silage effluent contained?		
120	Is there suitable secure slurry storage; are warning signs placed?		
121	Is there tidy dung storage?		
122	Is rubbish, containers etc regularly disposed off?		
123	Are chemicals disposed of according to GDF code of practice?		
Security, protection			
124	Are animal hazards secure with warning signs?		
125	Are chemical hazards secure with warning signs?		
126	Do physical hazards eg fragile roofs have warning signs?		
127	Is there sufficient and suitable gates/fencing around the farm?		
128	Is foot path access indicated?		
General appearance			
129	Are unsightly features screened?		
130	Are verges and ditches maintained?		
131	Are natural features cared for?		
132	Are there shrubs, trees, and/or flowers planted?		
133	Are historical features maintained?		

Section–C scores	Scores YES	Scores NO
Milk storage area	...	
Milking parlour	...	
Health & Hygiene	...	
Safety	...	
Documentation	...	
Farm surroundings	...	
Security & Protection	...	
General appearance	...	
Total score section C	...	

Evaluating high quality management

Total score section A	...
Total score section B	...
Total score section C	...
Overall total	...

There are, in total, 133 items to be scored. The total score divided by 133 and multiplied by 100 gives the final percentage score. You can use this score to assign final percentages to a predetermined category. For example: **EXCELLENT; AVERAGE; ACTION NEEDED.**

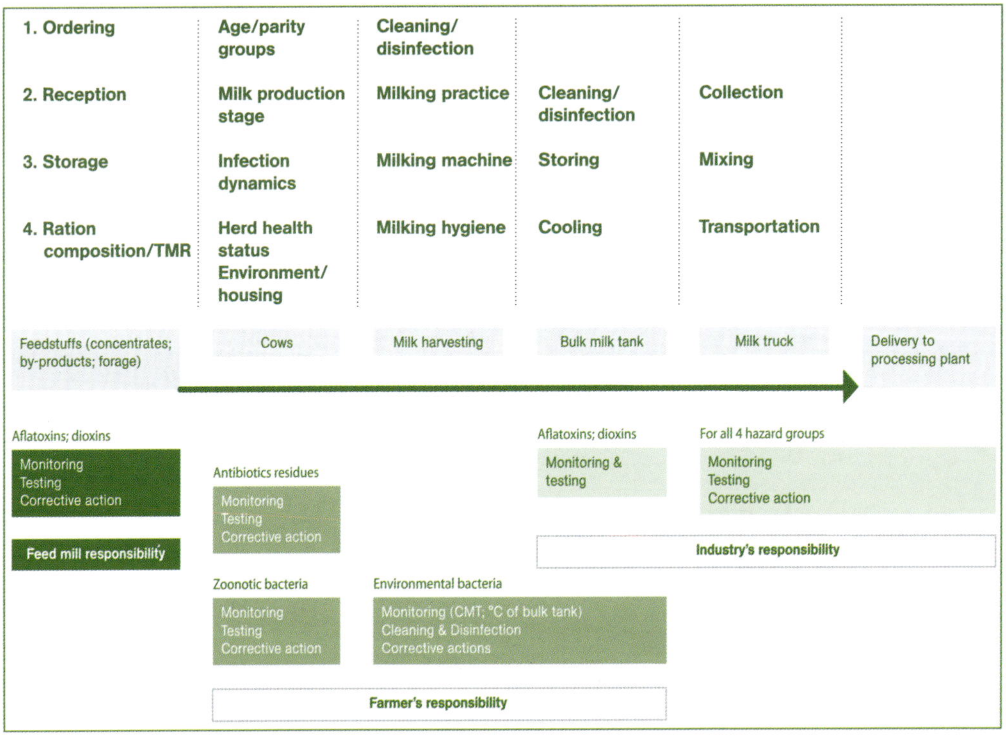

Figure 16.3 HACCP based quality assurance of raw material production: a flow diagram

Quality risk management based on HACCP principles

After Khandke, 1999

Table 16.2 Quality risk management based on HACCP principles - implementation table

Process step	Look at the diagram of the production process
Risks	What can possibly go wrong in the production process?
Prevention & Control measures	What has been done to prevent risks from occurring?
CCP yes or no?	See the formal criteria for CCP definition
What is measured?	What factors must be controlled to reduce or eliminate the risks?
How is it measured?	What methods or tests must be applied (sample size!)? Which senses must be used? How can you minimize the probability of failures?
Where is it measured?	Check the various production process steps where risks occur
How often is it measured?	Optimize frequency (predictive value; economics)
Who is measuring?	Identify the responsible person(s) and specify his/their task
Set standards & tolerance limits	For each CCP (if not a true CCP, a point of particular attention, POPA, and then "target values" must be defined)
Corrective measures	What must be done when test results are off limits? What happens to the product (e.g. milk)? Describe how the process can be adjusted What action(s) to take to limit the probability of repeated occurrence
Who will take corrective action?	Farmer, veterinarian, or someone else How and where are the QRM records kept?

CCP= critical control point; QRM= quality risk management programme

Quality control parameters on Dutch dairy milk (2009)

In addition to the regular checks on milk fat, milk protein, cell count, antibiotic residues and freeze point, the Qlip/NZO laboratories (under COKZ) can also monitor:

- Aflatoxin
- Heavy metals (Cadmium; Lead (Pb))
- Radio-activity
- Pesticides
- Chloroform
- Polychlorobifenyl (PCB)
- Dioxin
- Polycyclic aromates
- Anthelmintics (residues)
- Melamine

Factors affecting the omega 3 content of milk

My Campina Magazine, January 2008, The Netherlands

Positive effects

- increased access to grazing
- frequent rotation of grazing animals around different plots
- Short grass
- Grass with omega-3 > 3.5 mg/kg
- Grass with organic phosphate matter > 630 g/kg DM
- Provision of concentrates containing linseed
- Provision of concentrates without hard fat
- Provision of good quality silages

Negative effects

- restricted grazing
- over-grazing or prolonged grazing of one plot
- Long grass
- Grass with omega-3 < 3.5 mg/kg
- Grass with organic phosphate matter < 620 g/kg DM
- Poor weather conditions
- Concentrates with hard fats
- Maize silage or grass silage with a high DM level
- Wilted grass silage

Lactose in milk

Bernd Pieper, NL, 2006

Table 16.3 Lactose levels in milk from dairy cows

Level	Unit	Remark
Lactose too low	(< 4.50%)	Problems with high cell counts If cell counts are not high → check the ration → Possibly a decrease in propionic acid due to a poor protein to starch ratio (in that case, often the milk urea is increased!)
Lactose and urea are both low	→	Insufficient DMI
Lactose is elevated	(> 4.65%)	This is desirable and it demonstrates a well-balanced diet and/or healthy cows

- Cows in **early lactation** show a **lactose level higher** than that at the end of lactation.
- **After the first calving**, the lactating heifers show a **lactose level higher** than the adult cows.
- A **high lactose content + a low milk fat** content may indicate ruminal **acidosis**.

Action → increase the fibre in the ration and decrease the quantity of concentrates.

Preventing contamination of silage by *Clostridium spp*

Preventing contamination of silage by *Clostridium spp* (including silages of good quality)

Clostridium spp (or butyric acid bacteria) in silage cause economic losses, either by penalties for quality failure, or by loss of DM (up to 15%). In the latter case, the digestibility, feed intake (about 1.5 kg per cow per day) and the metabolisable energy also decrease. When the temperature of the silage is 5°C higher than the ambient temperature, the silage is at risk of developing *Clostridium spp*. These high temperatures can be found at depths of 50 to 100 cm from the silage face.

A recent study (Anonymous, 2009) has shown that, despite fresh silage analysis showing very low *Clostridial* bacteria count, the silage can become heavily contaminated once the clamp is opened. This phenomenon is caused by an aerobic instability which begins with yeasts and continues with moulds. This may happen **even in silage of good quality** and has been demonstrated in maize silage and sorghum silage.

After harvest, at ensiling, there can be around one hundred spores, however, after opening and feeding (exposure to air), the number of spores increases up to millions per gram silage after five days (in the forenamed study after 5 days !).

Treatment of silage with *L. buchneri* (NCIMB 40788; anti-mould product) could stabilise the silage for at least 300 hours (or 5 days) maintaining the pH at a level below 5. The spores of *Clostridium* did not develop under air exposure. However, there are significant DM losses often associated with application of this micro-organism.

In normal situations, ensiling maize reduces the pH rapidly (rapid acidification) and spores cannot develop. But at the moment of feeding (opening, cutting and feeding) there is ample exposure of the silage to air and a reactivation of aerobic fermentation occurs, starting with yeasts and followed by moulds. The pH increases and the temperature increases up to 45°C. The increase in pH causes the *Clostridium* spores to develop and so contamination begins.

Another way to protect the maize silage against the losses due to *Clostridium spp* is to use **poly-ethylene tissue** to cover the silage clamp.

However, recently it was determined that **black-on-white oxygen-barrier-films** (125 microns) were

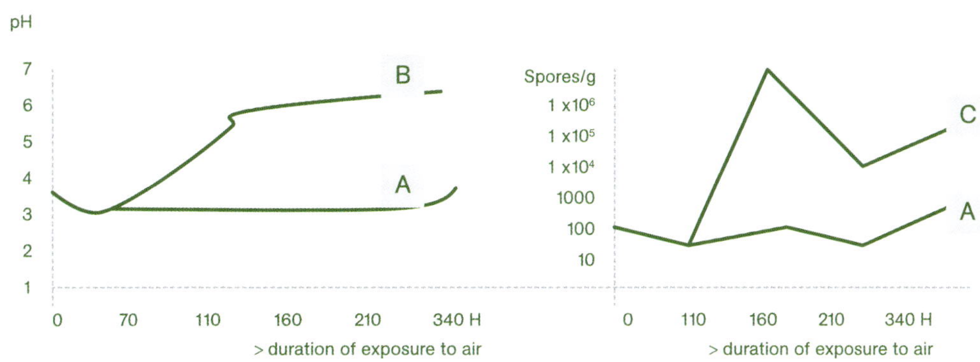

Figure 16.4 The formation of *Clostridium* spores in maize silage following exposure to air
A= with product *L. buchneri* ; B= pH in maize silage without this product
C= number of *Clostridium* spores without treatment

Preventing contamination of silage by *Clostridium spp*

Preventing contamination of silage by *Clostridium spp* (including silages of good quality)

Clostridium spp (or butyric acid bacteria) in silage cause economic losses, either by penalties for quality failure, or by loss of DM (up to 15%). In the latter case, the digestibility, feed intake (about 1.5 kg per cow per day) and the metabolisable energy also decrease. When the temperature of the silage is 5°C higher than the ambient temperature, the silage is at risk of developing *Clostridium spp*. These high temperatures can be found at depths of 50 to 100 cm from the silage face.

A recent study (Anonymous, 2009) has shown that, despite fresh silage analysis showing very low *Clostridial* bacteria count, the silage can become heavily contaminated once the clamp is opened. This phenomenon is caused by an aerobic instability which begins with yeasts and continues with moulds. This may happen **even in silage of good quality** and has been demonstrated in maize silage and sorghum silage.

After harvest, at ensiling, there can be around one hundred spores, however, after opening and feeding (exposure to air), the number of spores increases up to millions per gram silage after five days (in the forenamed study after 5 days !).

Treatment of silage with *L. buchneri* (NCIMB 40788; anti-mould product) could stabilise the silage for at least 300 hours (or 5 days) maintaining the pH at a level below 5. The spores of *Clostridium* did not develop under air exposure. However, there are significant DM losses often associated with application of this micro-organism.

In normal situations, ensiling maize reduces the pH rapidly (rapid acidification) and spores cannot develop. But at the moment of feeding (opening, cutting and feeding) there is ample exposure of the silage to air and a reactivation of aerobic fermentation occurs, starting with yeasts and followed by moulds. The pH increases and the temperature increases up to 45°C. The increase in pH causes the *Clostridium* spores to develop and so contamination begins.

Another way to protect the maize silage against the losses due to *Clostridium spp* is to use **poly-ethylene tissue** to cover the silage clamp.

However, recently it was determined that **black-on-white oxygen-barrier-films** (125 microns) were far more effective than poly-ethylene tissue (Borreani *et al.*, 2007).

The silages most susceptible to losses in DM, digestibility and aeration are:

- maize silage (contains already yeasts in the field)
- silage with high sugar contents and high DM content
- bales of silage with high DM content
- maize silage in a TMR wagon.

Maize silage should not contain more than 3 x log10 spores per gram of silage. If it is possible to restrict the concentration of spores to below 5 x log10 per gram, then the concentration in the bulk milk tank will be lower than 3 x log10 spores per litre of milk.

Useful reading

Anonymous 2009. Preventing Clostridia contamination at feed out – even in good quality silage. International Dairy Topics 8 (5) p.11

Borreani G, Tabacco E, Cavallarin L. 2007. A new oxygen barrier film reduces aerobic detoriation in farm-scale corn silage. J. Dairy Sci. 90 (10) 4701-4706

Borreani G, Tabacco E. 2008. Low permeability to oxygen of a new barrier film prevents butyric acid bacteria spore formation in farm corn silage. J. Dairy Sci. 91 (11) 4272-4281

Holzer M *et al.* 2003. The role of Lactobacillus buchneri in forage preservation. Trends in Biotechnology 21 (6) 282-287

Vissers *et al.* 2007. J. Dairy Sci. 90 (7) 3278-3285.

Urea concentration

Fermentable protein → **Metabolized into NH_3**
+ fermentable energy ↓
Microbial protein

Under normal feeding conditions, the urea level in milk is around 20 to 30 mg/L. This level increases when the fermentable protein level in the ration is in excess; the same is applicable when there is a lack of energy.

At high urea levels, a loss of protein nitrogen and increased liver stress can be observed. The same occurs when the ratio between protein and energy in the ration is high (e.g. with high grass rations).

The urea level can be reduced by lowering the protein content of the ration (less fertilizer; add beet, potatoes or maize) and/or stabilizing the supply of soluble nitrogen to the rumen microbes. The starch content can be increased by adding more energy to the ration (e.g. more maize). If the protein content has to be increased, beet, potatoes or maize may be added.

If the ratio between fermentable protein and starch is too low, proteins may be added to the ration, which may increase the urea level in milk.

If both milk protein percentage and milk urea are too low, protein can be added to the ration (and maybe also more energy) through products like soy, rapeseed or brewers grains.

Milk urea and milk protein levels are associated with certain reproductive performance results. Table 16.4 summarises these effects (Labbé & Ennuyer, 2010).

Table 16.4 Interpretation of milk protein levels

Milk protein content	Potential action	Potential action	Potential action
High milk protein level (e.g. > 3.50%)	Increase fermentable protein in the ration	Optimal fermentable protein content is present in ration	Lower the fermentable protein content in the ration
Low milk protein level (< 2.90%)	Increase protein content in the ration. And maybe starch content too?	Increase protein content in ration and also the starch content.	Maybe increase protein content in ration?? Increase starch content in ration anyway.
	Low milk urea	Milk urea level okay	Milk urea increased
	< 20 mg/dL	20 – 30 mg/dL lait	> 30 mg/dL

(ACM Animal Nutrition, Zwolle, the Netherlands, 2010)
Note that the threshold value for milk protein contents may vary according to region or country.

Urea concentration

Table 16.5 Associations between varying milk urea, milk protein levels and reproductive problems

Urea content low and milk protein content normal or elevated	Urea level normal and protein level elevated	Urea level increased and milk protein level elevated
= Too little nitrogen in the ration = Lowered fibre digestion? (milk fat content may decrease too)	= Excess energy in the ration = May lead to fat cow syndrome at the end of lactation and a subsequent difficult calving	= Excess energy and protein in the ration = Risk of fattening at end of lactation and hence problems at calving
	Urea and milk protein normal	Urea level elevated and milk protein normal
	= Good balance between ration energy and protein levels	= Too much nitrogen in ration = Reproductive problems
Urea low and milk protein low	Urea normal, but milk protein low	Urea elevated and milk protein lowered
= Too little energy and protein in ration = Risk of reproductive problems (e.g. heats not observed)	= Too little energy in the ration = Cow is losing body reserves or is unable to restore reserves	= Too much nitrogen, but too little energy in the ration = Risk of sub-fertility and early embryonic death

Milk harvesting hazards, risks & monitoring list

Table 16.6 Example of a hazards, risks,& monitoring list for milk harvesting in a HACCP-based quality risk management.

Identified Hazard	Probability of occurrence	Prevention measures	CCP or POPA	Critical limits / Targets	CCP monitoring	Corrective Actions	Records	Verification
MH.3 Contamin-ation of milk from milkers gloves	High	Use of gloves; or apply strict hands hygiene after milking each cow	CCP. 4	By sampling gloves or hands and bacterial score should be < 1	Samples taken from hands/ gloves 1x/ week at milking	Clean hands more thoroughly. Wear gloves, and disinfect them more often	See protocol of GDF, section milk harvest. Monitoring records	Check records 1x/ month
MH.4 Improper washing	High	Wash teat ends, dry afterwards with single paper towel	CCP. 5	Seconds spent at activity (60-90)	Check water temp (34°C). Check time spent 1x/week	Repeat or adjust teat & udder preparation	See protocol of GDF, section milk harvest. Monitoring records	Verify monitoring records 1x/month
MH.5 Cleaning water	No	Guarantee water quality	POPA 6	No microbes no chemical contamin-ation	Routine quality checks 2x/yr	Change water supply system compon-ents or source	Monitoring records	Check monitoring records 2x/yr

(MH= microbiological hazard; CCP= critical control point; POPA= point of particular attention; GDF= good dairy farming code of practice)

See also "Applying HACCP-based quality risk management on dairy farms" by Noordhuizen *et al.*, 2008, Wageningen Academic Publishers, Wageningen, the Netherlands

Reproduction

> "Reproductive performance has always been a key component of herd fertility schemes and herd health & productivity management. In this chapter, we limit ourselves to the interpretation of herd reproductive data, monthly and yearly performance data. Moreover, a protocol is introduced for analysing reproductive problems in the herd. Finally, some more specific aspects are addressed, such as *Neospora* abortions and the relationship between milking and oestrus/ovulation in dairy cows."

Interpreting a fertility-health wall card

A fertility-health wall card comprises the (daily) events for each individual cow. The starting point for each cow on the card is the date of calving, cows are listed in chronological order of calving. In addition to the calving date, calving history, dates of heats detected, insemination dates, reproductive disorders and treatment if any, date of pregnancy diagnosis and result, expected calving date, dry off date and treatment if any, clinical diseases and treatment if any, as well as date of culling and reason are listed by the farmer.

The following example looks at data from 32 cows over a selected period, there are, in fact, 33 cows on the card but some are counted twice! e.g. # 35, because she calved twice in the selected analysis period). The "🡐" below indicates a potential problem area that should be given special attention during subsequent farm visits.

Calvings

- majority in the period November–March but there are calvings throughout the year
- 18 female calves + 15 male calves (1x twins)
- Three stillborn calves out of 33 cows calved=9%
- Retained placenta: five out of the 32 cows calved=16 % 🡐

Intervals of reproduction

- Calving to 1° heat = 35 days (four out of nine cows at 14 days and two cows at more than 80 days)
- Calving to first AI=73 days on average (four cows at more than 80 days)
- Calving Interval (expected) = 374 on average (median is 372 days)

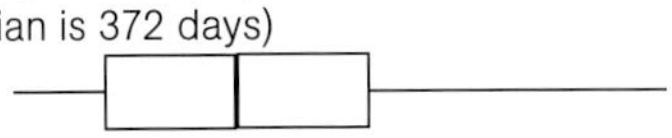

328 350 372 387 455

Mastitis

- Five cows out of 32=16 %
- Eight cases out of 32=25 %
- Repeat cases = two out of eight =25%🡐

Lameness

- 10 cows out of 32=33 %🡐
- Diagnoses are variable and not definitive

Other disorders

- Miscellaneous: ketosis; diarrhoea; milk fever 3x; 2x digestive problems (but no diagnosis)

Culling

- Five cows out of 32=17 % (= too few??) 🡐
- Reasons: 3x production (low production due to genetic background of heifers is usually a common cause; low production in older cows is commonly caused by something else, such as disease e.g. chronic mastitis coupled with lameness), 1x poor fertility, 1x mastitis

Average length of dry period is 60 days on average

	Plan of action:
1	Treat the clinical mastitis cases appropriately (antibiotics; milking practice; duration) and evaluate the treatment effects
2	Improve the diagnoses of claw lesions + start herd claw trimming
3	Work to reduce calf mortality (infections? sires? housing? hygiene? feed?)
4	Revise strategy for culling (mastitis; lameness; fertility?)
5	Address incidence of retained placenta if incidence remains high or increases further (disease resistance? ketosis?).

Suggestion:

Next page are two parts of a Herd Fertility-Health wall card containing data collected in the field. Try to conduct your own analysis of these data and subsequently design an action plan based on the results.

Monthly herd performance

Year= 2008 - 2009 **Farm Code=**

Date of visit (dd-mm)	31-10	30-11	24-12	31-01	28-02	....	....	Ref value
# cows present Lact / Dry	76/19	69/16	68/14	81/9	79/18			17% dry
L milk per day	30	28	28	30	32			
Mean milk fat %	41	39	40	41	41			
Mean milk protein%	34	39	33	33	34			
Ratio milk fat to protein	1.20	1.0	1.21	1.24	1.21			
% cows with fat < 3.00	1	2	4	3	5			
% cows with protein < 3.00	3	2	4	8	11			
Urea level in bulk tank (mg/dL)	-	-	-	-	-			
Somatic cell count (SCC) max (x 1000cells/ml)	398	411	453	822	612			< 250
SCC min (x 1000cells/ml)	78	88	92	81	67			< 100
% new cows SCC > 250 (x 1000cells/ml)	6	9	12	19	9			< 10%
% cows with SCC > 250 (x 1000 cells/ml)	8	11	13	21	22			< 15%
Bacterial Count (x 1000/ml)	7	26	43	6	13			< 50
# Clinical mastitis cases	3	4	2	5	2			< 25%
# Abortions	0	1	0	0	0			< 4%
# Retained placenta	0	1	1	2	0			< 10%
# Metritis	1	2	0	1	3			< 10%
# Abomasal displacement	0	0	0	0	2			< 5%
# Ketosis	0	0	0	1	2			< 5%
# Acidosis	1	0	0	0	1			< 5%
# Lameness cases	0	3	2	4	3			< 25%
# Stillborn calves	0	0	2	2	1			< 5%

Monthly herd performance

Year= 2008 - 2009 **Farm Code=**

Date of visit (dd-mm)	31-10	30-11	24-12	31-01	28-02	....	....	Ref value
# Calves dead > 3 days age	0	0	1	1	1			< 5%
# Calves with diarrhoea	0	0	1	3	4			< 10%
# Calves with respiratory problem	0	0	0	1	0			< 10%
# Cows culled	0	4	1	2	1			< 25%
Reason: Production			1		1			< 22%
Reproduction		2						< 30%
Udder/teats		1		1				< 15%
Claws/legs				1				< 10%
Other		1						< 23%
% pregnant per AI	54	50	46	51	53			> 50%
# AI/cow inseminated	1.6	1.8	1.7	1.6	1.8			< 1.7
Calving Interval (actual)	378	369	379	381	383			365-385
Calving Interval (expected)	381	379	389	391	392			365-385
% anoestrus in number of cows examined	10	12	17	21	23			< 10%

Interpretation of the above example

There were more lactating cows present in January and February (new calvings).The 17% of dry cows is valid for herd calving intervals of 365 days and a dry period of 60 days duration. If the dry period length decreases and/or the calving interval decreases, this dry cow percentage may increase.

The **milk production** per cow per day increased from January to February, possibly due to the new calvings.

The **milk fat** percentage decreased in November (transition from grazing to barn poorly managed?) but the figures are acceptable. The same is true for **milk protein**, with the exception of November where the **ratio of milk fat to protein is zero and undesirable.** Around December to February, the number of cows with milk protein levels < 3.00 increased; is there a lack of energy in the ration? Ketosis? **This should be followed-up**.

There is a large variation between cows regarding the individual **somatic cell count** levels. The number of mastitis cases indicates the same: there is something going on in the herd. The cows with a SCC > 250.000/ml and the mean SCC indicate a mastitis situation that could potentially explode. **This should be followed-up**

Monthly herd performance

The milk quality (microbiology) was always under the threshold of 50.000cells/ml but there were moments of risk in November and December. **This should be followed-up**

Abortions: no problem

Retained placenta: little problem

Endometritis: increases 🡐 **to be followed-up**

Abomasal displacement: no problem

Ketosis: no problem

Acidosis: no problem

Lameness: appears prevalent in the herd. Diagnoses?? 🡐 **to be followed-up**

Calves stillborn: five cases in three months 🡐 **to be followed up**

Calves dead after three days age: three cases in three months 🡐 **to be followed-up**

Calves diarrhoea: number increasing 🡐 **to be followed -up**

Calves respiratory disease: no problem

Culled cows: the total rate is rather low. Revisit the farmer's culling strategy!

Reproduction: the pregnancy rate is good. The number of inseminations per cow is acceptable. The intervals do not highlight any problems.

The number of cows examined for lack of heat behaviour increased 🡐 to be followed-up

The next step is to compose a Plan of Action, with interventions/advice for the short-term (a maximum of five priority items) and those for the mid-long term (see elsewhere in the series). The issues to be selected should derive from those marked with a "🡐".

Protocol for the analysis of reproductive problems

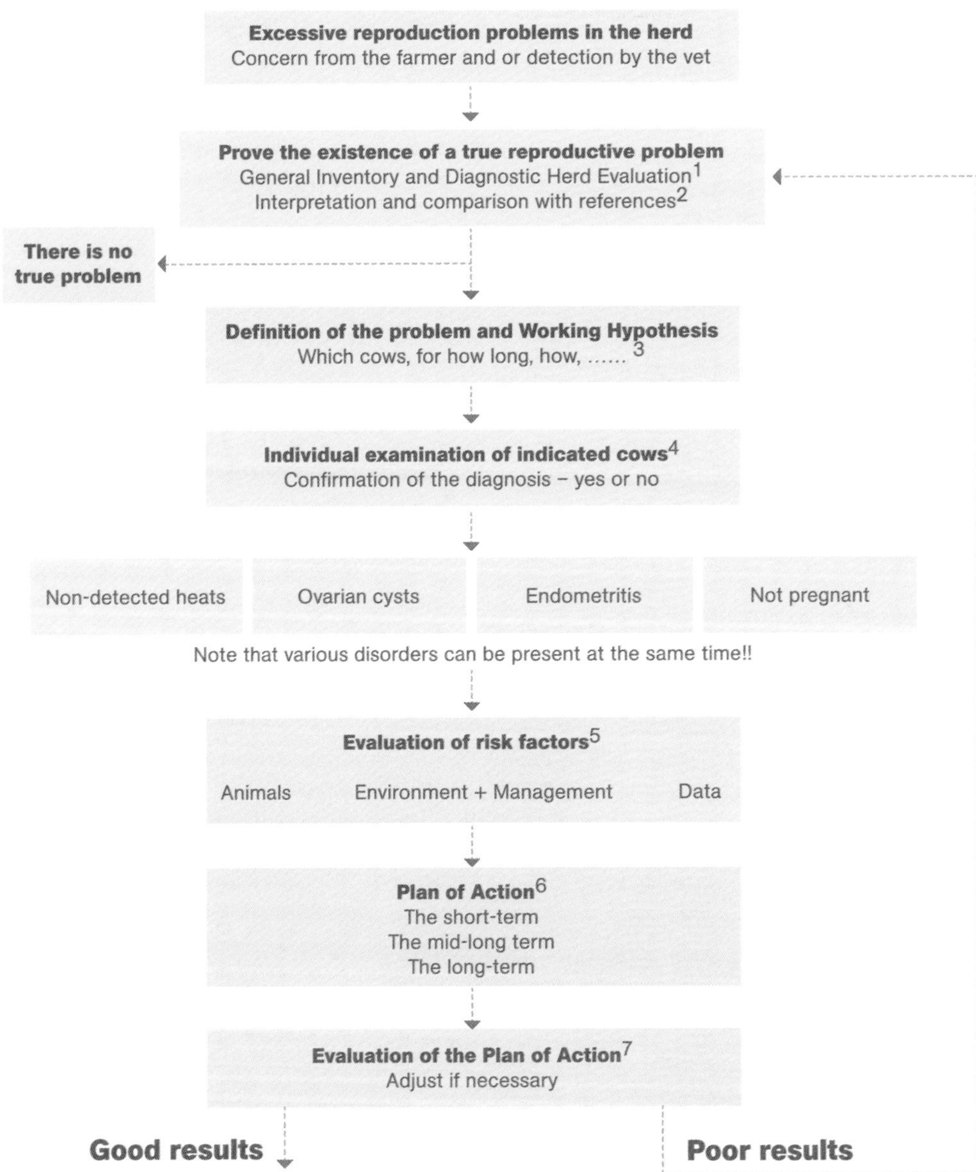

Figure 17.1 Protocol for the analysis of reproductive problems

Yearly reproductive performance data

An example year report concerning the reproductive performance of a dairy farm is given below. In practice, you would be asked to interpret the parameters with regard to their value and interdependence.

- List the reference values for your region in the column next to the farm figures.
- What are the greatest problems on this farm and why?
- What plan of actions do you propose to improve the situation?

See Table 17.1 for performance data

Interpretation
At herd level

- The **milk production** is good but at the same time you have to pay attention to potential problems of negative energy balance (and ketosis) in such herds.
- There were, on average, 84 cows present but there have been 94 calvings this year; have cows been culled?
- There are 27 heifers (parity 1) versus 69 adult cows. A percentage of 28% heifers is somewhat higher than the standard mean of 25%; were there forced culls?
- The mean **Calving Interval** (expected) is adequate for this herd (unless the farmer has reached this level by culling many cows). Unfortunately, it is not possible to compare this interval with that of last year to see whether there has been an improvement.
- The **interval from calving to first AI** is 84 days and somewhat long (reference is 70 days), especially when bearing in mind the large range in this parameter. The latter can be partly explained by the **interval from calving to last AI** of 116 days on average. There are cows that have had more than one AI (see also the **interval from first AI to conception** of 33 days!!). As this interval is longer than the interval from calving to conception (115 days), it means that there are cows inseminated but still not pregnant and possibly culled.
- Obviously, there are few inseminations before 50 days lactation (2/3), which partly explains the delay in the interval from calving to first AI, as well as calving to conception.
- Only **83 cows with a calving date have been inseminated**.
- The first 15 day lactation heats have not been observed, although there must have been some at least! Even between 16 and 30 days lactation only a few heats have been observed (10%) - this figure should have been between 40 to 50%. During the period 31 to 50 days, this percentage is only 17%, while between 60 and 70% is the reference value. In the AI period (50 to 75 days) the percentage is 37%, so there is a problem with heat detection, which contributes to abnormal mean intervals.
- The **number of first AI** is 72 out of 84 cows present, so there are cows that are not inseminated again and possibly have been culled.
- The **AI Efficacy Figure** is 1.82, which is higher than the reference value of 1.50. This may be related to the low heat detection rate and/or problems of cystic ovarian disease, endometritis, sire fertility or insemination technique. These parameters should be checked.
- The **pregnancy rate after first AI** of 54 % is acceptable but should be improved.
- The mean **culling rate** of 33% (28 cows) shows that there are, in fact, too many cows culled compared with the reference of 25%. This may indicate a forced culling due to e.g. reproductive problems. The latter is confirmed by the number of cows culled for reasons of reproduction (11 cows). Finally, there are, without doubt, udder health problems in the herd. See sheets in the section of mastitis for further analysis.

At group or individual level

- The data for each cow (milk recording; fertility data) should be available. However, these data are not always available on the dairy farm! If there is a real reproductive problem, it is highly recommended to complete such records in order to be able to carry out any analysis and/or interpretation. A good example is the use of a Fertility-Health wall card (and, at a later stage, a computer). See in this series for an example.

Yearly reproductive performance data

Table 17.1 Annual reproductive performance report

Average milk production per cow per year		8500 L
Average number of cows present		84
Number of cows/primipares with a calving date		69 / 27
Number of cows/heifers inseminated		61 / 22
Number of inseminations before 50 days lactation		2 / 3
Interval calving to 1st insemination		84 days
Interval calving to last insemination		116
Interval calving to conception		115
Calving Interval (expected)		395
% of heats detected	0 – 15 days after calving	0
	16 – 30 days	10
	31 – 50 days	17
	51 – 75 days	37
Number of inseminations in total		130
Number of first inseminations		72
Interval from 1st insemination-conception		33
Insemination Efficacy Figure		1.82
% of cows pregnant after 1 insemination		54%
Mortality of neonatal calves		0
Calves and maiden heifers culled		0
Cows culled	Mortality	1
	Metabolic disorders	1
	Calving problems	1
	Udder disorders	7
	Claw disorders	1
	Subfertility	11
	Other disorders/diseases	1
	Old age	0
	Non-health reasons	0
	Too low milk yield (genetics)	5
	In total 28 (=33%)	

Yearly reproductive performance data

- On this card you put all cows in chronological order of their calving date. All events can then be recorded: calving history, dates of heat, AI, reproductive disorders, pregnancy diagnosis, treatments and culling. See elsewhere in this series for an example.
- Such a card allows the easy detection of problem cows and heat-detection performance.

Conclusions and suggestions

The following conclusions can be drawn and a plan of action determined:

- There is a problem of heat detection and/or a problem of cows not showing their heat due to a severe negative energy balance around and after calving or many cases of cysts or endometritis. These must be checked during subsequent farm visits for a period of at least six to 12 months. If the heat detection is improved or the negative energy balance alleviated, the following issue will be easier to deal with.
- The farmer should start to inseminate much earlier. This means at the first heat detected after 50 days. This is only feasible if heat detection in the period from calving to 50 days is improved!

  The inseminated cows must be better monitored for return to heat: checks at three, six and nine weeks after AI should be made. Coloured cattle markers can be used to identify those cows. The place to mark is the pelvis or the tail base.

Note: As already stated, the clinical and sub-clinical lameness cases should not be neglected because they can impact on the expression of heats. Ketosis can also have a negative effect on follicle growth.

Final remarks

In the area of reproduction, the term mean is often used to derive reference values. Means are easy to calculate but, statistically speaking, it is not correct for the intervals named above. These intervals do not have a 'normal (Gauss) frequency distribution' but are 'skewed', usually to the right. This can be seen in a box and whisker plot by a long tail on the right side and a short one on the left. For those cases, the 'median' should be calculated in place of the mean. The median is the value that lies in the middle of all observations put in chronological or numerical value, the lowest value left, the highest value right. An example is given here in terms of a box-plot (box-and whiskers):

Now we have five parameters with which to evaluate this interval:

1. Median
2. Area of the left box
3. Area of the right box
4. Left tail
5. Right tail

Moreover, these parameters are statistically justified and can easily be calculated manually. Year Reports on reproduction differ from farm to farm. It is necessary to become accustomed to those used in your own region. However, the approach to data interpretation always remains the same!

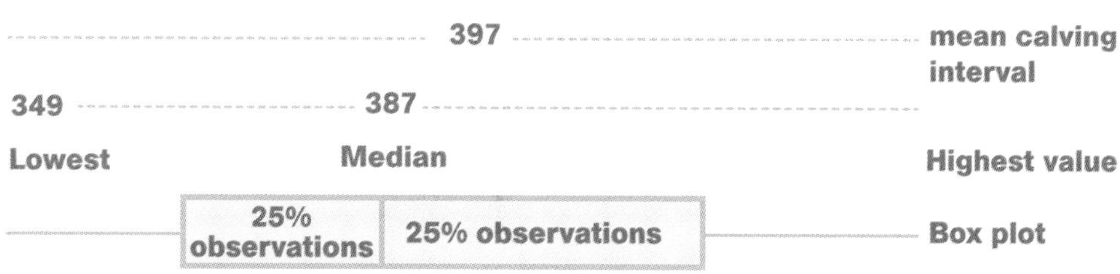

Figure 17.2 Example 'box and whisker' plot

Abortions due to *Neospora caninum*

This parasite is closely related to toxoplasma. Infected cows do not show specific clinical signs but often require more inseminations per pregnancy have more abortions or mummified foetuses. However, around 80% of affected cows appear to give birth to healthy calves that, in turn, become carriers.

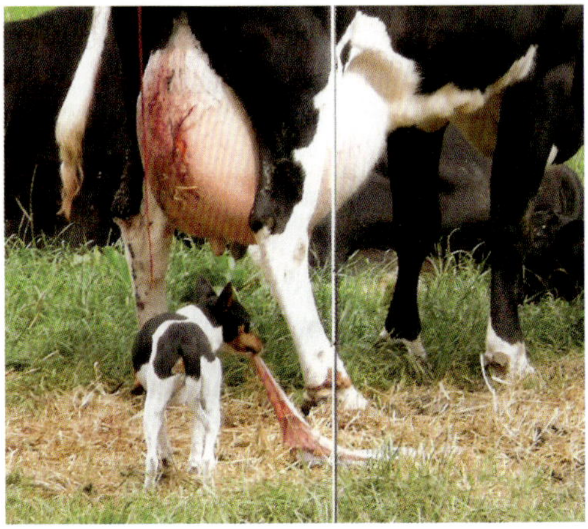

Transmission is mainly from dam to calf (intrauterine) and less so from dog to cow. Embryos become infected after implantation in the uterus. Therefore, to control an endemic infection, the 'carrier' animals should be identified and culled.

A dog or coyote can be infected by eating the placenta or ingesting the vaginal discharge of infected cows. The fox has never been identified as an intermediate host.

Procedure for the control of *Neospora*:

A Examine a foetus in a diagnostic laboratory
B Take blood samples from calves and dams of the same family
C Take blood samples from other cows after they aborted for identification of the presence of *Neospora* in the herd
D Take blood samples from all cattle older than six months for Identification of the presence of *Neospora*

Over time, the positively-testing cattle should be eliminated from the herd (or apply this after the transplantation of the embryos).

- Exclude the dogs from the premises.
- Install a rigid rodent control programme.
- Implement a **Biosecurity Programme**

Number of daily milkings in relation to ovulation/oestrus

Stevenson *et al.*, 1997 (LPS 50)

It is thought that the milking process may have an impact on the ovulatory process as observed in the case of lactation anoestrus in horses. Therefore, a series of farm field studies were designed to determine the impact of daily milking frequency. The results are summarised below. It appears that milking three times a day yielded the shortest interval from calving to first ovulation and oestrus.

Table 17.2 Milkings in relation to ovulation/oestrus

Daily milkings	Days to first ovulation	Days to first oestrus
Exp.1		
3x daily	26	46
6x daily	37	76
3x + 3x suckling	48	48
Exp.2		
3x daily	22	45
6x daily (3 weeks)	30	61
6x daily (6 weeks)	39	75

Table 17.3 Overview of some major casual agents for abortion in dairy and beef cattle (Belgium)

Infection type	Causal agent	Months of gestation when affected (XX= unknown)	% abortion in dairy	% abortion in beef
Virus	BVD		4.4	3.8
	IBR		XX	XX
Bacterial	A pyogenes		5.2	6.3
	E.coli		2.6	4.2
	Listeria		0.7	1.4
	B. licheniformis		0.6	1.0
	Q-fever		4.9	3.6
Parasitic	Neospora		14.3	8.1
Yeast & fungi	Fungi		3.5	3.5
	A fumigatus		2.4	2.4

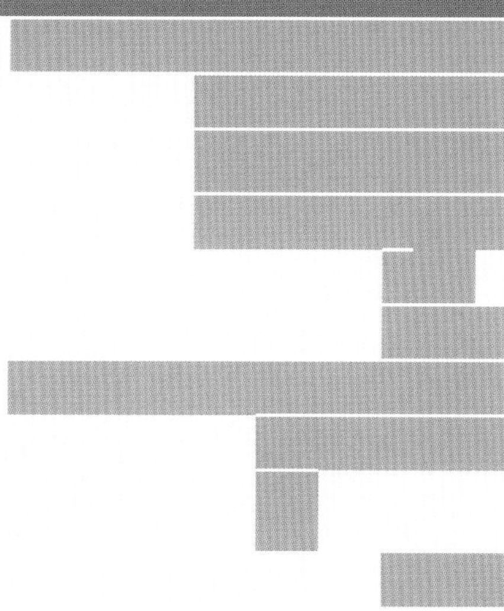

Source: DGZ, Flanders, Belgium; in Veeteelt, October issue 2011

Analysis of an abortion problem

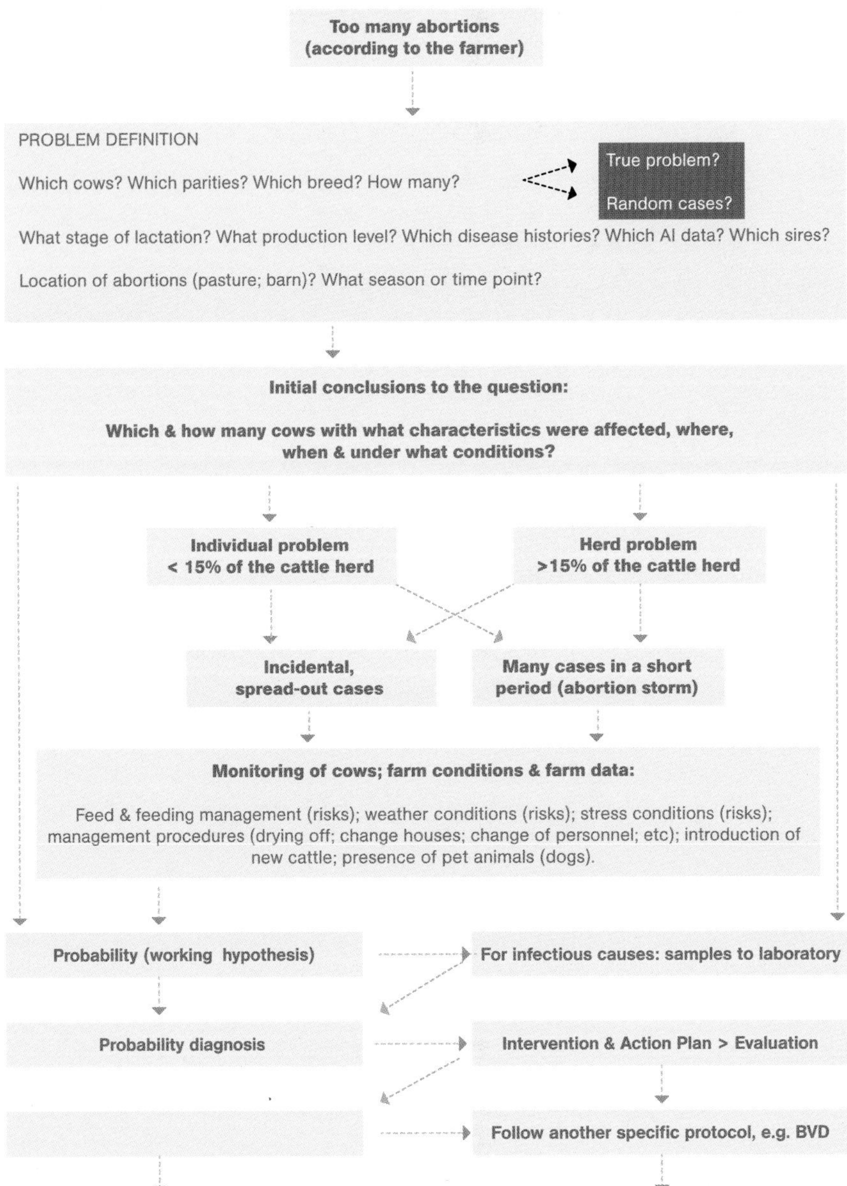

Figure 17.3 Protocol for analysis of an abortion

Hints & tips for farm visits

For herd fertility schemes in HHPM

Take several different coloured cattle markers with you (red; blue; green; violet)

Reproductive examination comprises:

A pre-breeding checks at 35-45 days in milk (DIM).
B check cows not seen in heat, or cows seen only once.
C check cows with an irregular oestrous cycle.
D check cows with reproductive problems.
E pregnancy checks. Mark cows with a G.

For cows marked A, O or I, as well as cows with a negative pregnancy check, it is desirable to determine the stage of the cycle that cow is in: either exactly which day of the cycle they are in (D 12 or D 7) or a more general estimate e.g 'early stage of cycle (D 5 to D 10)' or mid-cycle (D10 – D 12)'. Note that there is always a certain variation present of ~ 2 days around the 'predicted' stage. Physiological cycles run from 19 to 24 days! Cows that are predicted close to a new oestrus may be coloured with a specific colour to facilitate detection by the farmer.

Activities associated with reproduction

Since herd fertility work involves many rectal palpations, it would be highly desirable to carry out other observations at the same time. Moreover, the farmer is available at the time to record findings. These other observations may include: body condition score; rumen fill score; faecal consistency score; undigested fiber in faeces score, as well as teat-end callosity score, hygiene score, deviant posture score. This takes very little additional time and provides much useful information!

Other activities

If, during the execution of a herd fertility inspection, other tests/samples are to be carried out (CMT testing, or sampling milk for bacteriology) use a separate marker to identify these animals. For example, M for milk sampling; B for blood sampling.

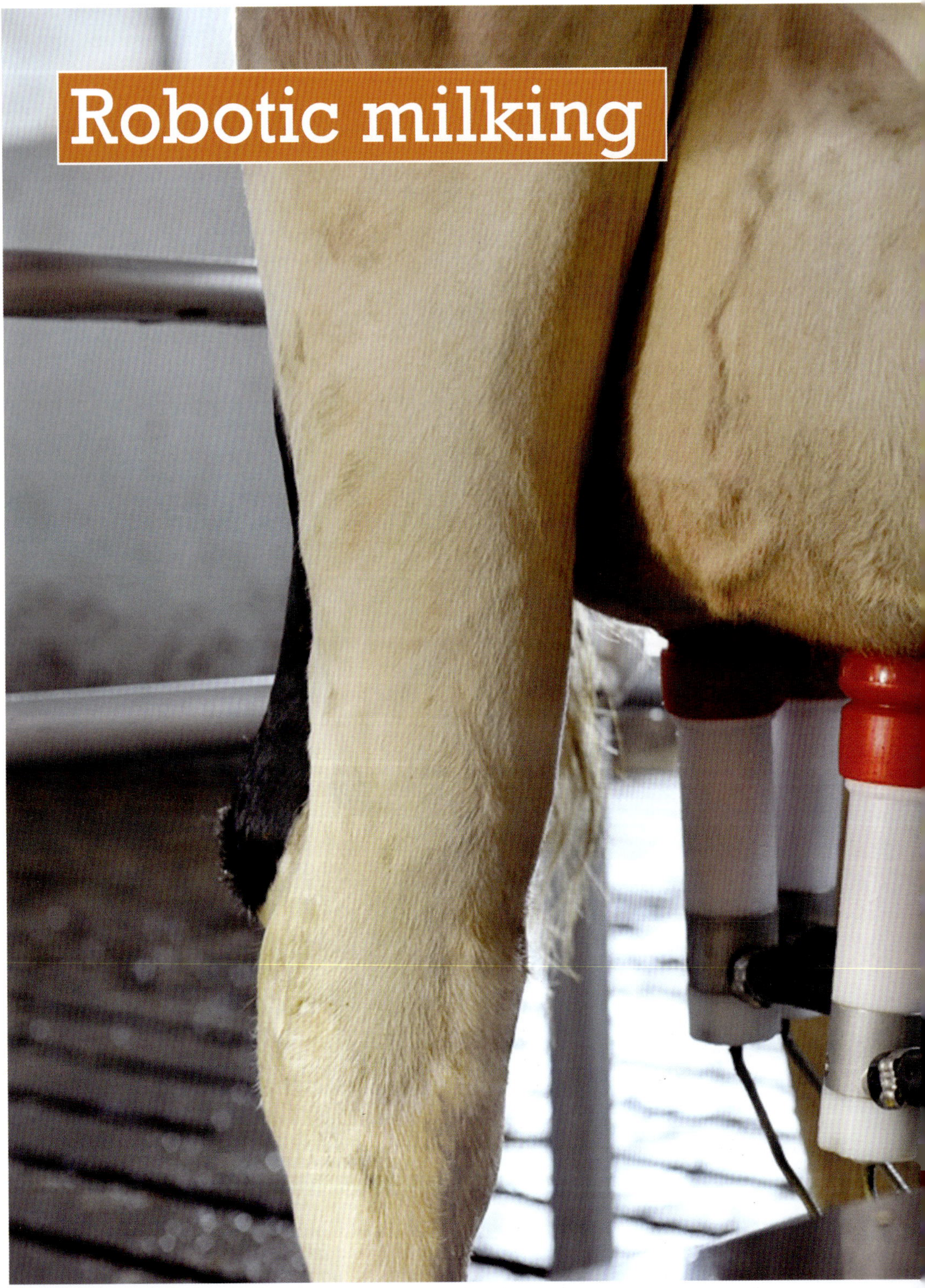
Robotic milking

> "This chapter starts with aspects to consider before installing a milking robot on the farm. Additionally, robot milking and udder health, the use of data from the robot and a protocol for using the robot are also dealt with, together with key success factors for robotic milking."

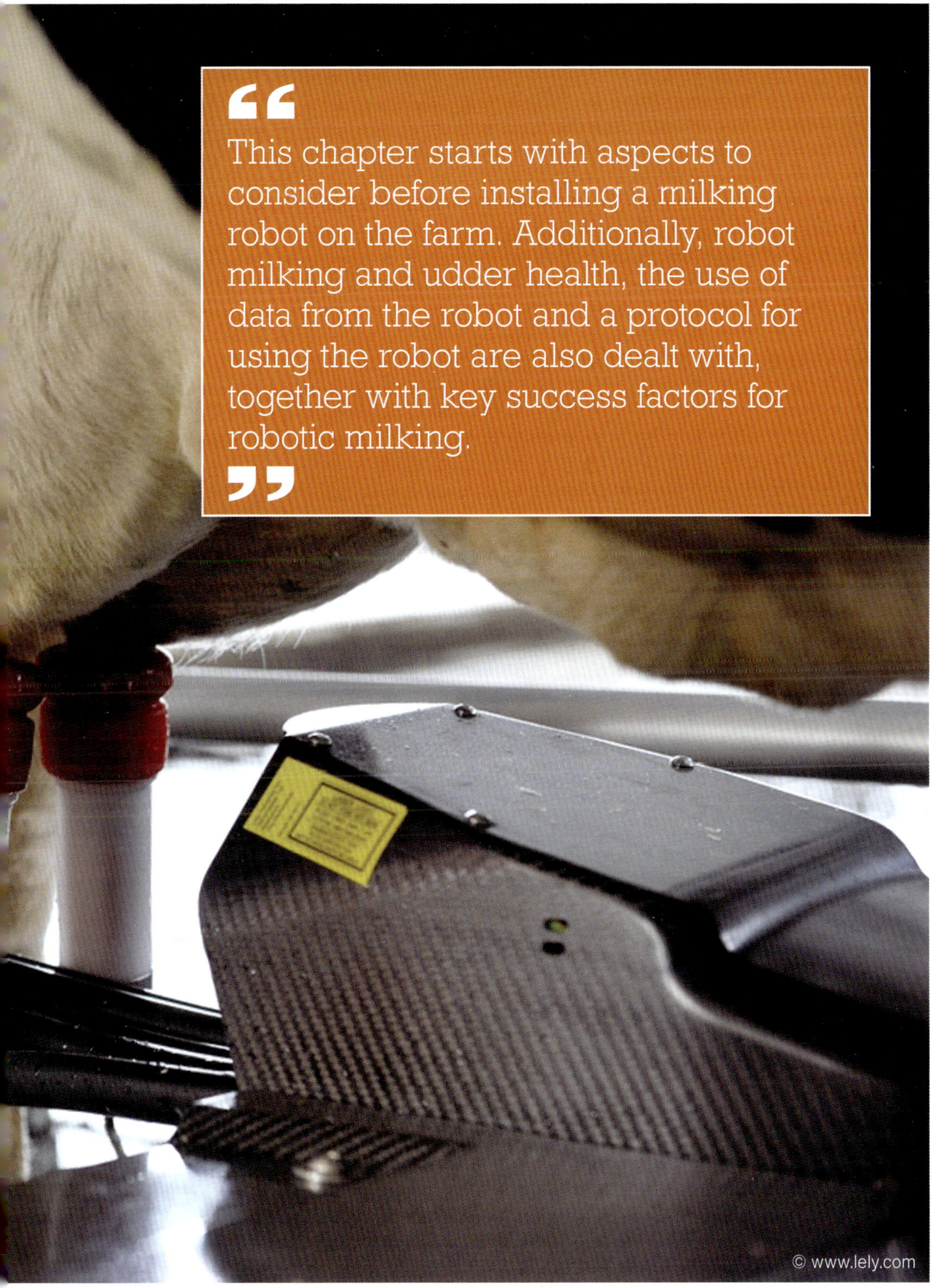

© www.lely.com

Points to consider when installing a milking robot

How to manage cows that require attention when milking with a robot

Currently, there are about 10.000 robot farms in Europe. The most influential reasons for farmers to change from a conventional to robotic milking system are:

- More flexible working hours (milkings are not fixed anymore)
- Reduction in total work load
- Increased independence (more time available to do other things in life)

Not all robotic farms are successful in achieving the factors above but, in general, farmers on a robotic system are satisfied with their decision to change.

> If you are trying to decide whether to convert to a robot, the following actions are important to carry out:
>
> - Detailed analysis of your current production system
> - Detailed definition of your new demands/requirements.

It is a completely different method of production and management.

With a robot, the number of milking places is limited, time becomes critical (for conventional milking it is the other way around) and the operation is virtually 24/7.

Any **events that could disturb the routines surrounding the robot** will reduce the capacity its function. Cows that need additional care from the farmer and demand a lot of time.

- Examples would be:
- Lame cows
- Cows with metabolic disorders (acidosis; ketosis; milk fever)
- Cows with reproductive problems
- Cows and heifers at the start of lactation
- Cows being treated for disease or other issues
- Cows with a lesion (hock lesion; other)
- Cows with mastitis
- Weak cows (subordinate animals; timid; 1st lactation heifers)

> *The time devoted to these cows should never be underestimated* **(the last 20% of cows in a herd are cause of 80% of the total daily work)**

A kind of management by exception focuses specifically on these cows and the prevalence of these cows in a herd varies from 8 to 15%.

Lame cows

A robot farm is more susceptible to lameness problems than a conventional one because cows must visit (walk to) the robot on a regular basis. Moreover, a lame cow often kicks the milking claw and the robot requires several attempts to attach it. Therefore, claw health must be maintained at a high level, so that the visit frequency will not be affected. Additionally, lame cows may well block the entrance to the robot for others. It is important that any exercise areas are clean and dry and routine, functional herd claw trimming must be in place (twice a year for all cows and first lactation heifers).

Reproductive problems

When operating a robot system, heats must be detected quickly and efficiently. As the farmer often doesn't have the time to observe the cows frequently, it is necessary to reserve at least two time points per day for heat detection. Automatic heat detectors can be used but a much simpler way is to mark cows due for heat with coloured cattle markers on the pelvic area or tail base. This can also apply to cows to be checked at three, six and nine weeks after AI or previous heat.

At the start of lactation

This group is highly sensitive to all kinds of different stressors; these cows need to be taken care of prudently. Their ligaments are not yet fully recovered, which renders them susceptible to disorders such

Changing from conventional milking to the robot

Abstract of a paper by Christian Mueller, WestfaliaSurge, Germany (Internatl Dairy Topics vol 8 no 3, 2009)

as locomotory. This group must be housed close to the robot to reduce the distance and time taken to reach it. This not only encourages the animal to visit the robot but also reduces the risk of lameness. For the initial milkings, it is wise to be present and assist in attachment of the claws if necessary or for monitoring udder health and/or milk fever. First-lactation heifers represent about 30% of animals in the herd; they need to learn the robot routines and may need help in doing so.

Cows under treatment

It is preferable to have a pen specifically available for drying off, treating sick animals or other activity. This pen must be situated after the robot **(post-robot selection area)**.

Mastitis

Mastitis treatments often take a great deal oftime. Furthermore, there is always the risk of infection of herd mates or the robot. Ideally, mastitic cows should be separated from other cows not too far away from the robot. This means they can be monitored more easily, treated more appropriately; hence the infection risk can be reduced.

Weak cows

Sometimes these cows are not free to enter the robot because other, more dominant cows, prevent them. These cows will be highlighted on the milking list generated by the robot (not milked). This number will increase when the robot's capacity is reached. These cows need to be manually taken to the robot by the farmer at least once a day. This could be combined with when he cleans the cubicles, for example.

Questions to ask to limit any negative impact on production.

- What should be done to ensure all cows are milked on a regular, daily basis?
- How can these animals be observed and monitored when they are not seen in the parlour two or three times a day?
- How can these cows be treated without having to be handled one at a time?
- How could these cows be manually milked from time to time?
- How can these activities be integrated into a robotic system?

These questions can be answered by designing the housing (centre of milking + calving pen + treatment/observation pen) around the robot to reduce distances between key areas. The cows requiring attention will, for the most part, be in this centre! Moreover, an area of selection before or after the robot (for inseminations; check-up) could be created.

➔ good housing design is critical for good organisation and management!!

Experiences with a ROBOT

The milking of cows with a robot is about more than simply technique. There is also the emotional adaption.

Once the robot is installed, it will in general take two to three weeks before the cows become adapted; it will take at least two to three months for the farmer to adapt and nearly a year before the farmer feels comfortable with it (Figure 18.1.). It is for this reason that farmers who decide to buy a robot, must be intensively guided and coached before the purchase and the first month after the installation, followed by less intensive help over the next 12 months.

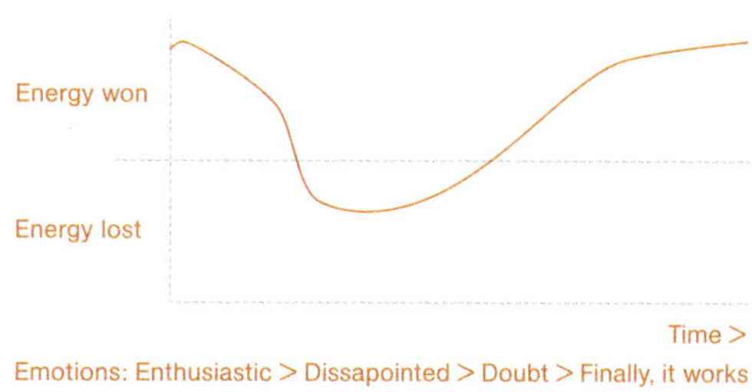

Figure 18.1. The different stages of emotions when a milking robot is installed (Hulsen, 2009) (SAC Vol.4, N°.7, 2009)

Udder health & robotic milking

Adapted after UGCN; K.deKoning & H.Miltenburg, Veeteelt 1, March 2009, the Netherlands)

Globally, there are about 10,000 farms with a robot (1500 in Holland).

The activities to promote udder health are basically the same as for conventional farms but generally more monitoring and care is required.

Before installation of the robot

During the months before installation of the robot, the farmer is often very busy with the construction and hence has less time for cow care. This is one of the causes of an increased somatic cell count, or even mastitis, on these farms. At the same time, it is possible that maintenance of the original milking machine is neglected, leading to the same problems.

Housing & hygiene

These two factors become even more important on robot farms. The robot does not always clean the dirty teats adequately and when environmental bacteria are subsequently found, the hygiene of the housing is often suspected. There are monitoring and scoring systems available to help the farmer in assessing the level of hygiene on his farm.

Poor preparation

This can be caused by poor calibration of the robot but also by factors, such as nervous cows, long hair on the udder or poor conformation of the teats.

During milking

Bacteria can be transmitted during the milking from cow to cow, especially when there is a high frequency of visits to the robot (up to 80 cows per robot or more). The cleaning of the robot between two cows is crucial (currently a steam cleaning procedure is employed). Hot water cleaning of the robot must be done three times a day and the equipment must be cleaned with high pressure, lukewarm water by the farmer daily.

Disinfection of teats after milking

Teat disinfection systems for robots are not always optimal. Poor calibration of the system, blocked tubes and an empty disinfectant container are just a few causes of inadequate teat disinfection. Poor pressure setting is also a problem: pressure too high + liquid disinfectant leads to a disinfection mist or pressure too low + fatty disinfectant can result in poor or no disinfection. So the farmer must check the system frequently. In order to test it, a paper towel can be wrapped around the teats before the disinfection begins and one can observe what happens.

Frequency of milking

Milking a cow more often leads to better health and more milk. However, there are prerequisites:

- the intervals between milkings must not be too short
- the volume of milk per milking must not be too low
- the milkings must be done in a routine manner

Intervals of less than six hrs lead to a rise in cell count and are a risk for splitting up of milk fat into fatty acids. They have also an effect on robot capacity. Cows with a low production level have a lower milking speed but their preparation and attachment take more time. Irregular intervals also increase cell counts. The general goal for frequency is between 2.5 and 3 times per cow per day (so a calibration for around 8 to 12 L milk per milking). At the start of lactation, there can be more flexibility around this setting.

There are numerous factors that have an effect on this setting. Cows, which are active and have easy access to the robot will visit more often. Cow activity is influenced by their health, the barn climate, claw

Udder health & robotic milking

After UGCN ; Kees de Koning ; Hans Miltenburg (Veeteelt,1 March 2009; the Netherlands

health, hyperthermia and nutrition. A basic ration that is too rich or a ration lacking fibre will lead to reduced activity and, hence, reduced visits to the robot. Long queues to enter the robot will also de-motivate cows to visit it. The same is valid for robot malfunction.

Attachments missed

The percentage of missed attachments must remain at an acceptably low level because they cause milk rests in the udder and a loss of milk into the cubicles.

Long hair on the udder, low milk yield, poor teat position, dirty robot detector and nervous cows are all factors causing an increase of this parameter.

Preventative maintenance and technical check-ups of the robot twice yearly help in keeping this percentage low. Moreover, the functioning of the robot can also be checked using parameters, such as mean visits per milking, timing of milkings and missed milkings as provided by the robot.

Detection and plans of actions

The conductivity of milk is the best means to detect (sub-)clinical mastitis. There are also other parameters that can be used to assess the situaiton (milk recording; level of daily milk yield; individual cell count; lowered feed intake; fever).

The surveillance of cows with high cell counts is as important as detecting them.

A plan of action must be designed at herd level, with different diagnoses, treatment indications and withdrawal time (**Herd Treatment Advisory Plan**, see elsewhere in this series).

The cows with chronically high cell counts often do not appear on the robot attention list; so these need to be monitored separately.

When the rate of high cell count cows increases above 15%, there is a problem at herd level. In this case, a control & prevention plan must be designed by the vet and farmer, with detailed actions, control, preventative and risk management measures.

Chronically high cell count cows must be culled.

Remarks

Implement a Protocol for Robot Checking. This protocol comprises the working instructions on a daily, weekly and monthly basis to avoid problems as mentioned above and to increase work efficacy, especially when several people work on the farm.

Pregnant cows (> six months) often show an increased cell count; they do not want to visit the robot frequently and their milk yield decreases.

Many robot farms appear to **change the teat liners every three or four weeks**.

For drying-off, they apply a procedure of milking once a day for a set period before completely drying off.

> More monitoring of udder health is needed than on conventional farms

Effective use of information

Mike Kuijperij – Lely Benelux

ANALYSIS OF DATA

Step 1 Visit behaviour (pattern)
Milking → Lely Astronaut

Table 18.1 The following parameters can be calculated

	ROBOT 1	ROBOT 2
# of milkings	....	....
# of refused attachments	....	....
# of missed attachments	....	....
Milking speed		....
% free	....	....
# of cows	....	....
# of milkings per cow per day	....	....
(#milkings minus # missed divided by # cows)	Animals →	review of the herd
Kg milk per day	....	....

Note

- Number of visits on average - this depends on litres of milk per cow per day and the visit threshold set (often at 2.6).
- Number of refusals - depends on silages, cow activity, feed rests in the feed bunk, settings of the software. Can be 70% in the first year. Check the irregular intervals too
- % Free. The threshold is 10 to 15 % (to facilitate heifers entering in the robot)
- Milking speed - depends on the herd and level of production; ideally should be ≥ 2 (for example 2.3 kg milk per minute)

The parameters must always be set or reset! For example, from 0 to 40 days lactation, you could set a threshold of four visits to the robot per day. settings based on the weekly average and the daily production can also be established.

Step 2 Behaviour (pattern) of individual visits

If the list "Behaviour Visits" is not available, you can create it as follows:

- Go to Per Day and then to Lists.
- choose Compose a new list.
- choose Cow data.
- choose the following parameters
  - Number of lactation
  - Days in lactation
  - Production of milk per day
  - Production mean
  - Visit behaviour.
- press Enter
- choose as order of appearance Production of milk per day
- choose only those cows which are in lactation for the 4 parameters named under 4.
- give the name Visit Behaviour
- give the description CompVis.
- save this list by pressing F2.

Check

- The number and % of cows with < 25 kg milk per day →
  These cows must not be milked more than 2 to 2.5 times per day. A more frequent milking pattern in low producing cows may provoke the splitting of fatty acids in the milk. If the percentage is, for example, 45 this can indicate a herd coming to the end of lactation or problems at cow level.

- The number of cows with > 35 kg milk per day and < 2.5 milkings per day →
  These cows make too few visits to the robot (because > 4 would be desired). Consider

Effective use of information

checking rumen fill, rumination frequency and manure quality. If, out of 26 cows, there are four that make few visits, check the fibre content of the ration, for acidosis or the energy-density of the basal ration.

Step 4 Production of milk per day

Per day → Lists → ProdDay

Name of list: ProdDay					
Review:	Herd Data				
Group	Number of days in lactation	Number of cows milked	Production in total/day	Production expected	Production per cow/day
	122	59	1775	1788	30.1

Check the milk production per day versus the number of days in lactation for the herd.
Note that the culled cows and cows dried off or non-pregnant, which are still labelled as « in lactation » in the computer, will contribute to the calculation of « stage of lactation »!

Corrected production

If the cows at the start of lactation do not produce as expected, compose a list from five to 60 days lactation to better observe the performance of and variation in those cows.

Name of list: 5 to 60 days								
Review:	Cow data							
No. of the cow	Parity	Days in lactation	Feed rests	Milk per day	Milk in average	Charact of visits		
						milkings	refus	mis
451	1	25	3.48	24.8	24.3	2.0		
317	3	17	3.61	36.0	34.1	2.0		
347	2	9	0	40.6	38.7	3.1	2.7	0.1

Special attention for = The number of cows with too many feed rests
The number of cows with decreasing milk yield
The number of cows with < 2.5 milkings per day

Step 5 Milk fat and milk protein contents

Check

- The graph for litres of milk over time (going down? variation?)
- The level of milk production (kg) per cow per day (check stage of lactation)

Effective use of information

- Check the milk fat graphs
- Check the cows with a milk fat < 3.80 (must be < 10% of all cows. If this % increases → check for adequate rumination and acidosis; fibres in the ration; disease.
- Check the milk protein graphs
- Check cows with a milk protein < 3.00 (must be < 5% of all cows. If this % increases → check for NEB; ketosis; other disorders.
- Check the distribution of milk fat and protein over the lactation (to detect problem groups).

Step 6 Udder health and somatic cell count (SCC)

Check

- % of cows with SCC > 250.000 (threshold for defining a cow problem is < 15% of the herd; when > 15% there could be a herd problem)
- % of cows that show a SCC > 250.000/ml for the first time (threshold is set at < 10%)
- Look for chronic cases by using F4 and compose a Plan of Action together with the vet.
- The threshold for conductivity of the milk is at 60 to 75.

Advice for setting up the robot

You should choose the site for the robot(s) carefully

Easy access and good cow circulation is paramount. The robot should be positioned in a spacious and inviting area where cows feel at ease. It must allow sub-ordinate animals to feel comfortable enough to visit while, at the same time, be able to escape dominant animals. Also, the robots should be reasonably close to a drinking trough. The exit from the robot must be open and direct, encouraging the cow to leave.

Secondary routes

The arrangement of cubicles and allowance for shorter routes is very important. The isolation of certain cows for example, close to the robot, is required for AI, treatment, claw trimming and/or veterinary work but also to facilitate the adaption of the heifers to the robot at the start of lactation.

Preparation for using the robot

The ideal situation is to limit the number of cows per robot. It is preferable to choose the period where the majority of the cows are dried off. In some cases, a premature drying off may be necessary (from six months gestation onwards for certain cows).

Do not be afraid of computers or the data they generate!

Simply use the PC in the easiest and most efficient way you can – learn what the most important pieces of information are and how they relate to existing records and data. Emphasis must be on those areas that have the greatest impact on management. Remember, there is always a consultant available to guide and help you through the initial phases.

Setting up the robot

After PLM 395, November 2008

Determine new benchmarks

Adaptation to a robot has less effect on the cows (adapted in three weeks) than it has on the farmer. The farmer must be willing to change existing habits and adopt a new system of observing the cows. Follow-up with cows is more focused on individual animals. For example, more attention would be placed on a cow at peak milk yield that has not visited the robot for the last eight hours compared with a cow at the end of lactation: *management by exception*.

Feed a good ration

This is paramount for good animal health, welfare (cow comfort) and milk production. The quality of ration components, the balanced quantities and the feeding frequency over the last 24 hrs are important parameters.

Cow comfort

When a robot is in use, cows must be in good health. Lame cows or cows with poor body condition (BCS score of 2 or less in early lactation) will not stand to wait or fight to enter the robot or ; they prefer to stay lying down.

Stocking density of cows

Stocking density should be assessed in terms of cows milked per robot as opposed to number of cows present. A threshold of 65 cows per robot should not be exceeded (depending on housing facilities; milk production level; feeding system; etc). The parameter to monitor is "*the number of incomplete visits to the robot*".

Distribution of calving dates

The calving pattern must be such that it doesn't overload the robot.

Robot Alarm Signals for monitoring & surveillance

- Number of cows not visiting the robot (lead them in manually each morning)
- Abnormal visiting pattern or rhythm of cows (heat, mastitis, lameness, ….)
- Abnormalities can be identified by comparing the daily production averages, the average number of visits per cow per day, the average duration of milking, and the average production at the left side versus right side. These must be continually monitored.
- Feed refusals (maize, grass, concentrates) must be monitored and the quantity of refusals measured (especially in winter). Cleanliness of the feed bunk should be checked daily (emptied by cows or not).
- An increase in the electrical conductivity of the milk and a decrease in L milk produced may indicate the occurrence of mastitis.
- Changes in the behaviour of cows: for example a cow in heat (pedometer)
- The frequency of visits to the robot should, on average, be around 2 ½ to 3 times every 24 hrs. Any deviation must be monitored and followed up. The same is valid for cows refused access to the robot by the computer.
- Finally, the idle time during a 24 hr period should be considered: if idle time turns out to be between 10 and 20% of the day, one still has a margin for adapting the programming of the robot (frequency of milking; feed delivery by computer).

Protocol for milking with robots

UGCN—2008—NL

Table 18.2 Protocol for milking with robots

Frequency per day or week	Activities	Complete? (= X)
1. At least 2x per day (set a fixed time for these activities)	1. Check the attentions/warnings and alarms	
	2. Look for cows with long intervals and/or irregular milkings	
	3. Monitor the cows with new irregular intervals more closely; look for causes; correct	
	4. Check the new attentions for mastitis; use the Herd Treatment Advisory Plan for mastitis	
	5. Clean the robot and, if necessary, the waiting area too.	
	6. Main cleaning 3x/day; replace the filters	
	7. Listen for abnormal robot noise(s)	
	8. Check automatic concentrate dispenser for dosage	
	9. In the barn, observe the cows, their feed intake, the drinking places	
	10. Clean cubicles dry them; fill with fresh bedding; clean exercise areas	
2. Once daily (set day and hour)	1. Check robot performance	
	2. Check robot function; follow a cow throughout a whole milking cycle	
3. Once monthly (set day and hour)	1. Check the criteria for entrance and the dosage of the concentrate dispenser	
	2. Check the stock of cleaning products and the teat-dip	
	3. Check the cleaning system: temperature; product; tank	
	4. Check the lubrication of the vacuum pump	
	5. Clean the robot and the concentrate dispenser thoroughly	
	6. Put new bedding material in the cubicles	
4. Other fixed activities (record these in the Farm Diary)	1. Shear udder hair and tails: 4x per year	
	2. Routine herd claw trimming twice per year (heifers too!) or each month the cows at > 120 days lactation)	
	3. Use foot baths fortnightly with formalin OR other products five days/week every month	
5. ROBOT: specific items	Replace the teat liners and other rubber material	
	Check maintenance contract for robot and tank	
	Check the oil circulation system, compressor, vacuum pump, electricity grounding wires	
	Calibration of concentrate dispenser and milk container	
	Completely clean the dispenser; maintenance of fences	

Main guideline factors for robotic milking

Floris Pellikaan, Cees Jan Hollander & Wilma Fokker, Veeteelt,1-2 June 2010; the Netherlands

In 2010, a study carried out over 125 top-yielding, Dutch dairy farms with robotic milking investigated whether there were specific factors that contributed to their success.

Below are the results and some suggestions of factors influencing success of robotic milking.

Average level of milk production per cow	10.038 L milk
Average milk fat content	4, 30%
Average milk protein content	3,49%

Given the usage rate of the feed rack, it is important to avoid over-stocking as heifers will be prevented from entering the feed rack!! This also means that with 70 cows per robot, this issue is already happening! The majority of the high-producing farms have **no more than 55 or 60 cows per robot**. At the same time, they operate a more rapid milking routine (often more than 2.5 L milk/min) meaning that the robot should be idle for **at least 10% of the time.**

Behind the feed rack, these farms have, on average, **3.3 m of space** available; between the robot and any other obstacle there is an average 3.9 m space.

These farms use a **PMR (partially mixed ration)** and not a TMR (total mixed ration). The idea is that the basic ration is made to a lower energy level than a TMR, and concentrates are then supplied in the robot to attract the cows and make up the ration.

Table 18.3 Main guideline factors for robotic milking

Parameter	Minimum	Maximum	Average
Number of robots	1	4	1.7
Milkings per day per cow	2.2	3.4	2.8
Milking refusals per day per cow	0.1	11.1	2.3
Milk production per day per cow (kg)	25.9	41.0	31.9
Milking speed (kg/min)	2.1	3.2	2.5
Duration of visit (min)	5.18	9.11	7.05
Production per robot (kg)	950	2266	1587
% of idle time per robot	2.9%	42.2%	20.3%
Annual revenue (€)	2269	3001	2507
Use of cubicles	35.9%	132.9%	84.7%
Use of feed rack (%)	37.0%	214.3%	101%
Age of buildings (years)	1	40	16

Around 30% of the farmers grazed their animals.

Milk parameters for on-line measurement

Tijmen van Zessen, Veeteelt, 2010; the Netherlands

Table 18.4 Milk parameters used for on-line measurement in different robotic milking systems

		Herd Navigator DeLaval	VMS DeLaval	MQC Lely	Afilab Fullwood
Metabolism & production	Milk fat	–	–	+	+
	Milk protein	–	–	+	+
	Lactose	–	–	–	+
	Urea	+	–	–	–
	Ketosis (BHB)	+	–	–	–
Udder health	Cell count	–	+	+	+
	Conductivity	+	+	+	+
	Blood in the milk	+	+	+	+
	Colour	+	+	+	–
	LDH 1-2-3	+	–	–	–
Reproduction	Progesterone	+	–	–	–
	Activity**	+	+	+	+
Annual costs	TOTAL*	?	21/cow	5–10/cow	?
	MAINTENANCE*	?	4	?	10/unit
Cost of purchase*		?	85/cow	?	1600/unit

* = in euro

** = measured with a pedometer

The costs of sensors have been estimated at €125 to €150 per cow per year, while the benefits were estimated at €250 to €300 per cow per year resulting in a return on investment of 2:1.

Using sire genetics

Veeteelt, No. 27 – July 12, 2010, pp 32-34

A study by Annemieke Sprengers (Wageningen Agricultural University, LELY and CRV, 2010) has demonstrated that there are large differences between sires regarding their fitness for creating female offspring that are fit for robotic milking.

The study parameters/traits were:

- Speed of milking
- Duration of milking
- Visit behaviour to the robot
- Robot efficacy

Table 18.5 Sires and their fitness for creating female offspring that are fit for robotic milking

For example	Sire 1	Sire 2	Sire 3
More milk per minute (L)	0.49	0.22	0
Heritability	0.245	0.11	0
Availability of robot/24 hrs	20	20	20
More milk/ robot/day	294 kg	133 kg	0
More milk/ robot/year	107.310	48.595	0
More €/ robot/year	12.190	5.520	0
More cows/ robot	11	5	0

(sire 1 = Canvas) (sire 3 = reference)

The most important parameter appears to be speed of milking.

The correlation coefficient between robot efficacy and milking speed is 0.9. So indeed, the milking speed should be part of genetic selection of sires and dams, where robotic milking is concerned.

Sustainability

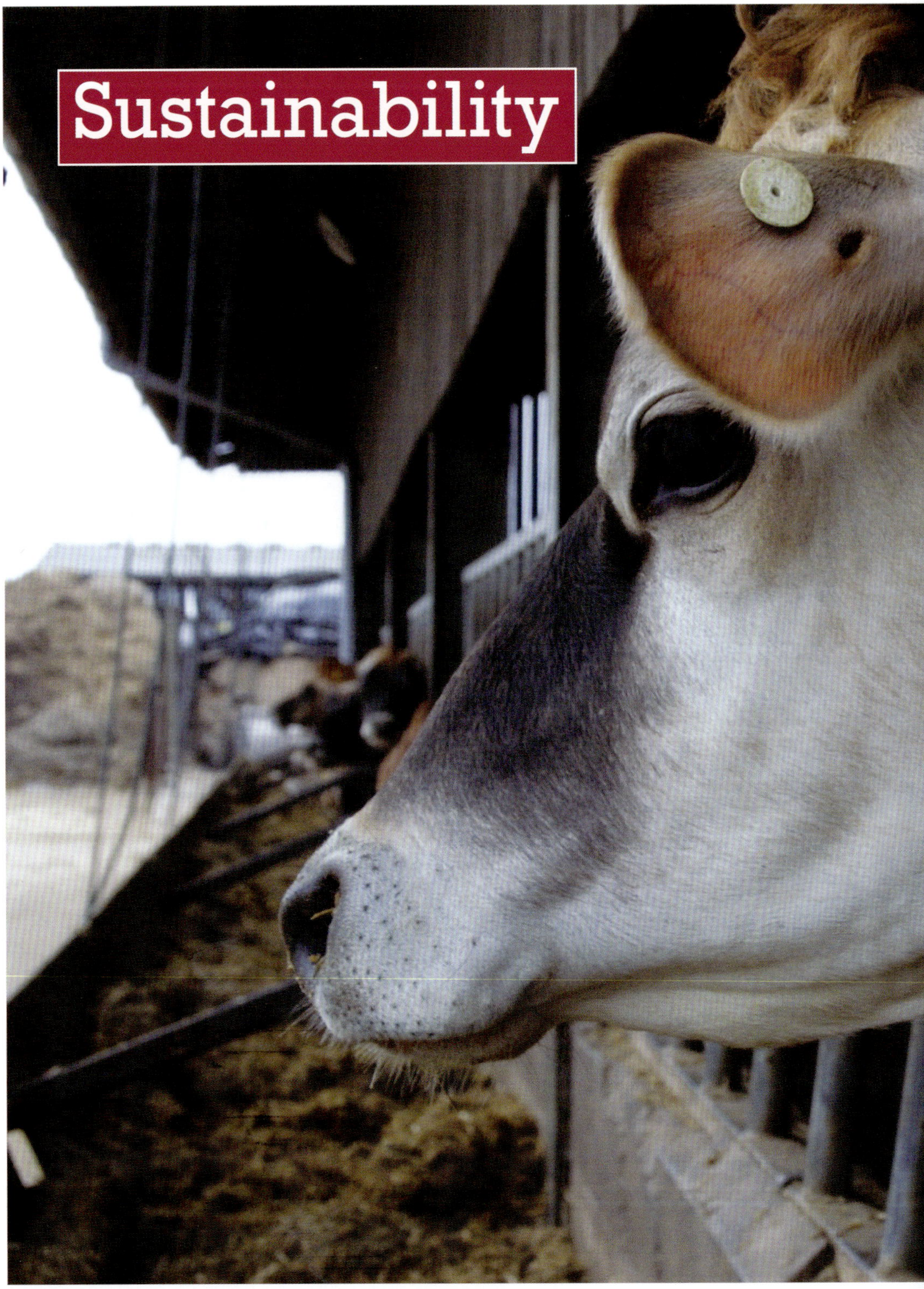

"Sustainability on the dairy farm is a hot topic and is, therefore, included in this book. This chapter presents the four pillars of the sustainability concept and subsequently deals with contributions from the food chain to climatic change. Finally, several approaches to improve sustainability on the dairy farms are presented."

Sustainability

Duurzaamheid, Bedrijfsduurzaamheidsindex, Farm Sustainability Index. KJ van Calker, P.Berentsen, P.Galama, G.Giessen, RBM Huirne. Veeteelt, November 2, 2005, the Netherlands

Sustainability comprises four components:

1 Economic issues ....profitability

2 Internal social issues......labour conditions

3 External social issues......acceptance by society

4 Ecological issues.............soil–air–water

For the first component, the main factor is the **family income,** values for which vary from € 18000 to € 74000.

The second component is related to the annual **work load**.

Public health, animal welfare, animal health and landscape value contribute to the third component. Application of a basic quality assurance programme would be valued at 0, while application of all possible quality control programmes would be valued at 100.

Component four is comprised of the **eutrophogenic potential, nitrate concentrations in ground water, use of water and its contamination, acidification potential, greenhouse gas potential and ecotoxicity potential**.

Different stakeholders/bodies place emphasis on different components (expressed as percentage of importance, 100% being the sum of all components):

Table 19.1 Stakeholder emphasis

Components → of sustainability	1	2	3	4
Dairy farmers	42	26	16	16
Consumers	13	13	38	38
Industry	33	17	28	22
Government	32	11	26	32

This requires much communication and discussion in order to bring balance to the focus on the different components.

Sustainability and the concerns of society

(Friesland–Campina, 2010)

Societal concerns:

- Around 20 % of cows are not grazed
- Around 30 % of the cows experience mastitis every year
- Around 25 % of the cows experience lameness every year

Question of the general public: "is the cow the real focus of the dairy farm(er)?"

Perceived as major dairy husbandry problems are:

- Increased level of CO_2, NH_3, CH_4
- Feeding of (imported) soy

Hence, the general public appears concerned with problems to do with energy and climate, phosphate and nitrate and the environment.

Food-chain contributions to climatic change

The food chain significantly contributes to greenhouse gas (GHG) effects and in the European Union, the food chain is estimated as a source of 31% of the total emissions (Smedman *et al.*, 2010). There exist various intervention measures to try to reduce emissions from the food chain.

The most effective way to assess the impact of such measures is to quantify the emissions according to the nutrient density of the product. The nutrient density/climate impact index is used to assess this and is calculated as the nutrient density of the product divided by the GHG emissions per gram of product (Smedman *et al.*, 2011).

Foodstuffs can be ranked according to their nutrient composition. The concentration of nutrients in a foodstuff can be compared with the recommended intake of such nutrients. This results in a score, which indicates the contribution of that foodstuff to the recommended intake.

The nutrient density to climate impact (NDCI) index

In a Swedish study (Smedman *et al.*, 2010), the NDCI index of vitamin D-enriched skimmed milk was compared with orange juice, beer, wine, soya-drink and others.

The nutrient content of these products have been defined using concentrations of protein, starch, fat and 18 vitamins and minerals. These concentrations were correlated with the Swedish recommended intake values. The emission of gases (expressed in g CO_2-equivalent according to the Kyoto protocol) was calculated per 100 g product. Taken into account were gases in emissions from farms, processing, packaging and transportation to the retailer. Other phases in the emission process (consumers, home transportation, preparation, waste etc) were omitted. Carbon sequestration (the process of carbon binding in organic matter in the soil) was also not included.

Although milk has a greater impact with regards to GHG emissions, its nutritive value are much greater compared with the other products, so much so that it has the most favourable NDCI index by far. The nutritive density of other products is very low, or non-existent, resulting in an undesirable NDCI index, irrespective of how low the GHG emissions are, additionally, the protein quality of cow milk is much higher than that of soy, a fact that is not taken into account in the NDCI index. (Blom, 2010)

Table 19.2 Nutrient density to climate impact (NCDI) index

Product	% of NNR in 100 g product	Number of nutrients > 5% of NNR	Nutrient density	Emissions of gas	NDCI index
Milk	126	9	53.8	99	0.54
Soda	7	0	0	109	0
Orange juice	90	4	17.2	61	0.28
Beer	18	0	0	101	0
Red wine	24	1	1.2	204	0.01
Mineral water	2	0	0	10	0
Soya drink	53	3	7.6	30	0.25
Oat drink	32	1	1.5	21	0.07

NNR= Nordic Nutrient Recommendations; nutrient density= % of NNR in 100 g product x number of nutrients > 5% of NNR divided by 21; gas emission= g CO_2 per 100 g product; NDCI= nutrient density divided by gas emission

Improving sustainability in dairy farming

Table 19.3 Farming domains that a dairy farmer can address to improve the sustainability status of their dairy farm

Farming domains	Details
Farm economics	It is a continuous challenge for farmers to keep the cost-price below farm returns in order to make a proper living.
Relationship with society	Public relations become more important for mutual understanding. Conservation of wildlife in pasture exploitation is an issue. Appropriate land use and the adequate definition of farm sites are others. For consumers, the proper husbandry methods, limited use of pesticides, a validated responsible use of antibiotics and hormones, animal health and welfare are of concern. Food availability, safety and quality are relevant.
Management planning (at the operational and tactical level)	Need for formal, operational protocols, addressing the different domains in a formal and organized way, taking into account the time factor (e.g. seasons and stadium of lactation/ rearing). Examples of General protocols: Farm Treatment Protocol, Farm Health Plan. More specific protocol: Fresh Calf Feeding protocol (Noordhuizen *et al.*, 2008).
Animal selection	Genetic improvement not only for "more milk" but rather for more milk under given conditions like health, welfare, longevity, or adaptability (heat; scarce feed conditions); is in fact a management activity (Oltenacu, 2009).
Animal nutrition	Optimizing feed conversion rates, emission levels and excretion (including waste material); optimizing forage quality and availability for good health and performance in every lactation stage, without the risk of contaminations (man; environment) and intoxications (e.g. mycotoxicosis from food or milk). Safeguards from the feed mills regarding limiting contaminations in their production process (Tielen, 2009).
Animal reproduction	Optimizing –under optimal feeding management and husbandry- the reproductive performance to maintain maximal productivity under given farm conditions.

See also the full paper by Noordhuizen, Oenema, Boersema and Cannas da Silva entitled "Veterinary advisory practice and Sustainable production on Dairy farms" published in Cattle Practice (UK) 2012.

Farming domains	Details
Herd health & production management programmes	The participation in a professional HHPM programme (which focuses on animal health, animal welfare, public health and food safety) appears warranted for obtaining professional advice and coaching. Monitoring/eradication and control of main cattle disease like BVD, IBR, *Neospora*, Johne's and Salmonella for herds is crucial with respect to an optimal resistance and a less challenged immune system. As result less anti-microbial and anti-parasitic products have to be used. License to produce and to market: consumer protection & consumer concerns.
Pasture exploitation, nutrient management & soil quality	Precision grassland farming will limit the waste of feed and optimize the use of manure and fertilizer. One should refrain from over-using chemical products. Limited use of (anti-parasitic) drugs will improve micro-flora in the soil. Animal manure is of higher quality than fertilizer and brings structure to the soil. Soil quality is a paramount issue.
Waste management (environmental) protection)	Refers to proper handling of waste material including chemicals, and to minimizing use of scarce resources such as fossil energy, phosphorus, fresh water. It also relates to minimizing the pollution of soil, air and waters.

Values of reference

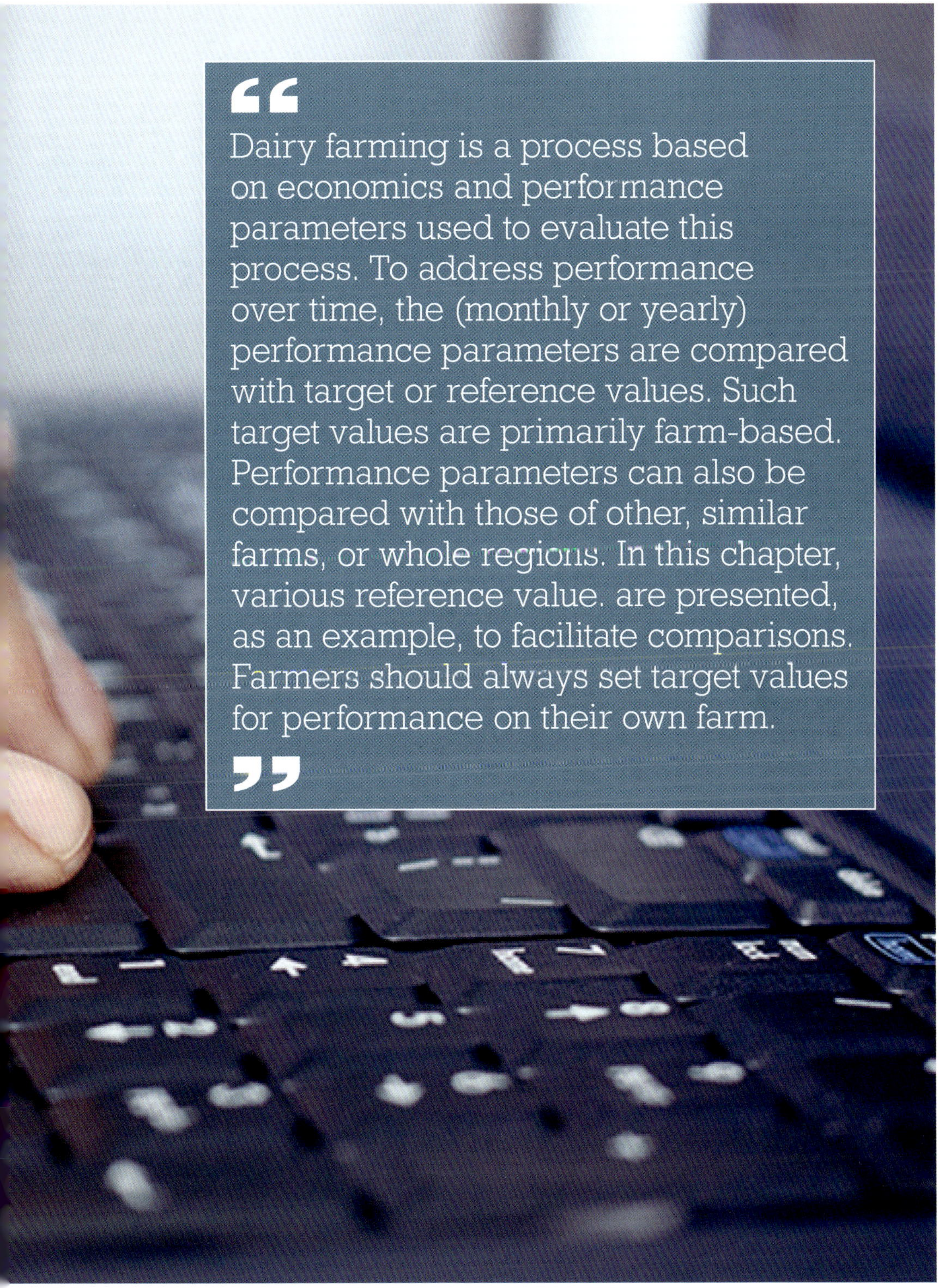

> Dairy farming is a process based on economics and performance parameters used to evaluate this process. To address performance over time, the (monthly or yearly) performance parameters are compared with target or reference values. Such target values are primarily farm-based. Performance parameters can also be compared with those of other, similar farms, or whole regions. In this chapter, various reference value. are presented, as an example, to facilitate comparisons. Farmers should always set target values for performance on their own farm.

Dairy production parameters

Table 20.1 Reference value for cows for important dairy production parameters- diagnostic evaluation

	At calving	Early lactation	Mid lactation	End lactation	At dry off
Body condition score (1-5)	3 ½ → 3	2 ½ → 3	3	3 → 3 ½	3 ½ → 4
Rumen Fill score (1-5)	3	3	3 ½	4	4 – 4 ½
Faecal consistency score (1-5)	2 - 3	2 - 3	3	3 - 4	4
Fibres in faeces score	1 - 2	1 - 2	1 - 2	1 - 2	1 - 2
Locomotion score (1-5)	Herd distribution →	> 85% at score 1 or 2	< 10% at score 3	< 3% at score 4	< 2% at score 5
Hock lesions*	< 15%				
Poor leg posture* (% of cows in herd)	< 15%				
Hygiene & cleanliness	Herd distribution →	> 85% of the cows at score 1 or 2	< 15% dirty cows (score 3 or 4)		< 5% dirty dry cows (score 3 or 4)
Ketosis *	Herd level	< 5 %			
Acidosis *	Herd level	< 2 %			
Digestive problems *	Herd level	< 5 %			
Frequency of cows ruminating in the herd	Herd level	> 85%			

* Reference & target values are dynamic and can vary according to production, breed, husbandry method etc

Environment & management

Table 20.2 Reference values for environment & management		
Housing	Feed rack	Inclination 10 cm
	Cubicles in loose house	125 cm (W) x 250 cm (L)
	Temperature of straw layer 10 cm deep (straw yard)	< 38°C
	Drinking space width	6 m (winter) – 9 m (summer) total width per 100 cows
	Exercise areas	Clean, dry, non-abrasive
	Manure scraping	4x per day; 6 – 8x per day (if claw health problems prevail)
Barn climate	Temperature	5 – 15 °C
	Luminosity	Can read newspaper
	Humidity	< 80%
	Moulds	none
	Ventilation	Smoke-test: smoke should disappear in 2-3 min; presence of cobwebs
Other management	Preventive trimming of the herd	2x per year
	Foot bath	3 days per month
	Foot bath size (W*L*D)	100 x 300 x 20
	Behaviour	Calm, unstressed

Farm data

Table 20.3 Reference values for farm data

Milk recording	Variation in milk yield	< 8% between 2 recordings
	Milk fat	> 4.0%
	Milk fat to protein ratio	> 1.2 to < 1.5
	Milk fat:protein > 1.5	Ketosis?
	Milk fat:protein < 1.2 (1.0)	Acidosis?
	Milk protein	> 3.0%
	Cell count of bulk milk tank	< 200.000/ml
	Bacterial count of bulk tank	< 10.000/ml
Reproduction	Calving interval	< 400 (385) days*
	Interval from calving – 1° heat	< 35 days*
	Interval from calving – 1° AI	50 – 70 days*
	Interval from calving – n° AI	< 100 days*
	Interval from calving –conception	< 85 days*
	Interval from 1°AI - pregnancy	< 12 days*
	% heats detected at 50-70 days	> 85%
	% pregnant after 1° AI	> 50%
	Age at first calving for heifers	24-25 month (breed!)
	% Abortions	< 4%
	% Retained placenta	< 12%
	% Cows with suboestrus	< 15%
	% Ovarian cysts < 70 days	< 5%
	% Endometritis < 70 days	< 15%

* = values will vary dependent on milk production level, farm economic conditions, farm goals etc.

UK141901
300192

Veterinary public health

“Dairy farms producing milk, meat and dairy products have a great responsibility toward society. This responsibility concerns veterinary public health and food safety elements. Public health hazards as related to food safety are summarised in this chapter. Subsequently, an overview of zoonoses is given in a checklist format, as well as the main characteristics of zoonoses. Methods for the improvement of food safety are also included.”

Food safety control

B Maunsell & DJ Bolton, Teagasc, Dublin, Ireland, 2004, the EU-RAIN project

The following issues are associated with food safety control on dairy farms: traceability; feed; water; cattle houses; cattle; health & disease prevention; use of detergents, chemicals and disinfectants; milk quality; animal waste material; biosecurity plans.

- Farmers should fully comply with animal **traceability** programmes (herd ID; cattle ID; cattle movement recording) as required by the authorities.
- **Feed** on dairy farms originates from the farm itself or is purchased elsewhere (e.g. concentrates, forages, industrial by-products). Feed manufacturers usually have a Good Manufacturing Code of Practice (GMP) in place in order to manage the risks of contamination and transfer of infectious agents. However, there may be a slight risk of chemical residues, e.g. in feedstuffs contaminated with fungi, which may produce mycotoxins.
- Poor silage quality (grass; maize) may be an indirect risk for food safety (*E.coli O 157; Listeria monocytogenes*).
- Cattle should have free and easy access to clean and drinking **water**. Private wells and surface water streams should be tested for microbiological and chemical contamination. Water troughs in pasture should be covered to protect against contamination by birds. Animal waste should not contaminate drinking water. The water troughs should be inspected and cleaned at least once a week.
- Cattle **houses** should be built and maintained according to the standards for cow comfort. Cattle pens should be regularly cleaned and disinfected (e.g. once a month or bi-monthly). A separate calving pen should be available and cleaned/disinfected after each calving. Waste material must be separated from livestock. The cattle houses should be rodent-/bird-/dogs-/cats-/wildlife-proof. The houses should be dry, well-ventilated and provide sufficient light. Overstocking should be avoided. Cattle overwintering outdoors should have a dry shelter.
- Cattle should be **clean** in order to prevent contamination of udder/teats and hence milk. Culled cattle for slaughter should also be as clean as possible. Before winter the cattle should be clipped either completely or at least on their back, legs and udder.
- There may exist several formal **disease control** programmes (e.g. IBR; BVD) to which a farmer can adhere. But there are also veterinary **herd health & production management** advisory programmes and **quality risk management** programmes to which the farmer can also subscribe. These programmes are carried out during regular (e.g. monthly) farm visits. Proper treatment of cattle can be carried out following **Herd Treatment Advisory Plans**, as proscribed by the vet and regularly updated. There are Good Antibiotics Application Code of Practice guidelines, which are both technical instructions and an aid in safeguarding food safety.
- Detergents, disinfectants and **chemicals** should be properly stored (dry, clean, ventilated) and used according to manufacturers' instructions. All surfaces, where milk can come into contact with chemical products, must be thoroughly rinsed with water.
- **Milk quality** can be safeguarded on the farm by applying high standard hygiene measures among cows, the milking parlour and milking machine and personnel. The milking machine should be properly functioning and maintained twice a year. Bulk milk temperature must be kept at 4°C. An udder health control programme as part of herd health & production management should be operational.
- **Waste material** should be stored away from livestock and feed storage. Slurry containing pathogens (e.g. *E.coli; Salmonella spp; Listeria spp; Campylobacter spp*) should not be spread on fields for grazing. If it is spread on these

Food safety control

fields, grazing should be delayed for at least 6 months. Waste should not contaminate surface water. On crop fields, waste should be spread following harvest.

Biosecurity plans should be in place to eliminate or manage the risks of introducing diseases or spreading them on the farms. Such plans should include staff instructions; foot bath application; restricted access to the farm; risk management instructions. Biosecurity can be integrated into quality risk management programmes following the HACCP principles (hazard analysis & critical control points). See also the sheets in the section Biosecurity in this series.

Preventative measures to safeguard feed from food safety problems:

- ✓ Feed should only be purchased from reputable suppliers who have a safety control system in operation
- ✓ Feed should only be used according to manufacturer's recommendations.
- ✓ Any medicated feed should be handled with great care.
- ✓ Feed must never contain prohibited ingredients, such as pesticides.
- ✓ Feed should be properly stored (clean, dry and bird/vermin-proof).
- ✓ Silage should be produced to the highest quality possible.
- ✓ Additives to improve ensiling should be considered under certain plant stress conditions.
- ✓ Ensure good consolidation and sealing of the clamp - aids in improving silage quality.
- ✓ If home-grown cereal crops are fed to cattle, withdrawal periods regarding pesticide residues should be adhered to.

See also "*Herd health & production management in dairy practice*" by Brand *et al.*, 2001 and "*Applying HACCP-based quality risk management on dairy farms*" by Noordhuizen *et al.*, 2008. Both are available at www.contextbookshop.com.

Public health hazards & food safety

Pearson and Dutson, 1996

Table 21.1 Potential health hazards regarding food safety-overview	
Antibiotics etc	Antibiotics, Sulfonamides
Metals	Arsenic, Lead, Cadmium, Mercury, Vanadium
Bacterial	*Acrobacter spp., Campylobacter jejuni, Clostridium perfringens, Escherichia coli, E. coli verocytotoxigenic* $O_{157}H_7$ *Listeria monocytogenes, Bacillus cereus, Salmonella spp., Staphylocoques aureus, Yersinia enterocolitica*
Parasites >< Protozoa	*Cryptosporidium parvum, Taenia spp., Toxoplasma gondii, Giardia spp.?*
Viral	Norwalk (norovirus)

Checklist for zoonoses

Species present	Total number	Animals older than 1 yr
Sheep		
Goat		
Cattle		
Horse/donkeys (state which)		
Poultry		
Swine		
Other species		

At the entrance to the farm there is a notice indicating that:	YES	NO
Pregnant women are especially susceptible to pathogens		
Young children are especially susceptible to pathogens		
An infection can be transmitted via stroking the animals but also via manure or inhalation (but risks are low)		
Hands must be washed with soap and water after having touched the animals		
Hands must be washed with soap and water before eating		
The face or mouth must not be touched before hands have been washed		
Eating & drinking is not permitted in the fields or animal houses & is only allowed in the indicated picnic area		
Any instructions of the owner/manager/employees must be followed		

Note the answers to the following questions:	YES	NO
Animal houses and their environment are clean		
Animals are not covered with manure		
Feed is stored in secured, closed sites or clamps		
Feed refusals are removed daily		
Dustbins are emptied daily		
Visitors have no free access to manure		
During the lambing season, straw bedding is renewed daily		

The hand washing facility comprises:		
Water from the public system		
Soap and/or disinfectant		
Disposable towels/hot air hand drying machine		
For personnel and veterinarians there are always available:		

Checklist for zoonoses

Clean boots	
Clean overalls	
A brush for cleaning boots	

There is a quarantine area on the farm premisses	
A well-identified, closed & clean facility, away from the main farm, is available for carcasses & dead animals	
In this facility, all carcasses and other body tissues (e.g. placentas) are stored (large carcasses are covered by a tarpaulin and kept away from the farm)	
There is a fixed contract with a veterinarian or clinic	
The veterinarian makes a farm visit at least twice yearly	
In the case of disease or other health disorders the veterinarian is called	
Only the owner/manager, employees and veterinarian are allowed to assist at calving	
Pregnant women are not permitted to work with manure	
Pregnant women are not permitted to wash dirty farm overalls or farm boots	
Pregnant women are not permitted to assist at calving or other births or manage sick or aborted animals	
Pregnant women are requested to minimize any contact with animals	
A daily record of events (farm diary) is kept, including:	
Diseases	
Mortality	
Health interventions or prevention (treatments; vaccinations)	
Zootechnical interventions (claw trimming; hair clipping)	
Dates of veterinary visits with diagnoses and/or treatment plan	
There is a Herd Health Treatment Advisory Plan	
All arrivals and departures of animals are recorded	
A list of all animals present is updated frequently	
All births take place in or around the farm	
Aborted animals are separated from other animals and from visitors	
In the case of abortions, the veterinarian is called	
In the case of abortions, placenta and foetal tissues are sent to a laboratory for analysis/diagnosis	
Sheep and goats are vaccinated against Q fever	
Sheep and goats are vaccinated against chlamydophilosis (CL)	
Sheep and goats > 1yr old are checked for antibodies against CL at least once per year (blood)*	
The results of these checks have been negative for the last two years *	
Sheep and goats > 1 yr old are checked for antibodies for Q fever at least once a year (milk or blood) *	

Checklist for zoonoses

A farm visit for this purpose, including the checklist, will cost about □50 per year.

The results of these checks have been negative for the last two years *	
All goats > 1 yr old are checked each year for CL or are certified free from CL	
The results of these checks have been negative for the last two years	
All goats tested positive for CL have been culled	
Of the zoonoses listed below, the veterinarian has determined clinical or suspect cases during the last 12:	
VTEC (cattle; sheep; goats)	
Salmonellosis (all species)	
Campylobacteriosis (poultry; swine)	
Psittacosis (poultry)	
Chlamydophilosis (sheep; goats)	
Q fever (cattle; sheep; goats)	
Colibaccillosis ETEC (all species)	
Toxoplasmosis (sheep; goats)	
Yersiniosis (poultry; swine; sheep; goats)	
Mycoses (all species)	
Ecthyma (sheep; goats)	
Listeriosis (cattle; sheep; goats)	
Caseous Lymphadenitis CL (sheep; goats)	
Leptospirosis (cattle; sheep; goats)	
Tuberculosis -- avian (poultry; swine)	
Tuberculosis -- bovine (ruminants)	
REMARKS	

Name of owner/farmer:	Name of veterinarian:
City and ZIP/post- code:	Code:
Date:	Date:
Signature:	Signature:

Score 1 point for every '**YES**' answered. The maximum score is 100 points.

With a total **score of 60** points or higher, the farm can be designated as having "**appropriate management regarding zoonotic control**"

With a score of **60 points or less**, it is highly recommended to **take action** towards improving the public health situation on the farm.
The veterinarian can deliver a plasticised certificate stating the aforementioned designation with the logo of the veterinary clinic, as well as the year of validity.

Characteristics of major zoonotic diseases

Animal Health Service GD, Deventer, The Netherlands, 2003

Table 21.2 Characteristics of major zoonotic diseases

Zoonosis	Pathogen	Reported cases in man per year	Clinical signs in cattle
Brucellosis	*Brucella abortus*	< 100	Abortion; retained placenta; metritis; vaginitis
	Brucella melitensis	< 100	none
Campylobacteriosis	*Campylobacter jejuni*	> 300.000	Abortion and enteritis
Chlamydia	*Chlamydia psittaci*	< 100	Abortion sometimes
Caseous Lymfadenits CL	*Cb. pseudotuberculosis*	< 100	Abscesses in lymph nodes sometimes
Cryptosporidiosis	*Cryptosp. parvum*	100 - 1000	Enteritis in calves < 3 weeks old
Echinococcosis	*Echinococcus granulosis*	< 100	In liver and lung
Ecthyma	*Parapox virus*	< 100	none
Erysipelas	*Erysipelothrix rhusiopathiae*	100 - 1000	Sometimes arthritis in calf
Leptospirosis	*Leptospira hardjo*	< 100	Abortion; milk deviating; no signs
Listeriosis	*Listeria monocytogenes*	< 100	Abortion; meningitis

Characteristics of major zoonotic diseases

Table 21.2 continued Characteristics of major zoonotic diseases

Clinical signs in sheep and goats	Clinical symptoms man	Infection transfer	REMARKS
Abortions rare; other signs as cattle	Mild process; swelling of liver and spleen; arthritis	Direct contact; animal products; lochia	
Abortion; stillbirth; fever; diarrhoea; mastitis; arthritis	Malta-fever; spleen and liver swelling; flue-like; heart, circulation, respiratory and repro disorders; death is rare	Contact with (abortive) animals; raw milk consumption	Mainly in Mediterranean countries and outside Europe
Abortion and enteritis	Enteritis; mild process; severe in children and elderly people	Direct contact; food (poultry, pork, shellfish)	Note: barbeque-borne infections
Abortion; eyelid infection; arthritis	Flu-like; abortion in pregnant	Contact with aborting animals	Bird contacts: lung problems; arthritis
Abscesses in lymph nodes	Lymph node infection (arm-pit)	Contacts (mainly abattoir people or sheep shearers	Occupational disease
Enteritis in lambs < 3 weeks old	Gastro-enteritis	Contact with infected animals, dogs, cats, man	Transfer from man to man possible with certain strain
In liver and lung	In liver, lung, spine brains, bones	Material contaminated by dog (vegetables)	Dogs eat contaminated offal
Wart-like skin lesions	Skin swelling, fever sometimes; lymphnode swelling; painful	Direct contact via lesions	
Chronic arthritis, lameness after dip	Skin lesions, arthritis sometimes	Contact with contaminated dip; lesions	
Same signs as cattle	Headache; fever; flue like; liver or brain affected	Direct and indirect contact; also via urine and semen	Mass disease control programme
Abortion; meningitis	Skin-infection; meningitis; abortion in pregnant	Direct contact with aborting animal, milk or dairy products	Many people and animals are carrier

Characteristics of major zoonotic diseases

Table 21.2 continued Characteristics of major zoonotic diseases

Zoonosis	Pathogen	Reported cases in man per year	Clinical signs in cattle
Anthrax	*Bacillus antracis*	< 1	Sudden death; bleeding from orifices
Q fever	*Coxiella burnetii*	< 100	No signs often
Rabies	*Lyssa virus*	< 1	Brain; behavior changes; death
Salmonellosis	*Salmonella typhimurium*	< 100	High fever; diarrhea; death; often no signs
	Salmonella dublin	< 100	As previous;+ also abortion
	Other species (S.enteritidis)	> 100.000	Often no signs
Toxoplasmosis	*Toxoplasma gondii*	100 - 1000	No signs
Tuberculosis	*Mycobact.bovis*	< 100	Short, dry cough; typical lesions
Trichophyton	*Trichophyton verrucosum*	< 100	Round, asbestos-like lesions, mainly on head and neck
VTEC	*E. coli O157*	< 100	No signs
Yersiniosis (rodentiosis)	*Yersinia pseudotuberculosis*	100 - 1000	Enteritis; abortion + abscesses in organs sometimes

Characteristics of major zoonotic diseases

Table 21.2 continued Characteristics of major zoonotic diseases

Clinical signs in sheep and goats	Clinical symptoms man	Infection transfer	REMARKS
Sudden death; bleeding from orifices	Toxaemia, cramps, death	Contact and products (soil workers!)	Spores survive for centuries in bottom
Often no signs	Often no signs; flue like; lung-liver + heart problems	Inspiration; contact; wool, skin, raw milk	Excretion via milk, genital tract
As cattle	Brain + meningitis; water fear; death	Dog or cat bite	
As cattle; long process	Enteritis; skin lesions	Milk; direct contact; abortion products	Mass disease control program needed
As cattle; long process	Enteritis; signs are more severe; very sick	As previous	Mass disease control program needed
---	Enteritis	Food intoxication; certain food products (poultry)	
Abortion; stillbirth; weak lambs (!!)	Flue-like; abortion + malformations in pregnant	Contact with cat faeces; eating raw meat with cysts; contact with aborting animal	
---	Lung tuberculosis: tubercles in lungs and skin	Contact with infected people or animals, air, milk	Mass disease control program needed
No signs	Infection of hair, nails and skin (children!)	Contact with infected animals or material	Rather high prevalence
No signs	Severe diarrhoea, hemorrhagic colitis; HUS; brains; coma, death	Via faeces of carriers; faeces contaminated food and water	Verocythopathogenic toxines . Older people and children!!
As in cattle. In goat also mastitis	Enteral lymphnodes affected; also toxaemia	Contact infection	Mice involved? Especially in rodents and turkeys

Improving food safety

Aafter Valeeva, Huirne, Meuwissen & Oude Lansink, Agricultural Systems, 2005/2006

Table 21.3 Chemical and microbiological food safety hazards on a dairy farm

Management Practice	Chemical hazard	Microbiological hazard
Origin of purchased feed	X	X
Cattle quarantine		X
Identifying treated cows	X	X
Action for Salmonellosis & Johne's		X
Udder cleaning before milking		X
Cattle purchase		X
Monitoring programme		X
Calving pen		X
Action when in doubt about withdrawal time	X	
Contacts with neighbouring cattle pasture		X
Calf feed before weaning		X
Herd treatment plan	X	
Veterinary check of cattle		X
Animal ID system		X
Housing of cows		X
Separating calf from dam after birth		X
Manure supply source for pastures		X
Ground water level of pastures		X
Manure application		X
Water used for drenching cows	X	X
Outdoor drinking system at pasture	X	X
Drinking system in barns		X

Table 21.3 lists the managerial action points for the dairy farmer relating to the chemical and microbiological food safety hazards

Improving food safety

Management Practice	Chemical hazard	Microbiological hazard
Re-use of waste water for cleaning & disinfection	X	X
Personal hygiene and general hygiene on farm		X
Adequate cleaning & disinfection	X	X
Maintenance of equipment	X	X
Good dairy farming practice performance	X	X
Quality assurance system on the farm	X	X

Table 21.4 The different choices available for applying certain management practices, the food safety level achievable and the associated costs for the farm in € per ton of milk.

	On a farm with 50 dairy cows				
Number of control measures applied	2	5	11	22	24
Achieved chemical food safety level (points scored)	0	3	7	18	19
Achieved microbiological food safety level (points scored)	1	3	11	22	23
Additional cost(€ per ton milk)	0.04	0.10	2.63	27.54	32.17

	On a farm with 250 dairy cows				
Number of control measures applied	3	7	13	21	24
Achieved chemical food safety level (points scored)	2	5	9	18	19
Achieved microbiological food safety level (points scored)	3	7	12	21	23
Additional cost(€ per ton milk)	0.08	0.34	1.95	19.64	27.40

Improving food safety

After Valeeva, Huirne, Meuwissen & Oude Lansink, Agricultural Systems, 2005/2006

Management practices, such as cow ID, drug residue testing before delivery, purchase of cattle with known health status, a closed herd principle and extension of the withdrawal period do not incur great additional cost to the farm.

The **major contributions to additional farm costs** in the case of food safety improvement originate from the following factors:

- Veterinary herd health & production management advisory programmes (HHPM and QRM).
- Culling of cows, as part of a salmonellosis and Johne's disease control programme.
- A maximum application of Good Dairy Farming Codes of Practice.

Due to the herd scale differences, the larger farms spend somewhat less for additional costs per ton of milk than smaller farms.

When the costs are assigned to different farm domains (herd management; pasture management; water management; other issues), **herd management is always the area carrying the greatest cost**, where large investments into farm management are required. Pasture and water management contribute the lowest cost, irrespective of the herd size (50 versus 250 cows).

When costs are assigned to the different stake holders in the dairy chain (feed mill; dairy farm; processing plant), the **farmers carry the greatest costs**, followed by the processing plant (about half the costs of the farmers) and the feed mill (about 15% of the farmers' costs), especially when large investments are required. Below the 50% threshold of food safety improvement, the extra costs of improvement are about equally distributed among the stake holders.

Overall, an improvement of food safety up to 50% of the maximum level can be achieved with relatively low additional costs to the farm: €0.10 and €0.34 per ton milk for a 50 and 250 cow herd, respectively.

The highest improvement level can only be achieved when the farmer invest in the majority of the management measures highlighted in Figure 21.1. This yields an additional cost of about € 44 and € 40 per ton milk for a 50 and 250 cow herd, respectively.

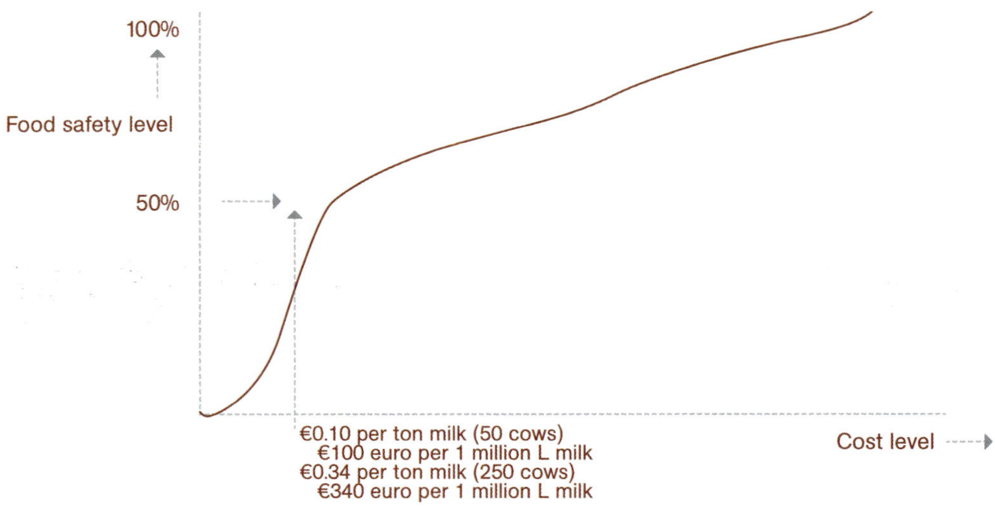

Figure 21.1 Food safety investments

Zoonoses - verocytotoxin producing *E. coli*

Verocytotoxin-producing *E. coli* can cause disease symptoms in man but doesn't always. Young children and elderly people (immune-compromised) are especially vulnerable. Symptoms include diarrhoea, colitis with bloody diarrhoea and renal failure (children!). Verocytotoxin-producing *E. coli* can be found in cattle, sheep, goats, dogs and rabbits.

Indirect infection transfer from animal to man may occur through:

- insufficiently heated meat products
- non-pasteurized milk and milk-products
- vegetables
- fruit juices
- salad
- drinking water
- swimming water

Direct infection transfer may occur through direct contact with ruminants and their faeces. In faeces, VTEC can survive for 10 weeks, depending on weather conditions (temperature; humidity).

Risk factors and control measures

Risk factors can be found in barns, manure, water and, to a lesser extent, in flies and feed. General risk factors that need management control are:

- newly-purchased cattle
- the presence of pigs on dairy farms
- on veal calf farms the presence of dogs
- the presence of other cattle farms within 1000 m
- flies in the barns
- wet bedding material in the cubicles of adult and young cattle.

A **Adult cattle:**
- closed farm principle
- clean and dry bedding
- balanced groups of young stock
- drinking troughs cleaned every two weeks
- chlorination of drinking water
- rodent control
- longer interval between manure spreading and grazing
- increasing the pH of the manure
- fly control
- Vaccination programme against VTEC
- Administration of probiotics
- Before slaughtering the (**beef) cattle**:
  * shortened period of feed withdrawal prior to slaughter
  * hay supplied during transport to abattoir and in the waiting area
  * cleaning/disinfection of boxes, waiting areas and transport means

B **For veal calf farms:**
- conduct a screening on VTEC
- slaughter calves from farms with a low infection level prior to delivery to the slaughterhouse that has calves with high(er) infection level (= logistic slaughter)

C **Personal hygiene:**
- wash hands after work, contact with animals, fence repair works, contact with manure, grazing/farm mud or soil, buildings, machinery and trucks

D **Young cattle:**
- closed farm principle
- clean and dry bedding
- clean animals
- balanced groups of young stock (= no mixing of ages)
- using farm boots/overalls
- using disinfection tubs

➲ Control measures that contribute to a better control of VTEC, may also help with control of Salmonellosis and Campylobacteriosis.

Water

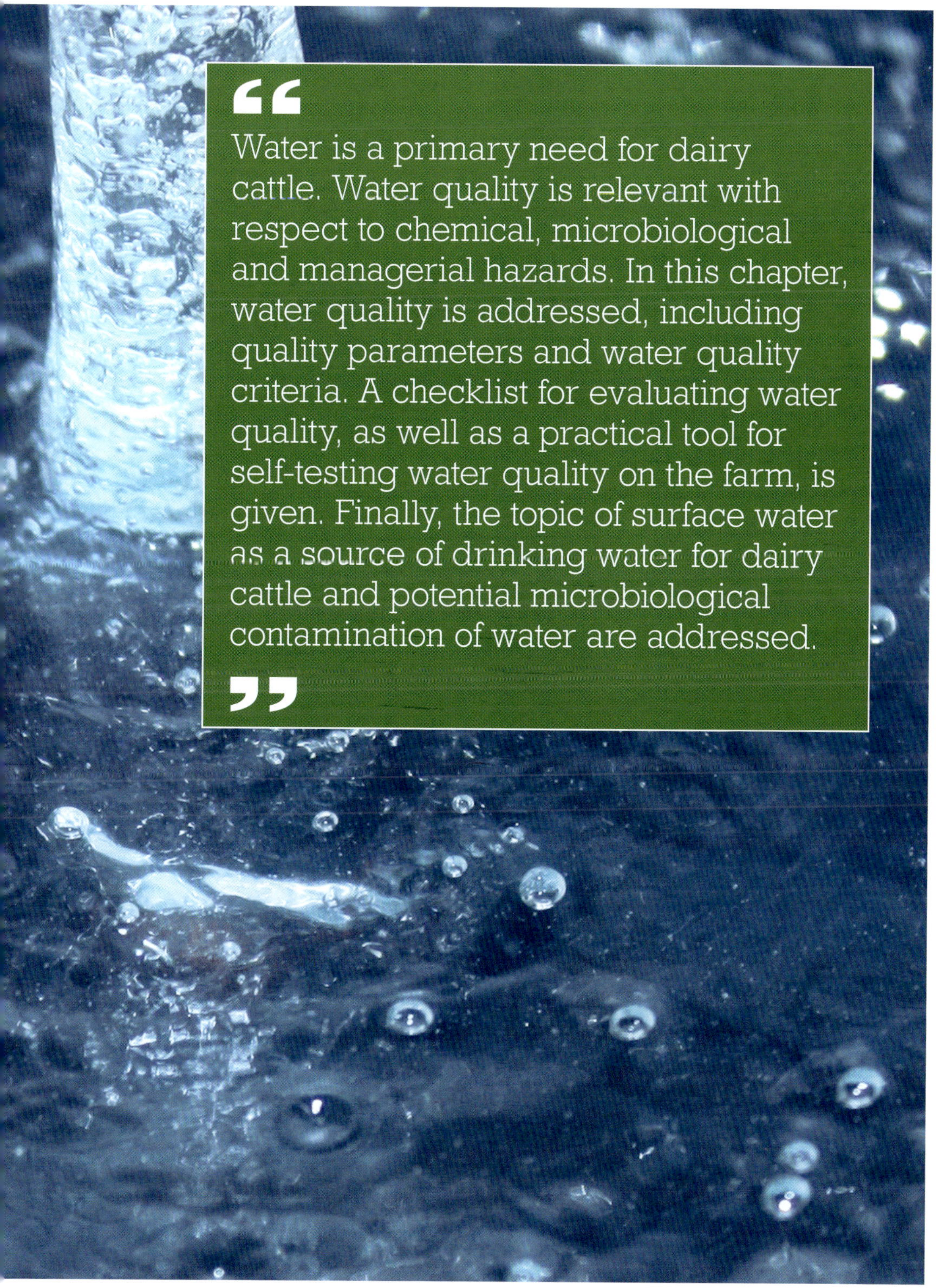

"Water is a primary need for dairy cattle. Water quality is relevant with respect to chemical, microbiological and managerial hazards. In this chapter, water quality is addressed, including quality parameters and water quality criteria. A checklist for evaluating water quality, as well as a practical tool for self-testing water quality on the farm, is given. Finally, the topic of surface water as a source of drinking water for dairy cattle and potential microbiological contamination of water are addressed."

Water quality

GD, Deventer, NL, 2007

A dairy cow drinks between 80 and 130 litres of water per day.

A cow will not drink contaminated or unpalatable water if she can avoid it. This phenomenon contributes to heat stress problems (especially when there is dirty or contaminated surface water at pasture!)

Practical checks to test water quality

- ✓ Take a clean and dry bottle
- ✓ Fill it with water from the cattle watering point
- ✓ Close the bottle and shake it well
- ✓ Leave the bottle to stand for four hours
- ✓ Open the bottle and smell the contents (good quality water has no smell)
- ✓ Check clarity of the water (good quality water has no colour and is clear)
- ✓ Check to see whether there are any deposits or iron flakes.

If all indicators are okay then there is > 85% probability that the water quality is good.

Polyethylene water tubes are dangerous because they are permeable to gases like ammonia or CO_2.

The ammonia causes biofilms in the tubes and changes the taste of water. During the summer when barns are not in use, this ammonia can be converted into nitrite, which is toxic.

Cows prefer open drinking places.

The depth of drinking places must be > 10 cm to avoid contamination.

Check chemical and microbiological water quality at least once a year.

Table 22.1 Reasons for poor water quality (not fit for cattle consumption)

	% of cases
H_2S	2
Bacteria	5
E. coli	4
Ammonia	17
Chlorides	7
Hardness	38
Ferrous Fe	27
Manganese	10
Natrium (sodium)	4
Nitrate	0.7
Nitrite	5
pH	1
Sulfate	5

Drinking water for cattle

GHM Counotte, 1999; GD Deventer, The Netherlands, 2006

Water quality must be evaluated using three parameters:

1	Palatability
2	Noxious substances
3	Availability

1. Palatability

The taste is influenced by Fe, salts and organic materials. Palatability is a relative term: when cattle are used to a certain type of water, they will continue to drink it (provided that the concentrations of chemicals etc are according to regulations).

2. Noxious substances

Examples of such products are nitrogenous compounds (ammonia; nitrite; nitrate; nitrosamines), organic material, sulphurous components, minerals, salts and certain bacteria. Ammonia can be found during processes of protein degradation under anaerobic conditions (animal tissue; wood; manure; dead leaves). The amines that are also produced during this process are toxic.

Nitrosamines can be found when the aforementioned process takes the inverse direction. The level of organic material can be assessed by observing the oxidation process. Elevated levels of organic material often go together with an elevated level of Fe and ammonia. Sulphurous components (H_2S and sulphates) indicate poor water quality and make the water inappropriate as drinking water for cattle. Elevated sulphate levels and elevated Mg levels will act as a laxative for cattle.

Water acidity (pH less than 6.5 or higher than 8.0) indicate contamination of biological importance. Excess Mn can cause an obstruction of the water pipes.

Surface water may also have a too high level of chloride (> 175 mg/L).

The number of bacteria present in the water is an indicator of hygiene - too high a level of coliforms can indicate faecal contamination.

3. Availability

Water must be able to run freely in the pipes without obstruction. Obstructions occur when concentrations of the following are too high:

- Fe
- Mn

Additionally, if the pH is too alkali and/or the water too hard, obstructions will also occur. Microfilms (biofilms) appear in the water pipes.

Source water can be used if the quality is okay. Most often, superficial sources are too highly contaminated. Water from deep in the ground is often of better (microbiological and chemical) quality. For smaller streams and ponds the same problem arises if cattle have had the opportunity to contaminate the waters.

A **stream of rain water** can be of good quality if there is no contamination.

Water quality of a stream is determined by:

- The quantity of slime/mud present
- The agricultural or industrial activities around the farm
- The presence of a current in the stream
- Possible contamination by septic tanks
- Biological processes (at high ambient temperatures)
- Soil type

Atrazine (a derivate of pesticides on maize cultures) may negatively affect water quality and transfer to milk. Diagnostic tests for detecting atrazine are available: ELISA test Envirogard kit (threshold for water= 0.025 µg/L; chromatography threshold = 0.05 µg/L). Peak concentrations of atrazine can be found in late spring and autumn.

Water quality Q & A

GD, Deventer, NL, 2007

	Yes-/-No
1. The water has a smell of rotten eggs and/or there are bubbles in the water	
If the answer is NO, continue to Question 2	
If the answer is YES, water quality is poor and inappropriate for cattle	
2. At pasture the colour of water is sometimes reddish, bluish or green	
If the answer is NO, continue to Question 3.	
If the answer is YES, the water quality is poor and inappropriate for cattle	
3. What is the colour of the water in the glass container?	
No colour + no sediment = 0 points; some colour + no sediment = 1 point; coloured water (yellow; brown) + sediment = 2 points	
4. What is the clarity of the water in the container? (first stir well and wait 5 minutes)	
Water is clear = 0 points; water is somewhat muddy = 1 point; water is very muddy = 2 points	
5. Is there sediment in the water?	
No, nothing = 0 points; some particles = 1 point; in the bottom of the glass container there is plenty of sediment = 2 points	
6. What is the odour of the water in the glass container?	
No odour = 0 points; some odour but no H2S = 1 point; 'rotten eggs' odour = 2 points	
7. Is there duck-weed in the water?	
0-25% = 0 points; 25-75% = 1 point; 75-100% = 2 points	
8. Is there water that comes from outside your premises? (see also Question 18)	
NO = 0 points;YES = 2 points	
9. What is the depth in the middle of the stream?	
From 0 to 20 cm = 0 points; from 20 to 50 cm = 1 point; > 50 cm = 2 points	
10. How much slime/mud is there in the middle of the stream?	
Nothing detectable = 0 points ; de 5 à 10 cm = 1 point ; > 20 cm = 2 points	
11. How is the current in the stream?	
The stream is often dry = 2 points; water does not flow = 2 points; there is a weak current = 0 points; there is a strong current = 1 point	
12. Are there septic tanks in the area (< 1000 m radius)?	
NO = 0 points; YES = 3 points	
13. Is it possible that household water enters the stream?	
YES, directly = 2 points; YES, but indirectly= 1 point; NO = 0 points	

Water quality Q & A

	Yes-/-No
14. Does rain water or cleaning water enter the stream?	
NO = 0 points; YES indirectly = 1 point; YES, directly = 2 points	
15. What is there on the other side of the stream?	
Industry or urbanisation = 2 points; pature = 1 point; agriculture = 2 points	
16. Is there an impact of salty water in the pastures?	
NO = 0 points; UNKNOWN = 2 points	
17. Have there been attempts to lower the ground water level?	
NO = 0 points; YES = 1 point	
total number of points noted (A) →	
The next three issues may serve to reduce the total score on questions 1–17	
18. The water coming from outside your premises is of good quality ?	
NO = 0 points; YES according to laboratoires = 2 points; YES according to yourself = 1 point	
19.The borders of the stream have vegetation (plants; bushes)?	
NO = 0 points; YES = 1 point	
20. Are there aquatic plants in the stream?	
NO = 0 points; YES = 1 point	
total number of points noted (B) →	
Final score on the quality of the water in the stream (A–B) =	…………..

Interpretation of test results:

0–8 points The quality of water seems acceptable, but this is not guaranteed. Be aware that the quality may change over time.

> 9 points It is doubtful whether the water is of sufficient quality to be used as drinking water for cattle. It is recommended to take a proper sample of 2 L of water in a clean, dry sterilized bottle and send it to a laboratory.

If you have answered YES to Questions 1 and/or 2, then the quality of the water is poor. Look for another source for drinking water for the cattle during grazing. Avoid cattle drinking from the poor water source (e.g. fence around it). Scoring 2 points may indicate respective sources of contamination of the water.

Quality of drinking water

The quality of drinking water is largely determined by heuristics and parameter values are not supported by scientific data or proof. Therefore, in many cases, we speak about reference values instead of standards or norms. An example is given in Table 22.3.

It has been stated that a water hardness of less than 4°D may cause problems in cattle (GD, 2011). This kind of water often has an unstable pH, which, in turn, may cause an increased corrosiveness of the water, affecting metal pipes and tubes. The phenomenon occurs more frequently when there are also lower levels of certain salts in the water (Ca and Mg). This kind of water is also less palatable for animals. The low hardness can be caused by (1) water de-hardening processes by reversed osmosis; (2) salt exchange processes. The latter causes an increased Na (salt) concentration in the water.

Table 22.3 Review of drinking water quality parameter values regarding surface water for cattle

Parameter	Reference values		
Bacteriology:	Suitable	Less suitable	Not suitable
Bacteria count (at 22°C)	< 100.000 cfu/ml	-	> 100.000 cfu/ml
E.coli count	< 100 cfu/ml	-	> 100 cfu/ml
Chemistry			
Ammonia	< 2.0	2.0-10.0	> 10.0 mg/l
Nitrate	< 100	100-200	> 200 mg/l
Nitrite	< 0.10	0.10-1.0	> 1.0 mg/l
Iron	< 0.5	0.5-10.0	> 10.0 mg/l
Sodium	< 800	800-1500	> 1500 mg/l
Chloride	< 250	250-2000	> 2000 mg/l
Sulfate	< 150	150-250	> 250 mg/l
Manganese	< 1.0	1.0-2.0	> 2.0 mg/l
H_2S	Absent	-	Present
pH	5.0-8.5	4-5; 8.5-9	< 4 & > 9
Hardness	< 20 °D; > 4°D	20-25 °D	> 25 °D
Organoleptics:	No colour; no odour; clear; no sediment; no iron-sediment		

Quality of drinking water

E. van Eenige, GHM Counotte,JPTM Noordhuizen. 2012. Drinking water for dairy cattle : a benefit or a microbiological risk ?

Contamination of drinking water at farm level may be chemical (iron; manganese) and/or microbiological (saprophytes; pathogenic bacteria and viruses, for example coliforms). *E.coli* bacteria may survive in drinking water systems, even under unfavourable conditions. *E.coli* often originates from faecal contamination of drinking troughs in pastures and/or barns. Infections may spread on a farm due to construction and design of the water distribution system. Cows around calving and in the early post-calving period are most susceptible to the negative effects of water contamination. In addition, high-producing cows consume around 120 L of water per day, especially in early lactation.

Even when the water quality has been confirmed at the entrance to the farm, this does not provide any guarantee that the quality remains unaffected further along the distribution system. On the contrary, a field study carried out in The Netherlands established that while 98% of samples at entrance showed water of "suitable quality", this figure decreased to 33% at the end points of the distribution system on farms, while another 33% was classified "less suitable" (field study on 130 dairy farms; 285 samples; Interact, 2006).

Some **features of the water distribution system** contribute to a decreased water quality: design and construction of the system; materials of tubes and pipelines (galvanized and stainless steel provoke biofilms with micro-organisms in the tubes; poly-ethylene shows permeability for gases); cleaning & disinfection procedures; water sources; design of water troughs. Large open water troughs in a distribution system according to the communicating vessels principles show the worst chemical and microbiological quality. Pressure-pipelines with anti-backflow-valves show less quality deviations than the previously described system. Without the communicating vessels principle, the open troughs are microbiologically similar to other models.

Cows prefer open drinking troughs above other models; they drink at a depth of 4 to 7 cm. The newest model of an open, rectangular drinking trough is funnel-shaped. This facilitates the easy removal of sediment material without spoiling much water. The temperature of the drinking water should ideally be around 10 to 15°C (important in summer: heat stress). The principle is to have 600 cm total width of drinking space for 100 cows; this width should be increased up to 900 cm in summer (heat stress periods).

Clean drinking troughs

Three methods were tested:

1. troughs were emptied, cleaned with water and refilled again
2. troughs were emptied, thoroughly brushed, water cleaned and refilled again
3. troughs were emptied, thoroughly brushed then a 5.25% Na-hypochlorite solution diluted 1 to 32 was put into water-filled troughs for 15 min. Afterwards, the trough was emptied and refilled with fresh water.

Results:

Methods 1 and 2 did not change the coliform bacteria count in any parameter (water, biofilm).

Method 3, however, reduced the coliform count by 99% in water and 99.9% in biofilm. The coliform count returned to the levels found before treatment only 24 hrs later. This was caused bythe simple fact that after one cow drank, the water was again contaminated (microbiologically and chemically).

This field study shows that the beneficial effects of treating water troughs chemically are of short duration only.

Quality of drinking water

(Interact AM, The Netherlands, 2006)

Table 22.4 Practical points for water supply	
1	If the drinking water is not tap water (for example well water or surface water), it is highly advisable to test the water (chemical and microbiological) quality at least once per year, because in well water often the levels of manganese and iron are higher than in tap water; while, pH is often lower and hardness higher than in tap water.
2	Distribution of drinking water should always be directed towards a drinking point, without backflow possibilities; if appropriate, a pressure system and valves must be installed to prevent such a backflow.
3	Drinking water systems for young stock should be separated from those for adult cattle; moreover, different age groups of young stock should preferably have their own water distribution system. When –in current systems- interventions are considered, it is advisable to conduct such intervention per age or production group. The best intervention would be in the communicating vessels systems because these yielded the worst microbiological results in all field tests.
4	Farmers can conduct a self-test on water quality, using clean dry bottles of 1-2 l. content. Samples are taken from the troughs and bottles left standing at room temperature for 2 or 4 hours. Then the results can be interpreted using parameters: odour; clarity; transparency; sedimentation (see the scoring card elsewhere in this series). When all parameters score "good", the probability that this water is of sufficient quality is high. When 2 or more parameters score "intermediate" or "bad", then water should be tested in the laboratory before using it as drinking water for cattle.
5	Drinking points must be easy to clean, especially when faecal contamination is detected. Mostly such cleaning is unfeasible due to construction features and the communicating nature of troughs, and therefore an alternative trough is proposed. Cleaning must be done at least monthly, preferably each week. Common cleaning regards, first, the closure of water flow and emptying the open trough; next, a thorough brushing with a detergent is conducted; thirdly, new water is allowed to enter the trough. Unfortunately, due to the high in-flow of water, sediment (iron- and manganese-deposition) is again flushed from pipeline into the trough. Therefore, low pressure (0.8 Bar) and air impact should be applied to pipelines to eliminate most of the sediment from the system. Water points in the pastures must not be forgotten or neglected.
6	When the coliform count in water samples exceeds the reference value for suitable water quality (see Table 22.3), this water often shows increased levels of iron, ammonia and nitrite too. The increased ammonia and nitrite levels may indicate faecal contamination. Increased ammonia levels in water samples, as a parameter for organic material, may also indicate backflow problems in the water distribution system and the creation of biofilms.
7	When well water is pumped up from (deep) underground, the maintenance and treatment of this water becomes paramount. Negligence regarding de-ironing and filtering, may put the water quality at risk. Best practice would be to design a proper maintenance protocol and stick to it. Additionally, it is recommended to conduct water quality testing more frequently than once a year.
8	The material and the position of pipelines used for the distribution of drinking water can play a substantial role in risk increase. For example, poly-ethylene has permeability for gasses (ammonia) which can negatively affect water quality and contributes to biofilm creation. The less susceptible materials for pipelines are glass, stainless steel, Teflon; better are concrete or PVC.
9	The palatability of the drinking water for cattle is influenced by its chemical composition (e.g. iron, nitrite, ammonia, calcium, organic matter) and hardness. Palatability is a relative concept; if cattle is used to water of less quality, then they still may continue to drink it. But large changes over time in the composition of water hamper such a "getting used to it"; water intake may drop and hence feed intake and milk production. Rumen acidosis too may be indirectly associated with poor microbiological water quality.

Surface water as a source of drinking water

	Question	Answer 1	Answer 2	Answer 3	Answer 4
1.	The stream has a nasty smell (H2S) or shows gas bubbles	NO See Question 2	YES Water is probably not fit for cattle; see Question 3		
2.	The water in the stream is sometimes red, green or blue	NO See Question 3	YES Water is probably not fit for cattle; see Question 3		
3.	Check the water quality with 1 L bottles for colour, cleanliness and sediment	See page 454			
4.	What is the clarity of the water?	Clear 0 points	Slightly cloudy 1 point	Not transparent 2 points	
5.	Is there sediment in the water sample?	There are no particles visible 0 points	There are some particles 1 point	The water is full of particles 2 points	
6.	How does the water sample smell?	No smell 0 points	There is smell but not H2S 1 point	Water smells of H2S (rotten egg) 2 points	
7.	Is there a cover of duck-weed on the stream?	0 to 25% 0 points	25 to 75% 1 point	75 to 100% 2 points	
8.	Is there inlet of water from elsewhere into the farm's streams?	NO 0 points	YES 2 points		
9.	What is the depth of the stream in the middle?	0 to 20 cm 0 points	20 to 50 cm 1 point	> 50 cm 2 points	
10.	How much mud is present in the middle?	None 0 points	5 to 10 cm 1 point	> 20 cm 2 points	
11.	Is there a current in the stream?	Regularly dry 2 points	Still water 2 points	Slowly flowing 0 points	Rapid flow 1 point
12.	Is there a nearby sewer over-flow?	NO 0 points	YES 3 points		

Surface water as a source of drinking water

	Question	Answer 1	Answer 2	Answer 3	Answer 4
13.	Could household disposal water enter the network of streams?	NO 0 points	YES but indirectly 1 point	YES, directly 2 points	
14.	Could water from the farm grounds enter the streams?	NO 0 points	YES, indirectly 1 point	YES, directly 2 points	
15.	What is on the other side of the stream?	Grassland 1 point	Urbanisation/ industry 2 points	Agricultural crops 2 points	Flowers/ horticulture 2 points
16.	Is the area affected by salty water?	NO 0 points	I do not know 2 points		
17.	Is there draining going on in the area?	NO 0 points	YES 1 point		
	The subtotal of points resulting from the questions 1 to 17 =				points

Score question 18 only if you have scored question 8!!					
18.	Is the inlet water of good quality?	NO 2 points	YES, my observation 1 point	YES, lab tested 0 points	
19.	Are the sides of the stream over-grown with plants?	NO 1 point	YES 0 point		
20.	Are there plants in the water?	NO 1 point	YES 0 point		
	The subtotal of points resulting from questions 18 to 20 =				
	Final score for this farm is (subtotal A + subtotal B) =				

A final score from 0 to 8 indicates a fair to good water quality (but no guarantee…)
Final score of more than 9 points indicates the need for a formal water quality testing at a laboratory.
When questions 1 and 2 have been answered with YES it is advised to change the water supply.
Any answer scoring two points, indicates a potential contamination issue and action is required.

Microbiological contamination of drinking water

Table 22.5 Schematic overview of the potential risk areas concerning the microbiological contamination of drinking water

Potential risk sites	Details of sites
Source	Mains water from public system
	Water ponds
	Wells
	Streams; rivers
Farm entrance	House
	Barns
	The same for house and barns
Type of distribution system on the farm	Design and construction
	Pipeline material
	System with or without pressure
	System with or without backflow valves
	Barn system yes/no separated from pasture system
Water drinking points	Design; construction
	Material of pipelines and water points
Other issues	Separation of age groups
	Cleaning & disinfection
Other management activities	Cleaning cows in milking parlour
	Flushing barn alleys
	Using sprinklers and alike in heat stress management

Major risk factors related to microbiological contamination of drinking water for dairy cattle.

Major microbiological contaminants are

- *E.coli*
- *Salmonella spp*
- *Mycobacterium paratuberculosis*
- *Yersinia enterocolitica*
- *Campylobacter jejuni*
- *Pseudomonas aeruginosa*
- *Norwalk virus*
- *Rota virus*
- *Calici virus*
- *Cryptosporidium parvum*
- *Giairdia lamblia*

Some of them may survive in water systems (e.g. *E.coli*), even under harsh conditions. **Contributory risk factors** for such contamination are the formation of biofilms (sometimes with a resistant exo-polysaccharide matrix) in the water distribution system, substrate for microorganisms (phosphates; nitrates) and sediment (iron; manganese; calcium), the design and construction and material of water systems for cattle (distribution systems of the communicating vessels principly are more prone to contamination than others; lack of backflow valves contributes to spread of pathogens throughout the whole system; lack of pressure in the system), the material of pipelines (polyethylene absorbs and leaks gasses; stainless steel and galvanized metal water troughs show lower quality water than plastic, PVC or concrete ones), drainage of rain water from agricultural land into well water or ponds.

Other risk factors are, for example, the (abundant) use of water to flush exercise areas and alleys; the formation of aerosols during cleaning and disinfecting procedures; the application of sprinklers and alike for controlling heat stress effects in cattle; high ambient temperatures; migrating birds and water fowl contaminating wells and ponds.

Consequences of poor water quality are: decreased water intake, decreased feed intake, loss of milk production, disturbed rumen function and subsequent health problems (acidosis; ketosis), food safety hazards.

Self-testing water quality

E.van Eenige, GHM Counotte, JPTM Noordhuizen

Self-testing water quality: Colour, transparency and sediment

Colour good
Transparency good

Colour poor
Transparency good

Colour poor
Transparency good

Colour bad
Transparency bad poor

Colour good
Transparency good
Sediment poor

Colour good
Transparency good
Sediment poor

Drinking water

Worksheets

Analysis of a mastitis production problem No 1

Analysis of a mastitis production problem - Example 1

This worksheet aims to train practitioners in their daily herd health work when dealing with day-to-day herd problems. A problem is postulated and then questions are asked. The practitioner is asked to answer these questions, preferably on a separate sheet of paper in order to be better able to evaluate the process. At the end of the exercise, the answers are provided. The main goal of this exercise is to **provide structure for analysing herd problems**.

Part I.

A farmer calls you one morning and tells you that his cows are not performing as expected in the beginning of lactation. He wants you to visit and propose a solution to his problem.

Question 1
What are your first **working hypotheses**?

Question 2
What elements would you include in your **work plan** to tackle this problem? (create a schematic overview)

Question 3
What are the **important questions** to raise in order to be able to define the problem more concisely?

Part II.

Which cows are involved?

- Parity > 4

What problems occurred at or around calving?

- Hardly any problems occurred

What problems occurred after calving?

- A few mastitis; no retained afterbirth

What occurred exactly and when?

- Up to 50 days after calving; from the end of the housing period

What was exactly was observed in those animals?

- Poor appetite; production loss in spite of good quality grass silage and corn silage

Question 4
Which cows or which cases would you want to see for a clinical examination and why?

Part III.

The **inventory** has given you the following results: *problems in cows with parity > 4 at the start of lactation (< 50 days after calving); *these cows show poor BCS and poor rumen fill score, despite feed quality being good. Feed is provided 3 times per day; *acute endometritis is found in two cases, with a poor rumen function, but no other signs; *problems do indeed seem to have started at the end of the housing period.

Question 5
What should you do now?

Part IV.

The results of the **overall herd inspection** are the following:

- BCS and rumen fill scores are normal and vary according to lactation stage
- there are no tripping cows or cows with bad posture
- no hock Lesions
- housing conditions, barn climate and hygiene are okay
- there are no concentrate refusals in the automatic distributor and or the feed bunk
- corn silage is of good quality, but grass silage is only acceptable, not optimal.

Analysis of a mastitis production problem No 1

Question 6
What should you do now and why?

Question 7
What is your **current probability diagnosis** and why?

Question 8
How can you **confirm** your probability diagnosis?

Question 9
Which **interventions** and what **advice** should be given?

Question 10

How can you make an **evaluation** and **which cows** would be included in that evaluation?

Question 11
What are the potential **consequences** of a negative energy balance or ketosis on the health of the cows and why?

Question 12
How could you limit the occurrence of those consequences?

Problems around calving (cesarean section, retained afterbirth, endometritis); low BCS at and after calving; negative energy balance with/without ketosis, with/without acidosis; abomasal displacement; mastitis cases; feeding/digestive problems; genetics in parity 1 animals; other conditions (virus infections, occasional diseases).

The 10 steps of the analysis procedure

(a) Anamnesis – definition of the problem; (b) Clinical examination of cases that are present; (c) Examination of the whole herd; (d) Examination of the environment; (e) Examination of the farm data; (f) Additional examinations (farm; laboratory); (g) Synthesis of issues detected and conclusions; (h) Probability Diagnosis; (i) Action Plan (the cases; short-term; mid-long term); (j) Evaluation during subsequent farm visits

Which cows? what problems at calving? what problems after calving? when did everything occur and since when? What exactly did one observe ?

Those cows which still show –according to the farmer– signs of the original problem (so acute cases) and in which one would expect a low BCS, low RF score about 3 to 4 weeks after calving and

Analysis of a mastitis production problem No 1

in which the rumen function can be well examined. Then, one can look for other signs and disorders (metritis; disease)

Inspection of herd mates showing no clinical signs : BCS, RF, faecal consistency and faecal content of undigested fibre; frequency of rumination in the herd (should be > 75 percent), claw lesions; an inspection of the cows' environment (housing; climate; feed)

Additional examinations may be on milk, urine, rumen fluid, manure, blood (at the farm) and blood samples for the laboratory

Primary ketosis due to lack of energy in the ration for fresh cows and hence a negative energy balance leading to a low BCS and a low RF score and then production loss; Secundary ketosis cannot be excluded in the case of severe endometritis

On the basis of sampling: milk, urine, blood. Also on the basis of evaluating the milk recording forms of several dates (decrease in kg milk, decrease in milk protein)

The cases: treat endometritis; in case of negative energy balance and/or ketosis treat with dexamethasone + propylene-glycol + ration adaptations At herd level: adapt the rations for the fresh cows and transition cows; create production groups for better surveillance and better feeding; reduce all forms of stress around calving; evaluate the management of silages during all housing period (answer the question: which best quality silage is to be given to which cows in which part of the housing period?)

During one or more subsequent farm visits (starting 1 month later) you should look for the cows that have calved since the last visit - they have been submitted to the new regime; check BCS and RF in those cows, as well as the milk recording data.

One could test for ketone bodies in milk or blood. At the same time, one should always check for other disease occurrences, partly to exclude secondary ketosis.

We know from research that ketotic cows are at a higher risk for mastitis due to immune-depression, as well as a lower reproductive performance (NEFA are toxic substances at follicle level)

Treatment of those cows should be according to the rules. The farmer should increase surveillance of those cows in order to detect any case of mastitis as early as possible. For example, by following closely the somatic cell count patterns of these cows over time.

Analysis of a mastitis production problem No 2

This document is aimed at self-training the practitioner to gain experience in matters of problem analysis. It comprises seven parts, each with questions. The user is asked to answer these questions, preferably on a separate sheet, and then compare his/her answers with the results presented at the end of this document.

Part 1

The farmer calls you one morning to tell you that he has recently experienced too many clinical mastitis cases in his herd. He asks you to visit and informs you that he has already prepared a document containing all the data on mastitis cases over the past year.

Question 1
What would be the different steps in your Plan of Analysis and why?

Part 2

Once you arrive on farm, the farmer informs you that most of the clinical cases occurred in the autumn and a only minor proportion in the summer. The younger cows especially were affected and mainly around calving. These cows were not clinically sick (no fever; no general disease signs). The average bulk milk somatic cell count is 100.000; individual cow cell counts are not available.

Question 2
What action should you now take?

Question 3
What is your current conclusion?

Part 3

On the basis of the available information, the following Tables could be constructed:

Table 23.1 Frequency distributions of mastitis cases versus lactation stage, season and quarters involved

		% of mastitis cases
Stage of lactation and clinical cases	0 – 5 days	69
	6 – 20	14
	21 – 100	6
	101 - 353	11
Season and clinical cases	Mid-September – end of December	42
	Start of January – end of March	21
	Start April – end of June	24
	End of June – mid September	13
Affected quarters and clinical cases	Left fore	24
	Right fore	16
	Left hind	34
	Right hind	26

Analysis of a mastitis production problem No 2

Table 23.2 Distribution of parities in a dairy herd of 100 cows or 50 cows respectively, and the mastitis incidence per parity group

Parity		100 cow herd	50 cow herd	% mastitis cases
Parity 1	35	35	18	38
Parity 2	20	20	10	22
Parity 3	20	20	10	18
Parity 4	15	15	7	12
Parity 5+	10	10	5	10

Table 23.3 The actual udder health parameters for this farm, as compared to reference values

	Actual	Reference
Bulk tank SCC	115000	<250000
% clinical mastitis	54	<25
% other teat/udder issues	4	<5
Culled for mastitis	4	<10
% SCC individ > 200	11	-
% repeat mastitis	7	<10
Peak in SCC (November – December)	380000	-

Question 4
What is your primary hypothesis regarding the causative micro-organisms?

Question 5
What should you do next?

Part 4

Milk samples have been taken from 20% of the herd (predominantly fresh cows) for laboratory culturing. The results will be available in a few days. In the meantime, you continue your analysis by inspecting the housing, the barn climate and the feed and feeding management. The outcome of this inspection does not yield useful information for the analysis. At the end of the day you make another visit for an inspection of the milking practices in the parlour.

Maintenance of the milking machine by the dealer is twice yearly, every six months. Despite this, you carry out an overall inspection when you are in the milking parlour. The parlour is clean, well-lit and well-ventilated. There is no automatic detachment of the cluster. One of the vacuum tubes shows a narrowing along it; the vacuum tubes to the milk containers are long; the teat liners are clean, flexible and in good condition; you do not find any deformities in the rubber material.

Question 6
Knowing that the laboratory results will be available in a few days, what should your next step be?

Analysis of a mastitis production problem No 2

Part 5

The results of the milking practices inspection are reviewed below:

Milking is carried by two people: father in the morning, son in the evening. Before milking starts, feed is provided in the feed bunk, with the feed rack closed (this then opens after milking). In the parlour, no concentrates are offered. For each milking five cleaning towels are used for cleaning teats; the towels are washed each week at 60°C. The milking principle is: prepare four cows in a row, extract the first milk jets, attach cluster. Mastitis cows remain with the rest of the group to be milked. After milking the teats are disinfected. During milking, cows are quiet and defecation is not observed. Milk interval study does not show any problem; milking is carried out properly; there are no milk rests in the udder. Teat ends do not show any callosity or blue discolouration rings at the teat basis.

Question 7
What would be your conclusions at this point: strong points (= positive) &weak points (= to be improved?

Question 8
What else could you do now?

Part 6

The farmer treated his mastitic cows with antibiotics in the affected quarter for generally two days; the milk of that quarter is discarded. Injections are not administered.

Question 9
What do you think about this treatment? And why?

Question 10
What should you do now?

Part 7

The results from the laboratory arrive (email or fax, or GSM, or posted). Out of all the samples:

- 60% concern staphylococci
- 5% concern Streptococcus dysgalactiae
- 6% concern mixed infections
- 30% negative results.

Question 11
What can you do with these results? What do you advise?

Question 12
What can you tell the farmer about the prognosis of treatment: should these cows be treated or culled?

Question 13
What should you do next?

Worksheet answers - example 2

ANSWER 1

According to the general plan of analysis, the following steps should be taken:

1 anamnesis of the farmer
2 definition of the problem
3 inspection of data
4 inspection of clinical cases present
5 inspection of the herd
6 inspection of the environment (machine, housing, climate, feed, management, hygiene)
7 synthesis of findings
8 additional examinations if required (samples of milk for bacteriology; milking practices; milking interval survey)
9 conclusions and action plan (short- and mid-long term), activities of other specialists (evaluation of machine, evaluation of rations...)
10 Evaluation of effect of interventions and advice given during subsequent farm visits.

ANSWER 2

Problem definition in detail and evaluation of data; frequency distributions are especially instructive.

ANSWER 3

It confirms exactly what the farmer has told you

ANSWER 4

Given the SCC values, the incidence and prevalence of mastitis, and stage of lactation, you should think about *Coliform* bacteria and/or *Staphylococci*

ANSWER 5

Milk sampling of mastitic cows and suspected cows (SCC) for bacteriology, especially in early lactation. CMT does not work well because of rather low SCC. Check clinical cases that are present (samples for culture); check cows at parturition and after calving; inspection of milking machine; inspection of milking practices.

ANSWER 6

Always checking the milking practices during the milking itself (**wet test**); if needed, set a new appointment with the farmer

ANSWER 7

See the + and the -. Changing the person who milks during the day may be a risk factor. Hygiene must be improved. Animals previously having mastitis, as well as the currently known infected cows should be milked last.

ANSWER 8

Evaluate the efficacy of antibiotic treatment by the farmer (days interval between treatment start and recovery; types of products used). It also needs discussion whether or not antibiotics by injection are indicated.

ANSWER 9

This treatment could be weak and insufficient (risk of remaining subclinical or latent infections; resistance). The type of antibiotic is very important in association with the bacteria found. As soon as the laboratory results are known, you must design a Herd Mastitis Treatment Advisory Plan, to be updated every three or six months. This Plan also comprises a paragraph on dry cow treatment.

ANSWER 10

Make a preliminary synthesis (clinical mastitis in younger cows in autumn especially early after calving, where *Coliforms* and or *Staphylococci* are expected; in the milking practices several issues should be improved, such as towels and cleaning; separate (former) mastitis cows from others; wear gloves during milking especially when *Staphylococci* are present; milking machine needs adjustment and make a preliminary action plan based on these items, while waiting for laboratory results to arrive. Discuss everything with the farmer

Answers - example 2

so he understands what is needed and why; this makes it easier for him to accept advice. Set a new appointment for evaluation and adjustment of action plan.

ANSWER 11

Continue to discuss with the farmer; emphasise the relevance of interventions and advice. The old cases of mastitis in the herd remain a source of contamination and infection for herd mates. Chronic cases should be culled than repeatedly treated. To lend support to the decision of whether to cull or treat, the farmer should subscribe to individual cell counting at the milk recording DHIA.

ANSWER 12

The prognosis of treatment is poor if the cow is old, in early lactation, has more than 1 infected quarter, is a chronic case and/or shows high SCCs.

ANSWER 13

Set a new appointment to evaluate the situation and the effect of advice and suggested interventions. Maybe the action plan has to be adjusted. Maybe new samples are to be taken from new cows after calving for culturing.

Epilogue & Acknowledgements

By reading through this book you most probably have found out that indeed this is not a textbook. For the different domains addressed in this book, several textbooks and other scientific publications have been published. That is why we have chosen another way of presentation, more practical and more easily and rapidly accessible. Nevertheless we consider it worthwhile to present here some links to practical textbooks and websites which are more or less directly related to the issues we have dealt with. .

Practical textbooks for further reading available at www.contextbookshop.com

Farm health and productivity management of dairy young stock rearing. Boersema JSC, Cannas da Silva J, Mee J, Noordhuizen JPTM. 2010.
Herd health and production management in dairy practice. Brand A, Noordhuizen JPTM, Schukken YH. 1996 & 2001.
Applying HACCP-based quality risk management on dairy farms. Noordhuizen JPTM, Cannas da Silva J, Boersema JSC, Vieira A. 2008.

Practical websites of interest

Monitoring sheets for farm visits; analysis; evaluation; advice. www.vacqa-international.com
Dairy herd health and productivity management. www.dairyherd.com
Issues regarding mastitis control programmes. www.mastitisnetwork.org
Mastitis and quality related topics www.milkquality.ca
Website of the World Health Organisation www.who.int
Veterinary issues www.medvetnet.org
EXCEL calculation devices for silage humps www.uwex.edu/ces/crops/uwforage/whatsnew.htm

Thanks

The author and Context would like to acknowledge the support of Dr Helen Warren from Alltech Inc in editing this book.

Photographs

We would also like to acknowledge the following contributors of photographs to this book;

Charles-Sainsbury Plaice
Ed van Eenige
Gilles LeSobre
John Mee
Luc Mounier
Jos Noordhuizen
Otlis Campimon
Veeteelt
Wesley Ewing
www.dairymaster.com
www.lely.com

Index

Index

Index

Index

I

J

K

L

M

Index

N

O

P

Index

Index

Index

Y

Z